Readings from *Conservation Biology*

EDITED BY DAVID EHRENFELD

The Social Dimension—Ethics, Policy, Law, Management, Development, Economics, Education

A Joint Publication of
The Society for Conservation Biology
and Blackwell Science, Inc.

BLACKWELL SCIENCE

EDITORIAL OFFICES:
238 Main Street, Cambridge, Massachusetts 02142, USA
Osney Mead, Oxford OX2 0EL, England
25 John Street, London WC1N 2BL, England
23 Ainslie Place, Edinburgh EH3 6AJ, Scotland
54 University Street, Carlton, Victoria 3053, Australia
Arnette Blackwell SA, 1 rue de Lille, 75007 Paris, France
Blackwell Wissenschafts-Verlag GmbH, Kurfürstendam 57, 10707 Berlin, Germany
Blackwell MZV, Feldgasse 13, A-1238 Vienna, Austria

DISTRIBUTORS:

USA
Blackwell Science, Inc.
238 Main Street
Cambridge, Massachusetts 02142
(Telephone orders: 800-215-1000 or 617-876-7000)

Canada
Oxford University Press
70 Wynford Drive
Don Mills, Ontario M3C 1J9
(Telephone orders: 416-441-2941)

Australia
Blackwell Science Pty Ltd
54 University Street
Carlton, Victoria 3053
(Telephone orders: 03-347-5552)

Outside North America and Australia
Blackwell Science, Ltd.
c/o Marston Book Services, Ltd.
P.O. Box 87
Oxford OX2 0DT
England
(Telephone orders: 44-865-791155)

Printed and bound by The Sheridan Press, Hanover, PA, on acid-free recycled stock.

© 1995 by Blackwell Science, Inc. and The Society for Conservation Biology

Printed in the United States of America

95 96 97 98 5 4 3 2 1

All rights reserved. No part of this book may be reproduced in any form or by any electronic or mechanical means, including information storage and retrieval systems, without permission in writing from the publisher, except by a reviewer who may quote brief passages in a review.

Library of Congress Cataloging in Publication Data

The social dimension: ethics, policy, law, management, development, economics, education/edited by David Ehrenfeld. p.   cm.—
(Readings from Conservation biology)
   Includes bibliographical references.
   ISBN 0-86542-455-1
   1. Nature conservation—Social aspects. 2. Biological diversity conservation—Social aspects. 3. Nature conservation—Economic aspects. 4. Biological diversity conservation—Economic aspects. 5. Nature conservation—Moral and ethical aspects. 6. Biological diversity conservation—Moral and ethical aspects. I. Ehrenfeld, David. II. Society for Conservation Biology. III. Series.
QH75.S585 1994
333.95'16—dc20   94-46715
                                                                CIP

The Social Dimension—Ethics, Policy, Law, Management, Development, Economics, Education

Contents

1 **Values and Perceptions of Invertebrates**
STEPHEN R. KELLERT

12 **Forest-Dwelling Native Amazonians and the Conservation of Biodiversity: Interests in Common or in Collision?**
KENT H. REDFORD AND ALLYN MACLEAN STEARMAN

20 **Indigenous Peoples and Conservation**
JANIS B. ALCORN

23 **On Common Ground? Response to Alcorn**
KENT H. REDFORD AND ALLYN MACLEAN STEARMAN

25 **Indigenous Reserves and Nature Conservation in Amazonian Forests**
CARLOS A. PERES

28 **Scientists as Advocates: The Point Reyes Bird Observatory and Gill Netting in Central California**
JAMES E. SALZMAN

39 **The Sweetwater Rattlesnake Round-Up: A Case Study in Environmental Ethics**
JACK WEIR

51 **The Olympic Goat Controversy: A Perspective**
VICTOR B. SCHEFFER

55 **Response to Scheffer**
CATHY SUE ANUNSEN AND ROGER ANUNSEN

59 **Reply to the Anunsens**
VICTOR B. SCHEFFER

60 **Assessing Extinction Threats: Toward a Reevaluation of IUCN Threatened Species Categories**
GEORGINA M. MACE AND RUSSELL LANDE

70 **Extant Unless Proven Extinct? Or, Extinct Unless Proven Extant?**
JARED M. DIAMOND

73 **Extant Unless Proven Extinct: The International Legal Precedent**
F. WAYNE KING

76 **What Exactly Is an Endangered Species? An Analysis of the U.S. Endangered Species List: 1985–1991**
DAVID S. WILCOVE, MARGARET MCMILLAN, AND KEITH C. WINSTON

83 **Habitat Protection Under the Endangered Species Act**
JOHN G. SIDLE AND DAVID B. BOWMAN

86 **Six Biological Reasons Why the Endangered Species Act Doesn't Work—And What to Do About It**
DANIEL J. ROHLF

96 **Response to: "Six Biological Reasons Why the Endangered Species Act Doesn't Work and What to Do About It"**
MICHAEL O'CONNELL

100 **Response to O'Connell**
DANIEL J. ROHLF

102 **Rejoinder to Rohlf and O'Connell: Biodiversity as a Regulatory Criterion**
JOSEPH P. DUDLEY

105 **On Reauthorization of the Endangered Species Act**
DENNIS MURPHY, DAVID WILCOVE, REED NOSS, JOHN HARTE, CARL SAFINA, JANE LUBCHENCO, TERRY ROOT, VICTOR SHER, LES KAUFMAN, MICHAEL BEAN, AND STUART PIMM

108 **Biological Integrity and the Goal of Environmental Legislation: Lessons for Conservation Biology**
JAMES R. KARR

115 **The Question of Management**
DAVID W. ORR

117 **Techno-Arrogance and Halfway Technologies: Salmon Hatcheries on the Pactific Coast of North America**
GARY K. MEFFE

122 **"Costs" and Short-Term Survivorship of Hornless Black Rhinos**
JOEL BERGER, CAROL CUNNINGHAM, A. ARCHIE GAWUSEB, AND MALAN LINDEQUE

127 **Lion–Human Conflict in the Gir Forest, India**
VASANT K. SABERWAL, JAMES P. GIBBS, RAVI CHELLAM, AND A. J. T. JOHNSINGH

134 **Ecotourism: New Partners, New Relationships**
JOAN GIANNECCHINI

138 **The Limits to Caring: Sustainable Living and the Loss of Biodiversity**
JOHN G. ROBINSON

147 **Limits to Caring: A Response**
MARTIN HOLDGATE AND DAVID A. MUNRO

150 **"Believing What You Know Ain't So": Response to Holdgate and Munro**
JOHN G. ROBINSON

152 **Natural Capital and Sustainable Development**
ROBERT COSTANZA AND HERMAN E. DALY

162 **Water, Endangered Fishes, and Development Perspectives in Arid Lands of Mexico**
SALVADOR CONTRERAS-B. AND M. LOURDES LOZANO-V.

171 **Brazilian Crocodilian Tears**
R. EUGENE TURNER

172 **Livestock Breeds and Their Conservation: A Global Overview**
STEPHEN J. G. HALL AND JOHN RUANE

183 **Can Extractive Reserves Save the Rain Forests? An Ecological and Socioeconomic Comparison of Nontimber Forest Product Extraction Systems in Petén, Guatemala, and West Kalimantan, Indonesia**
NICK SALAFSKY, BARBARA L. DUGELBY, AND JOHN W. TERBORGH

197 **Assessing the Economic Value of Traditional Medicines from Tropical Rain Forests**
MICHAEL J. BALICK AND ROBERT MENDELSOHN

200 **The Business of Conservation**
DAVID EHRENFELD

203 **The Role of Foreign Debt in Deforestation in Latin America**
RAYMOND E. GULLISON AND ELIZABETH C. LOSOS

211 **Financial Considerations of Reserve Design in Countries with High Primate Diversity**
JOSÉ MARCIO AYRES, RICHARD E. BODMER, AND RUSSELL A. MITTERMEIER

217 **Cheese, Tourists, and Red Pandas in the Nepal Himalayas**
PRALAD B. YONZON AND MALCOM L. HUNTER, JR.

224 **Biological Diversity, Agriculture, and the Liberal Arts**
DAVID W. ORR

227 **Professionalism and the Human Prospect**
DAVID W. ORR

230 **Architecture as Pedagogy**
DAVID W. ORR

233 **The Virtue of Conservation Education**
DAVID W. ORR

Preface

Conservation must rest upon a foundation of good science, but science cannot by itself bring about the protection of species and habitats. This collection of papers from *Conservation Biology* demonstrates the critical importance of the social context in which all conservation challenges occur and are resolved. From its first issue, in 1987, *Conservation Biology* has published pioneering articles relating conservation science to its sister subjects of ethics, policy, management, law, development, economics, and education. For this volume, I have selected 43 of the best of these articles from the 30 issues of *Conservation Biology* published at the time of this writing. Some of these papers have already received much publicity, having been reported in major North American newspapers. But all are readable and stimulating, and all transcend the narrow details of their particular subject matter.

Rarely is there only one viewpoint in conservation—conservation is anything but monolithic. For this reason, I have included a number of controversies (on the conservation role of native peoples, on the ethics of removing mountain goats from a national park, on the Endangered Species Act, on sustainable development, and on profit-making conservation), each of which involves an article, its rebuttal, and perhaps a counter-rebuttal. I believe that such controversy is a sign that conservation biology is a growing and vital discipline; the tolerance of disagreement is one of its finest characteristics.

Together, the papers in this book were chosen to demonstrate in an enjoyable and memorable way the necessity and the fascination of bringing science and society together in a productive dialogue. I hope and expect that this collection will remain of lasting relevance even as the particular topics change with the ceaseless flow of events.

The volume begins with a group of papers about attitudes and ethics. Stephen Kellert looks at public attitudes towards invertebrates, among the least valued of all animals. What is it about invertebrates that makes them so generally—and in most cases inappropriately—hated, and is there anything that can change this attitude? Kent Redford and Allyn Stearman consider the conservation goals of groups representing indigenous Amazonian peoples, and find that they are not always the same as those of conservation biologists. Redford's and Stearman's views are discussed and sometimes challenged by Janis Alcorn, and the debate is reevaluated by Carlos Peres. James Salzman looks at a perennial problem for conservation biologists: to what extent should scientists be environmental advocates—can advocacy be compatible with scientific responsibility? Capitalizing on the widespread fear and loathing of rattlesnakes, the Sweetwater (Texas) Rattlesnake Round-Up has become an economically significant local festival, but nobody knows how many rattlesnakes are left. Philosopher Jack Weir stirs a complex brew of ethics, sociology, economics, biology, and conservation, and comes up with some recommendations. Mountain goats were introduced into what is now Olympic National Park in the 1920s and are destroying native vegetation and disrupting the regional ecosystem. They must be removed. Alternatively, the goats may be native to the Olympic Peninsula, and their adverse impacts on ecosystems have not been proven. Wildlife biologist Victor Scheffer and animal rights advocates Cathy and Roger Anunsen exchange very different opinions about the goats of the Olympic forests.

The second group of papers takes up the question of defining what exactly "threatened" and "endangered" mean—an interesting mix of biology and policy. Georgina Mace and Russell Lande, in a landmark analysis, reassess the IUCN's categories of threat, and find them inadequate. Jared Diamond takes up the surprisingly tricky question of when do we declare a species extinct; and Wayne King puts this issue into an international legal context. And in a widely publicized paper, Wilcove et al. use hard data to evaluate two major public criticisms of the U.S. Endangered Species Act: that it is overly concerned with subspecies and populations, and that its criteria for protection are inconsistently applied. They find neither criticism justified.

Conservation law is a rapidly growing subject, and much of it is centered around the Endangered Species Act. John Sidle and David Bowman begin our examination of the ESA by looking at the extent to which the original act, as amended, can be construed to confer protection on habitats, which is increasingly seen as the key to conservation of species. Daniel Rohlf finds that the Endangered Species Act hasn't worked very well because it is poorly framed with respect to biological realities such as ecosystem function, patch dynamics, and the probabilistic nature of threats to the persistence of species. Michael O'Connell responds that the legislation is not at fault; it is the implementation that is ineffective. Joseph Dudley, supporting O'Connell, questions whether a focus on biodiversity rather than single species can be translated into a regulatory context. Murphy et al. (the "al." comprises a distinguished group of conservation biologists and lawyers) present a list of twelve widely accepted reasons why a strong Endangered Species Act is necessary, and suggest modifications of the existing act that will make it more effective. Our discussion of conservation law concludes with an overview, by James Karr, of the history of the difficult relationship among environmental legislation, implementation, and ecology—with some suggestions for changes.

David Orr opens our consideration of conservation management with what might be called the First Law of Management: know the difference between what can be managed and what cannot. Gary Meffe, writing about Pacific salmon hatcheries, brings up an important problem: are we doing more damage than good with management technologies that are incompletely worked out and poorly related to the biology of the threatened organisms? In this vein, is the dehorning of black rhinos a clever conservation strategy, or is it an African version of the Pacific salmon hatcheries? Berger et al. sum up the evidence. One of the most difficult tasks confronting conservation managers is to protect the welfare of both humans and an endangered species when that species is a dangerous predator living in close proximity to human settlements. Saberwal et al. study the case of the Asiatic lion in India's Gir Forest.

The largest group of papers falls in the category of economics and development. Although economists and ecologists generally find themselves on opposite sides of the fence, this is not always the case, as our selections show. Joan Giannecchini leads off with an informative glimpse of the promises and pitfalls of a young but powerful new industry, ecotourism. John Robinson follows with a scalpel-sharp examination of establishment conservation's new love—sustainable development. Robinson finds a multitude of warts, but Martin Holdgate and David Munro defend the relationship. Economists can be ecologically informed, as Robert Costanza and Herman Daly show in their discussion of "natural capital." Salvador Contreras-B. and Lourdes Lozano-V. use the example of the arid lands of northern Mexico to show that development is frequently far from sustainable. Eugene Turner proves that conservation biologists can be poets as he sings of the ecological dangers of the farming (for leather) of Nile crocodiles along the Brazilian Amazon. In one of the most thought-provoking papers in this collection, Stephen Hall and John Ruane uncover a direct relationship between the proportion of extinct breeds of domestic livestock and the per capita GNP among the countries of Europe. Salafsky et al. compare rain forests in Guatemala and Indonesia to see whether extractive reserves can help to protect them. As usual in the real world of ecological, political, and socio-economic constraints, the answer is not a simple one. Traditional medicines are a potentially significant variety of non-timber forest product, which can be harvested without destroying the ecosystem according to Michael Balick and Robert Mendelsohn. In my own comment, I agree but caution that it is unlikely that such harvesting schemes can be sustainable if they must supply the world market with goods. Raymond Gullison and Elizabeth Losos examine the role of foreign debt in Latin American deforestation and find it overstated. Ayres et al. look at the design of nature reserves for primates in developing countries and conclude that in some cases economic considerations can be factored in without compromising conservation objectives—in others, not. The last paper of this group, by Pralad Yonzon and Malcolm Hunter, Jr., considers how the economic parameters of cheese production and tourism in the Nepalese Himalayas might be safely juggled to help the few surviving red pandas of the district, rather than hurt them as is now the case.

Conservation education is the most seriously deficient part of the educational system of the United States and most of the developed world. In the four, brief essays that conclude this collection, David Orr provides suggestions for change that are at the same time visionary and practical.

This is the second of six volumes currently planned for this series. The first was *To Preserve Biodiversity—An Overview,* a general survey of the field of conservation biology. The others will be: *Wildlife and Forests; Plant Conservation; The Landscape Perspective;* and *Genes, Populations, and Species.* Overlap of papers between volumes is small—approximately 10 to 20 percent. Each volume is intended to stand by itself with a minimum of duplication. I have made every effort to select papers that will not age rapidly; nevertheless, the contents of each volume will be reviewed and revised at regular intervals.

I thank Jim Krosschell, Jane Humphreys, and Kathleen Grimes of Blackwell Science, Inc., for their assistance and for their effort to produce attractive, durable volumes printed on acid-free, recycled paper and published at an affordable cost. I am also grateful to the officers of the Society for Conservation Biology, especially Peter Brussard, Steven Humphrey, and Reed Noss, for their support.

David Ehrenfeld

Values and Perceptions of Invertebrates

STEPHEN R. KELLERT

Yale University
School of Forestry and Environmental Studies
205 Prospect Street
New Haven, CT 06511, U.S.A.

Abstract: *In this paper I explore the value of invertebrates to human society. I initially examine various ecological, utilitarian, scientific, and cultural benefits provided by invertebrate organisms. I then explore the extent of appreciation and understanding of these values among the American public. This assessment was based on a study of residents of the state of Connecticut, including randomly selected members of the general public, farmers, conservation organization members, and scientists. The general public and farmers were found to view most invertebrates with aversion, anxiety, fear, avoidance, and ignorance. Far more positive and knowledgeable attitudes toward invertebrates and their conservation were observed among scientists and, to a lesser extent, among conservation organization members. I finally examine the motivational basis for hostile attitudes toward invertebrates, particularly arthropods, among the general public. Important factors include the possibility of an innate learning disposition, the association of many invertebrates with disease and agricultural damage, differences in ecological scale between humans and invertebrates, the multiplicity of invertebrates, the apparent lack of a sense of identity and consciousness among invertebrates, the presumption of mindlessness among invertebrates, and the radical autonomy of invertebrates from human control.*

Valor y percepción de los invertebrados

Resumen: *Este trabajo explora los valores de los invertebrados para la sociedad. Inicialmente, examino varios beneficios ecológicos, utilitarios, científicos y culturales provistos por organismos invertebrados. Luego exploro el grado de apreciación y entendimiento de estos valores entre el público Americano. Esta evaluación fue basado en un estudio de residentes del Estado de Connecticut, incluyendo miembros del público general, granjeros, miembros de organizaciones conservacionistas y científicos elegidos al azar. Se encontró que el público general y los granjeros ven a la mayoría de los invertebrados con aversión, ansiedad, miedo, rechazo e ignorancia. Se observaron actitudes mucho más positivas y de mayor conocimiento entre los científicos y en menor grado entre los miembros de asociaciones conservacionistas. Finalmente examino las bases motivacionales para las actitudes hostiles hacia los invertebrados, particularmente antrópodos, en el público general. Importantes factores incluyen la posibilidad de una inclinación de aprendizaje innato, la asociacion de muchos invertebrados con enfermedades y daño agrícola, diferencias de escala ecológica entre humanos e invertebrados, la multiplicidad de invertebrados, la aparente falta de sentido de identidad y conciencia de los invertebrados, la presunción de negligencia entre los invertebrados, y la radical autonomía del control humano por parte de los invertebrados.*

Introduction

The world's current large-scale loss of biological diversity has been described as the silent crisis of our time (Eisner 1991), the death of birth (Rolston 1991), even the fourth horseman of a possible twenty-first-century environmental apocalypse—the three others being various forms of atmospheric degradation, resource depletion, and toxic pollutants (Wilson 1991). For the general public, the scope of this potential loss of biological diversity has been most dramatically expressed in the projected extinction of hundreds of thousands of species over the next few decades, based on current rates of habitat destruction, particularly in the moist tropical forests.

One often-unappreciated aspect of this potential mass extinction has been its concentration among invertebrate organisms. This faunal group represents more than

Paper submitted May 4, 1993; revised manuscript accepted July 22, 1993.

90% of the planet's estimated 10 million–plus animal species, mainly arthropods (Erwin 1983; Wilson 1992). Despite the catastrophic loss of so much invertebrate life, the general public seems largely unaware of its possible impact on human well-being. Although various publications (Kellert 1985; Prescott-Allen 1986; Wilson 1992) have cited compelling evidence of the diverse benefits of invertebrate organisms, these values remain largely unrecognized by the general public (Ehrenfeld 1976; Tangley 1984; Kellert 1993a). This paper will consider three important aspects of this subject. First, I will review some of the ecological, utilitarian, scientific, and cultural values of invertebrate animals, as set forth in the scholarly literature. Second, data will be presented from a study of the general public regarding commonly held perceptions of invertebrates and their conservation, as well as knowledge of invertebrate life. I will then explore the possible motivational basis for negative sentiments toward—principally—insects and spiders, and the implications of this information for developing a strategy of invertebrate conservation.

The Value of Invertebrates

Although invertebrate life comprises roughly 90% of all animal species, only an estimated one in 30 have been scientifically described (Wilson 1992). Not surprisingly, all but a handful of invertebrates are known to the general public (Pimentel 1975). Even invertebrate species officially listed as endangered or threatened (Perkins 1980; U.S. Fish and Wildlife Service 1992) constitute for most people a "rogue's gallery" of esoteric popular names such as the Oklawaha sponge, Holsinger's groundwater planarian, the Nickjack Cave isopod, the Kauai cave wolf spider, Keys scaly cricket, the American burying beetle, and the North Platte montane butterfly, to cite but a few.

The scholarly literature has documented considerable benefits humans derive from invertebrate organisms. Important *ecological* invertebrate values have been extensively described. Major categories of ecological benefits include ecosystem stabilization, energy and nutrient transfer, maintenance of trophic structures, plant pollination, plant protection, and the provision of major habitats for other organisms, among others.

Because of their sheer numbers, invertebrates perform essential roles in various ecological interactions, including herbivory, predation, parasitism, mutualism, and competition, to mention a few. While relationships are often unclear, it appears that these interactions generally enhance the stability of most ecosystems, and this resilience is generally weakened when large-scale reductions occur in the complexity, diversity, and redundancy of species richness and interactions (May, 1973; Goodman 1975; Orians 1975; Bormann & Likens 1979; Cairns 1980; Usher 1986). Various adverse ecological effects have often been associated with extensive invertebrate-species extirpations resulting from large-scale anthropogenic activities such as extensive forest clear-cutting or broad-based monocultural agriculture (Connell 1978; Myers 1979; Coe 1987; Pimentel 1991).

Some species play particularly important rates in the maintenance of biotic communities; they are sometimes referred to as "keystone" species. In moist tropical forests, high species diversity is often explained by the existence of many parallel, host-restricted plant-herbivore relationships and food-webs (Gilbert & Raven 1975; Bentley 1977). The perpetuation of these subsystems frequently depends on a few critical species performing essential roles such as pollination or seed dispersal (Dodson 1975).

The role of invertebrates in energy and nutrient transfers has also been widely documented (Krebs 1978). In most aquatic ecosystems, energy and nutrients assimilated by plants are often transferred to other consumer levels by invertebrates (for example, krill act as the link between phytoplankton and many marine mammal species; Paine 1974, 1980). In terrestrial ecosystems, insects and other invertebrates are often energy and nutrient sources for many species of vertebrates. Ants and earthworms frequently play critical roles in the decomposition of dead organic material (Crossley 1976; Edwards & Lofty 1977; Holldobler & Wilson 1990).

A constant stock of edible organisms is essential to support the next trophic level in any ecosystem. The bulk of this consumer-level biomass consists largely of invertebrate organisms (Gilbert 1980). In the United States, for example, the biomass of earthworms and arthropods is estimated at 1000 kg/ha, while the comparative biomass of human beings is just 18 kg/ha, and that of all other wild terrestrial vertebrates at about the same level (Pimentel et al. 1980).

Pollination and seed dispersal are two functions invertebrates perform that are essential for plant reproduction. Bees, wasps, butterflies, and flies comprise invertebrate groups that extensively pollinate plants (Bales 1985). In North America, some 5000 bee species are said to pollinate plants, while about 20,000 other species appear to depend on cross-pollination performed mainly by insects (Pimentel et al. 1980).

Parasitic and predatory insects often help prevent herbivorous insects and some vertebrate organisms from reaching "outbreak" levels in natural ecosystems (Anders et al. 1976; Bennett et al. 1976). As a consequence, in some monocultural agricultural situations, reduced diversity and abundance of predatory and parasitic invertebrates has resulted in herbivorous insects inflicting major crop damage (Pimentel 1961; Perkins 1982).

Certain invertebrate species provide habitats for many other organisms. Coral reefs are perhaps the most dramatic example, providing a wide range of niches for a diversity of plants and perhaps one-third of all fish species (Goreau et al. 1979).

A wide range of *utilitarian* benefits have been associated with invertebrate organisms, many of which are derivatives of the ecological functions described. These practical benefits include pest and weed control, nutrient circulation and soil quality, waste decomposition, pollination and seed dispersal, human food, industrial and medicinal products, monitoring of environmental quality, various fashion and decorative products, and others (Lindberg 1989).

Naturally-occurring invertebrates have helped control agricultural pest species, resulting in billions of dollars saved annually (Pimentel 1991). Introduced invertebrate organisms have proven successful in limiting agricultural pests (Laing & Hamal 1976; Hussey 1985). In addition, chemicals isolated from invertebrates have been used to repel pests and to attract potential problem animals into traps. Invertebrates have been employed to transmit diseases to pest populations (DeBach 1974).

Soil invertebrates, especially earthworms and ants, often benefit people by accelerating decomposition and increasing nutrient availability in commercial agricultural and forested ecosystems. These species can significantly assist in disintegrating fibrous tissues, decomposing sugars and cellulose, mixing organic materials with the soil, and increasing soil porosity, aeration, and drainage.

Modern societies annually produce huge amounts of organic waste. In the United States alone, the human population generates some 130 million tons of excreta annually, while livestock produces an additional 12 billion tons of manure (Pimentel 1975). Of this waste, 99% is thought to be decomposed by invertebrates (Pimentel et al. 1980), leaving one to wonder, if not for our "little friends," what humans would be "up to their eyeballs in."

The ecological importance of invertebrates in plant pollination has been noted and, not surprisingly, this capability has proven useful in many commercial applications (Levin 1983). In the United States, some 90 agricultural crops are cross-pollinated by insects (Pimentel et al. 1980).

Invertebrates are also part of the human diet. Marine invertebrates such as shrimps, clams, crayfish, oysters, lobsters, scallops, and squids are extensively used as human food, with the total American fish catch only slightly exceeding that of marine invertebrates in dollar value (Anonymous 1987; Lindberg 1989). Many non-Western countries utilize other invertebrates for food, such as locusts among Arab populations, or ants, termites, grasshoppers, and beetle grubs among some African societies. Honey is among the most important food by-products obtained from invertebrates. In the United States, more than four million bee colonies produce over 90 thousand tons of honey annually, and total world honey production is estimated at more than 800,000 tons (Anonymous 1982, 1983; Free 1982).

Another important commercial product obtained from invertebrates, dating back to prehistoric times, is silk (Anonymous 1987). Current annual production of silk is estimated at more than 30 million kilograms (Lindberg 1989). Pearls are also an important commercial use of invertebrates with a long historic tradition. In New Guinea, many villagers are currently engaged in the innovative commercialization of lacewing butterflies (Pyle 1976, 1981). Taiwan exports hundreds of millions of dollars of wild butterfly specimens annually, while the sale of various mollusc shells is a worldwide industry (Wells 1987).

Invertebrates are increasingly employed as indicators of environmental quality. A classic historic example is the melanistic change in genetic structure exhibited by several moth species in nineteenth-century England due to air pollution (Bishop & Cook 1981). More recently, certain amphipods and bivalues have been used to assess and monitor water pollution and heavy metal contamination.

Plant species have historically provided most of the natural pharmaceuticals for fighting disease. The actual and potential medicinal benefit of invertebrate species has been indicated by recent studies (Smith 1973; Eisner 1991). Some 500 marine invertebrates, for example, have been found to possess anticancer properties, and numerous heart, bone, urogenital, and other diseases have been identified as amenable to the biochemical properties of diverse invertebrates (Myers 1979).

The *scientific* value of invertebrates is widely evident, primarily due to the enormous variability in form and function of this faunal group. Current theory and methods in taxonomy and systematics have largely been derived from the study of invertebrates. In the anatomical sciences, the structural adaptations of invertebrates have been a major source of discovery and have led to many practical uses. The understanding of mutualistic relationships, such as mimicry and adaptation, has been fostered by the study of invertebrates. The experimental study of the *Drosophila* fruit fly has provided major revelations in genetics and natural selection. Invertebrates are extensively used in biological and ecological experiments due to their shorter generational times, greater numbers, and easier availability. Research on insects has increased knowledge of animal behavior, with the study of social insects contributing to the emergence of an entirely new field of study, sociobiology (Wilson 1975).

Less easy to document are the various *cultural* benefits associated with invertebrates. Certain butterfly and beetle species have been used throughout history for aesthetic and decorative purposes (Hogue 1987). Designs based on invertebrates have been employed in art, jewelry, fashion, and other decorative motifs. Hundreds of stamps display insects and other invertebrates (Southwood 1977). Modern film and photography have used invertebrates as popular subjects (Mertins 1986; Leskosky & Berenbaum 1988).

Musical composition has been influenced by invertebrates. As Frost (1959:64) noted, that "the songs of insects have inspired many is unquestionable ... Rimsky-Korsakov in his delightful 'Flight of the Bumblebee' reproduces the familiar hum of the bees ... Joseph Strauss caught an inspiration revealed in 'Dragonfly' ... In Japan, cicadas and crickets are placed in small cages, like birds, and their songs are considered agreeable." As Frost (1959:65) further reports, "We find insects mentioned frequently in the writings of poets and philosophers, and the folklore of nearly every country refers to them."

Invertebrates have been sources of spiritual inspiration. An eloquent, simple statement of this value is suggested by the ancient nursery rhyme: "Surely, wisdom is given to all living things, and the tiniest of creatures are teachers of kings"; and, in the Proverbs, it is said: "Go to the ant, you sluggard, consider her ways and be wise." Quammen (1984:25) noted that, in ancient Greece and Rome, "This link with the spiritual realm was applied to both groups within the Lepidoptera, moths as well as butterflies. Both ... were delicate enough to suggest a pure being, freed of the carnal envelope. Both were known to perform a magical metamorphosis."

Perceptions of Invertebrates

To what extent does the general public recognize, appreciate, or understand these diverse values provided by organisms who, Wilson suggests (1987), constitute "the little guys who run the world?" How do most people perceive and value invertebrates? What is their current level of knowledge of these organisms? Do they support the conservation and protection of invertebrates?

A limited attempt at answering these questions was provided by a study of the general public, farmers, conservation organization members, and university scientists. The primary methodology for this research was a closed-ended survey of 214 Connecticut residents. The survey focused on basic values regarding invertebrates, attitudes toward invertebrate-species conservation, assumptions regarding invertebrate sentience and intelligence, preference for types of invertebrate species, knowledge of invertebrates, behavioral contacts with invertebrates, and the demographic characteristics of respondents. The survey consisted of nearly 200 questions and was pretested prior to its final administration.

The general public sample consisted of 145 randomly selected Connecticut residents residing in the city of New Haven, the New Haven suburbs of Hamden and Bethany, and the rural Connecticut towns of Killingsworth, Union, and Norfolk. Special samples were randomly selected, including 24 farmers, 20 conservation organization members, and 25 scientists. Farmers were randomly selected from lists provided by officials of the Connecticut agricultural extension. Conservation organization members were randomly selected from membership lists of the Connecticut Audubon Society, Connecticut chapters of the National Wildlife Federation and The Nature Conservancy, and Connecticut members of the Humane Society of the United States. Scientists were randomly selected from the faculty of Yale University and were roughly divided into two groups consisting of biotic (e.g., biological and ecological) and abiotic (physical and geological) scientists. All surveys were personally administered and required an average of 45 minutes to complete. Of the initially selected respondents, 80% completed the survey. No significant difference occurred in refusal rates among the major sample groups.

A typology of basic attitudes toward animals (Kellert 1980) was adapted to assess attitudes toward invertebrates. One-sentence definitions of basic attitudes toward invertebrates are indicated in Table 1. Question-

Table 1. Basic attitudes toward invertebrates

Aesthetic:	Primary interest in the physical attractiveness and symbolic appeal of invertebrates.
Dominionistic:	Primary interest in the mastery and control of invertebrates.
Ecologistic:	Primary concern for interrelationships among invertebrates and other species, as well as between invertebrates and natural habitats.
Humanistic:	Primary orientation of strong emotional affection for invertebrate animals.
Moralistic:	Primary concern for the right and wrong treatment of invertebrates, with strong opposition to presumed cruelty toward invertebrate animals.
Naturalistic:	Primary interest in direct outdoor recreational contact and enjoyment of invertebrates.
Negativistic:	Primary orientation a fear, dislike, or indifference toward invertebrates.
Scientific:	Primary interest in the physical attributes, taxonomic classification, and biological functioning of invertebrates.
Utilitarian:	Primary interest in the practical value of invertebrates or the subordination of invertebrates for the material benefit of humans.

naire scales consisted of three to nine questions and measured all the basic attitude types, with the exception of the dominionistic, which was partially covered by the utilitarian scale. The scales were constructed based on cluster and factor analysis results. Correlational findings indicated that the scales were relatively independent of one another. A scale of knowledge of invertebrates was developed based on responses to 40 true/false knowledge questions. Most knowledge questions focused on insects, although questions also considered knowledge of spiders, worms, jellyfish, molluscs, and crustaceans. Questions covered a variety of topics, including biological characteristics, population status, taxonomy of invertebrates, invertebrates in agriculture and gardening, and human injury and disease.

Limited space precludes presenting most of the study results. Individual question findings are restricted to comparisons among the general public, conservation organization members, scientists, and farmers. Attitude and knowledge scale results include comparisons among these groups, as well as demographic variations among the general public based on age, gender, education, and income.

The general public largely expressed feelings of aversion, dislike, or fear toward most invertebrates, particularly insects and spiders. A large majority of the general public indicated a dislike of ants, bugs, beetles, ticks, cockroaches, and crabs; an aversion to insects in the home; a fear of stinging insects, spiders, and scorpions; a desire to eliminate mosquitoes, cockroaches, fleas, moths, and spiders; and a view of the octopus and cockroach as highly unattractive animals. Farmers generally expressed views similar to those of the general public. In contrast, scientists indicated far more appreciative attitudes toward invertebrates, while conservation organization members expressed attitudes more favorable than those of the general public and farmers but less sympathetic than those found among scientists.

A more positive view of invertebrates generally occurred when the taxa possessed aesthetic value (such as butterflies) or practical value (such as bees). Also, scientists and, to a lesser extent, conservation organization members indicated a greater interest in direct recreational contact with invertebrates than did the general public or farmers.

In a limited way, the survey explored views regarding the affective and cognitive capacities of invertebrates, mainly insects. The majority of all sample groups regarded invertebrates as capable of experiencing pain, but only a small minority perceived these animals as possessing the capacities for affection, conscious decision making, or future thinking. The most striking exception was the tendency of farmers to believe bees possessed the capacity for rational decision making and future action.

Most of the general public and farmers disapproved of major expenditures or economic sacrifices on behalf of protecting endangered invertebrates, mainly spiders and molluscs. Scientists and, to a lesser degree, conservation organization members were far more likely to support the conservation and protection of invertebrates. Only among conservation organization members did a majority endorse the notion of expending a great deal of "money protecting endangered spiders."

Space limitations preclude more than a cursory review of the knowledge results. A summary indication of overall knowledge is provided by the results presented in Table 2, which groups questions into a number of broad knowledge categories. The general public possessed greatest knowledge of invertebrates in relation to agriculture and gardening, basic biological characteristics, injury and disease, and—taxonomically—toward butterflies and moths (lepidoptera). They revealed the least knowledge of taxonomic differences among invertebrates, marine invertebrates, and—taxonomically—toward bees, cockroaches, grasshoppers, termites, and beetles.

A number of illustrative knowledge question results are presented in Table 3. More than 60% of respondents knew that insects lacked backbones, although relatively few thought of corals as invertebrates. Only a minority had much concept of the overall number of insect species. Greater practical knowledge of invertebrates was suggested by responses to questions regarding silk and the role of insects in fruit farming. A wide variation in taxonomic understanding of invertebrates was indicated by public understanding of types of crustaceans, although few revealed an understanding of the relation between caterpillars, earthworms, and beetles, or between snails in relation to spiders or turtles, or the notion of an octopus as a kind of fish, or of a spider as a type of insect. Limited knowledge of invertebrates was

Table 2. Mean invertebrate knowledge score for varying categories of true/false questions.*

Knowledge Category	Mean Score
Invertebrates in agriculture and gardening (7)	67
Biological characteristics of invertebrates (12)	50
Invertebrates related to human injury and disease (3)	50
Lepidoptera (6)	49
Invertebrates generally regarded as unattractive (11)	45
Dragonflies, fleas, and aphids (6)	45
Spiders (3)	41
Population and endangered status of invertebrates (5)	39
Octopus, sponge, and coral (3)	34
Bees (3)	33
Taxonomy of invertebrates (14)	28.5
Cockroaches, grasshoppers, and termites (3)	27
Beetles (5)	23

* 0–100 scoring range; number of questions in parentheses.

Table 3. Percentage of correct answers among general public on selected illustrative knowledge questions.

Question	% Correct
Most insects have backbones.	77
Coral reefs are made by plants.	61
The number of insect species is 10,000; 50,000; 100,000; 1 million and more.	62
Silk is an artificial fabric invented in the 1930s.	88
Insects visiting flowers are unnecessary in modern fruit farming.	69.5
Caterpillars are more closely related to earthworms than to beetles.	30.5
An octopus is a type of fish.	27
Snails are more closely related to turtles than to spiders.	12
Shrimp, crabs, and lobsters are all crustaceans.	83
Spiders are not insects.	23
A sponge is a large species of sea weed.	40
Insects maintain a constant body temperature similar to birds.	28
The snail darter is an endangered butterfly.	34
There are more beetles than any other type of insect.	21
All insects are arthropods.	14
A cockroach is a kind of beetle.	11
Coleoptera is a group of insects that includes ants.	14

Table 4. Scale scores of knowledge of invertebrates.

Main Groups	Mean Scores
General Public	0.53
Environmental Group Members	0.70
Farmers	0.61
Scientists	0.90
$F = 57.8; p = <0.0001$	
Age	
18–35 years	0.53
36–45 years	0.56
46–65 years	0.55
66+ years	0.47
$F = 2.2; p = 0.09$	
Gender	
Male	0.55
Female	0.51
$F = 2.4; p = 0.12$	
Education	
<High School	0.46
College	0.56
$F = 17.6; p = <0.001$	
Income	
<$20,000 annually	0.53
$20,000–$35,000	0.52
>$35,000	0.57
$F = 1.5; p = 0.23$	

further suggested by responses to questions concerning the body temperature of insects, the overall number of beetle species, the notion of insects as arthropods, the relation of cockroaches and beetles or of coleoptera and ants, and the suggestion of the snail darter as an endangered butterfly.

Individual knowledge questions were used to create a scale of knowledge of invertebrates. As Table 4 reveals, scientists had significantly higher invertebrate knowledge scores than did any other group, particularly in contrast to the general public. While the knowledge scores of conservation organization members were significantly lower than those of scientists, they were substantially higher than those found among farmers or the general public. Among demographic groups, the college-educated had significantly higher scores than did persons with less than a high school education. At a less rigorous 0.10 confidence level, elderly respondents had significantly lower knowledge scores than those found among other age groups.

As previously noted, seven scales of basic attitudes toward invertebrates were developed. The comparative frequency of the scales among the general public is indicated in Figure 1. These results suggest that feelings of fear, dislike, and indifference toward invertebrates were the most frequently encountered views of invertebrates among the general public. In contrast, the most infrequently encountered attitudes toward invertebrates included affection, ethical concern, or scientific curiosity.

Relatively more positive attitudes toward invertebrates were encountered when the animals possessed aesthetic value (butterflies), utilitarian value (shrimp), ecological value (bees), or outdoor recreational value (mollusc shells).

Attitude scale scores among the major sample groups are revealed in Table 5, with highly significant differences occurring on every scale. Scientists expressed by far the greatest overall appreciation, concern, and inter-

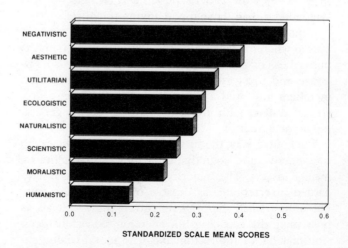

Figure 1. Basic attitudes of general public toward invertebrates. See text for explanation of attitude types.

Table 5. Scale scores of attitudes toward invertebrates among general public, members, farmers, and scientists.

	General Public	Members	Farmers	Scientists	F	p
Aesthetic	3.2	4.1	3.3	5.5	10.3	0.0001
Ecologistic	6.0	7.6	6.5	10.6	13.3	0.0001
Humanistic	1.1	1.5	0.5	3.6	23.3	0.0001
Moralistic	2.6	4.5	2.3	5.3	11.6	0.0001
Naturalistic	2.9	4.3	2.2	5.5	13.1	0.0001
Negativistic	8.9	5.2	8.1	2.9	17.4	0.0001
Scientistic	1.5	2.6	1.3	3.5	14.9	0.0001
Utilitarian	3.1	0.9	4.6	0.2	19.1	0.0001

est in invertebrates, especially compared with farmers and the general public. Farmers generally revealed more pragmatic, antagonistic, and emotionally detached attitudes, tending largely to view invertebrates as either threats or sources of material gain. Conservation organization members expressed relatively appreciative and protectionist attitudes toward invertebrates, although to a lesser degree than did scientists.

Basic attitude scale scores were compared among major demographic groups of the general public. Educational variations were highly significant, with the college-educated expressing significantly greater ecological and protectionist concern, aesthetic appreciation, and scientific curiosity toward invertebrates than did respondents with a high school education or less. Less educated respondents also indicated greater fear and utilitarian attitudes toward invertebrates.

Age-group differences were also substantial. Elderly respondents expressed significantly greater fear and exploitative attitudes toward invertebrates, as well as less ecological and ethical concern, than did other age groups. Among the sexes, female respondents were significantly more likely to express fear and less protectionist attitudes toward invertebrates than were males. Finally, scientists were divided into two groups, based on their classification as either biotic (biological or ecological) or abiotic (physical or geological) scientists. Biotic scientists were significantly more knowledgeable, more aesthetically and emotionally interested, and less fearful of invertebrates than were abiotic scientists.

Discussion

Despite the initial description of the considerable benefits derived from invertebrates, we learned that—at least among the general public—most invertebrates are viewed with attitudes of fear, antipathy, and aversion. These sentiments likely constitute a formidable challenge to developing an effective strategy of invertebrate species conservation.

It will be important to engender greater appreciation and concern among the general public toward the spineless kingdom. In order to accomplish this, it will be necessary to achieve a better understanding of the basis for hostile attitudes toward invertebrates, especially arthropods. The remainder of the paper will identify some potentially critical factors associated with public anxiety, aversion, and antipathy toward invertebrate organisms. Far more careful study and analysis will be necessary, however, before firm conclusions can be drawn.

A limited review of the scholarly literature largely corroborates the finding of generally negative public attitudes toward invertebrates, especially insects and spiders. Hardy reported (1988:64) that "public sentiment rarely favors insects." He suggested a bell-shaped curve of human attitudes toward arthropods, with a small minority of people taking pleasure in insects and spiders, a somewhat larger minority expressing sentiments of indifference, with the great majority harboring feelings of apprehension, fear, and outright phobia. He also reported higher levels of anxiety toward invertebrates among children and females (Marks 1969; Agras 1985).

As noted, few respondents in this study viewed invertebrates as capable of pain, rational consciousness, or planned action. Despite these sentiments, Lockwood (1987) offered considerable evidence—based on criteria of communication, problem-solving, and learning—to suggest that some insects may be viewed as capable of exhibiting consciousness, although many entomologists differ with this conclusion (Wigglesworth 1980). Lockwood admits (1987:83) that, for most people, "there seems to be an overall aversion to recognizing insects as organisms deserving of moral consideration."

Various factors have been offered to explain the consistent pattern of human aversion, avoidance, and disdain for most insects and spiders. One speculative assertion hypothesizes an innate fear of potentially dangerous insects, generalized to include other invertebrates (Adams 1981; Bennett-Levy & Martean 1984). Fear and avoidance of many arthropods has been suggested as conferring evolutionary advantages resulting in their statistically greater frequency among the general population. Spontaneous and often unprovoked expression of this fear may represent a biologically "prepared" learning tendency, a kind of conditioned response occurring with minimal stimulus. The psychologist Öhman (1986) has suggested that "animal fear originates in a predatory defense system whose function is to allow animals to avoid and escape predators ... It is appropriate to speak about biologically prepared learning ... likely to require only minimal input ... to result in very persistent responses that are not easily extinguished." Limited corroboration of this innate aversion toward arthropods has been indicated in studies where the occurrence of invertebrates, especially spiders, provoked withdrawal responses among vertebrate neonates in the absence of overt and obvious threats (Ulrich 1993).

A more conventional motivation for avoiding invertebrates is the presumed connection between many arthropods and human disease (Cheng 1973; Bay et al. 1976; McNeill 1976). Contemporary standards of hygiene, the desire for sterile home environments, and modern theories of disease transmission may exacerbate these fears, particularly in highly urbanized, industrial societies (Busvine 1966; Cloudsley-Thompson 1976). An entire specialty in modern medicine has been developed around the role of invertebrates in human disease transmission and epidemics (Horsfall 1962; Leclerq 1969; Frankie et al. 1981).

An additional practical basis for hostility to invertebrates, described earlier, is their frequent association with crop depredations and other forms of agricultural damage (Pimentel 1975, 1991). While only a small percentage of invertebrates are associated with this loss, the presumption may be one of "guilt by association."

Another explanation for human avoidance of invertebrates is the possible alienation from creatures so morphologically and behaviorally unlike our own species. Invertebrates represent radically different survival strategies, most dramatically expressed in very different ecological, spatial, temporal, and morphological scales. Their very multiplicity may constitute a major alienating element, as Samways suggests (1990:7): "With insects ... there are so many of them. Also ... they are lilliputians in our much larger world."

This disparity in numbers and size has been emphasized by Hillman (1991), who suggested that it can profoundly threaten the human species' cherished notions of individuality and independence. The idea, for example, of a beehive including 50,000 organisms, or an ant colony consisting of a half million ants, or an acre of soil containing 65 million insects, or the beetle order numbering more than one million species, may represent a fundamental challenge to human assumptions of personal identity and individuality. As Hillman remarked (1991), "Imagining insects numerically threatens the individualized fantasy of a unique and unitary human being. Their very numbers indicate insignificance of us as individuals."

Speculating on "why we hate bugs" (1991), Hillman offers additional psychological insights for human aversion and fear of arthropods. Reviewing a long history of prejudicial attitudes toward arthropods, Hillman (1991) ruefully remarked, "what we call the progress of Western civilization from the ant's eye level is but the forward stride of the great exterminator."

He suggests that a basis for human dislike of many invertebrates is their perceived "monstrosity" from an anthropocentric perspective. It is not uncommon, he points out, to associate insects and spiders with metaphors of madness, many common terms for insanity employing common arthropod names. As Hillman notes (1991), "Bug-eyed, spidery, worm, roach, blood sucker, louse, going buggy, locked-up in the bughouse—these are all terms of contempt supposedly characterizing inhuman traits ... To become an insect is to become a mindless creature without the warm blood of feeling."

A further explanation offered by Hillman originates in the radical "autonomy" of most invertebrates from human will and control. A particularly disturbing aspect is the willingness of many arthropods to invade human space and habitations in unexpected ways.

Finally, Hillman suggests that the "mystery" associated with most invertebrates may be a disturbing element. While such "other-worldliness" can provoke curiosity and even wonder in some people, the more typical human response to the unknown is that of fear and disdain. For most people, invertebrates remain largely alien and unfathomable.

The conservation of invertebrates will necessitate a far greater understanding of why we so consistently react with hostility and antagonism toward these organisms, particularly insects and spiders. To reverse the current trend toward the increasing impoverishment of the planet's biological diversity, we will need to acquire a more appreciative attitude toward the biological matrix of so-called "lower" life forms represented by the invertebrates.

It is unlikely that very many people will develop an affection or affinity for these animals, but it may be that a more compelling depiction of the extraordinary contributions to human welfare and survival made by invertebrates will do much to dampen prevailing negative attitudes. An encouraging finding of the research reported here, and elsewhere (Kellert 1993b), is the association of greater education with increased appreciation, concern, and knowledge of biological diversity and its conservation. This result suggests that more ambitious educational programs may greatly assist in the enhanced recognition of the positive values of invertebrates.

Although this will be less simple and straightforward, we will need to cultivate our sense of communality with all living organisms. While this must start with creatures we can empathize with readily—the larger, charismatic vertebrates—eventually we will need to extend our appreciation to the grandeur found within all living organisms. As Hillman suggested (1991), "we must start [with animals] not in their splendor—the horned stag, the yellow lion and the great bear, or even old faithful 'spot'—but with those we fear the worse—the bugs."

Acknowledgments

The author would particularly like to thank Dr. Timo Törmala for his excellent assistance in detailing the diverse values of invertebrates, and both Dr. Törmala and Professor Joyce Berry of Colorado State University for their invaluable assistance with research. Additional thanks are due Professor Holmes Rolston, III, and an anonymous reviewer for their excellent editorial advice.

Literature Cited

Adams, K. A. 1981. Arachnophobia: Love American style. Journal of Psychoanalytic Anthropology **4**:157–197.

Agras, 1985. Panic. W. H. Freeman, New York.

Anders, L. A., et al. 1976. Biological control of weeds. Pages 481–500 in C. B. Huggaker and P. S. Messenger, editors. Theory and practice of biological control. Academic Press, New York.

Anonymous. 1987. More than a numbers game. Bulletin of the International Union for the Conservation of Nature and Natural Resources **18**:7–9.

Anonymous. 1983. Wild bees do better at crop pollination. International Wildlife **13**:28.

Anonymous. 1982. World honey crop reports. Report No. 15—Bee World **21**:105–109.

Bales, G. L. 1985. The honeybee's environmental role. Annual Bee Journal **125**:234–235.

Bormann, F. H., and G. Likens. 1979. Pattern and process in a forested ecosystem. Springer-Verlag, New York.

Bay, E. C., et al. 1976. Biological control of medical and veterinary pests. Pages 457–480 in C. B. Huggaker and P. S. Messenger, editors. Theory and practice of biological control. Academic Press, New York.

Bennett, F. D., et al. 1976. Biological control of pests of tropical fruit and nuts. Pages 359–396 in C. B. Huggaker and P. S. Messenger, editors. Theory and practice of biological control. Academic Press, New York.

Bennett-Levy, J., and T. Martean. 1984. Fear of animals: What is prepared? British Journal of Psychology **75**:37–42.

Bentley, B. L. 1977. Extrafloral nectaries and protection by pugnacious bodyguards. Annual Review of Ecological Systems **8**:407–428.

Bishop, J. A., and L. M. Cook. 1981. Industrial melanism and the urban environment. Advanced Ecological Research **11**:373–404.

Busvine, J. R. 1966. Insects and hygiene. Methuen and Co, London, England.

Cairns, J. 1980. The recovery process in damaged ecosystems. Ann Arbor Science, Ann Arbor, Michigan.

Cheng, T. C. 1973. General parasitology. Academic Press, New York.

Cloudsley-Thompson, J. L. 1976. Insects and history. St. Martins Press, New York.

Coe, M. 1987. Unforeseen effects of control. Nature **327**:367.

Connell, J. H. 1978. Diversity in tropical rain forests and coral reefs. Science **199**:1315–1320.

Crossley, D. A. 1976. The role of terrestrial saprophagous arthropods in forest soils: Current status of concepts. Pages 49–56 in W. J. Mattson, editor. The role of arthropods in forest ecosystems. Springer-Verlag, New York.

DeBach, P. 1974. Biological control by natural enemies. Cambridge University Press, London, England.

Dodson, C. H. 1975. Coevolution of orchids and bees. Pages 91–99 in L. Gilbert and P. M. Raven, editors. Coevolution of plants and animals. University of Texas Press, Austin, Texas.

Edwards, C. A., and J. R. Lofty. 1977. Biology of earthworms. Chapman and Hall, London, England.

Ehrenfeld, D. W. 1976. The conservation of non-resources. American Scientist **64**:648–656.

Eisner, T. 1991. Chemical prospecting: A proposal for action. Pages 196–204 in F. H. Bormann and S. R. Kellert, editors. The broken circle. Yale University Press, New Haven, Connecticut.

Erwin, T. L. 1983. Tropical forest canopies: The last biotic frontier. Bulletin of the Entomological Society of America **29**:14–19.

Frankie, G. W., et al. 1981. A study of attitudes and practices of pest control operators toward pests and pesticides in selected urban areas of California, Texas, and New Jersey. Pages 9–32 in Proceedings of the Urban Integrated Pest Management Workshop. National Cooperative Extension, Dallas, Texas.

Free, J. B. 1982. Bees and mankind. Allen & Unwin, London, England.

Frost, S. W. 1959. Insect life and insect natural history. Harper and Row, New York.

Gilbert, L. E. 1980. Food web organization and conservation of tropical diversity. Pages 11–33 in M. F. Soule and B. A. Wilcox, editors. Conservation biology. Sirans Association, Sunderland, Massachusetts.

Gilbert, L. E., and P. M. Raven. 1975. Coevolution of plants and animals. University of Texas Press, Austin, Texas.

Goodman, D. 1975. Diversity, stability and maturity in natural ecosystems. Pages 139–150 in W. H. van Dobben and R. H. Lowe-McConnell, editors. Unifying concepts in ecology. Junk B.V. Publishers, The Hague, The Netherlands.

Goreau, T. F., et al. 1979. Corals and coral reefs. Scientific American **24**:124–137.

Hardy, T. N. 1988. Entomophobia: The case for Miss Muffet. Bulletin of the American Entomological Society **34**:64–69.

Hillman, J. 1991. Going bugs. Spring Audio, Gracie Station, New York.

Hogue, C. L. 1987. Cultural entomology. Annual Review of Entomological **32**:181–199.

Holldobler, B., and E. O. Wilson. 1990. Ants. Harvard University Press, Cambridge, Massachusetts.

Horsfall, W. R. 1962. Medical entomology: Arthropods and human disease. Ronald Press, New York.

Hussey, N. W. 1985. Biological control—a commercial evaluation. Biological Control News and Information **6**:93–99.

Kellert, S. R. 1980. Contemporary valuers of wildlife in American Society. Pages 31–60 in W. W. Shaw and I. Zube, editors. Wildlife values. Institutions Series Report. U.S. Forest Service, Ft. Collins, Colorado.

Kellert, S. R. 1985. Social and perceptual factors in endangered species management. Journal of Wildlife Management **49**:528–536.

Kellert, S. R. 1993a. The biological basis for human values of nature. Pages 42–69 in S. R. Kellert and E. O. Wilson, editor. The biophilia hypothesis. Island Press, Washington, D.C.

Kellert, S. R. 1993b. Attitudes, knowledge, and behavior toward wildlife among the industrial superpowers: United States, Japan, and Germany. Journal of Social Issues **49**:53–69.

Krebs, C. J. 1978. Ecology. Harper & Row, New York.

Laing, J. F., and J. Hamal. 1976. Biological control of insect pest weeds by imported parasites, predators and pathogens. Pages 686–744 in C. B. Huffaker and P. S. Messenger, editors. Theory and practice of biological control. Academic Press, New York.

Leclerq, M. 1969. Entomological parasitology—the relations between entomology and medical science. Pergamon Press, New York.

Leskosky, R. J., and M. R. Berenbaum. 1988. Insects in animated films. Bulletin of the Entomological Society of America **34**:55–63.

Levin, M. D. 1983. Value of bee pollination to US agriculture. Bulletin of the Entomological Society of America **29**:50–51.

Lindberg, C. 1989. The economic value of insects. Traffic Bulletin **10**:32–36.

Lockwood, J. A. 1987. The moral standing of insects and the ethics of extinction. Florida Entomology **70**:70–89.

Marks, I. M. 1969. Fears and phobias. Academic Press, New York.

May, R. M. 1973. Stability and complexity in model ecosystems. Princeton University Press, Princeton, New Jersey.

McNeill, W. H. 1976. Plagues and peoples. Anchor Press and Doubleday, Garden City, New York.

Mertins, J. W. 1986. Arthropods on the screen. Bulletin of the Entomological Society of America **32**:85–90.

Myers, N. 1979. The sinking ark. Pergamon Press, New York.

Öhman, A. 1986. Face the beast and fear the face: Animal and social fears as prototypes for evolutionary analyses of emotion. Psychophysiology **23**:123–145.

Orians, G. H. 1975. Diversity, stability and maturity in natural ecosystems. Pages 139–150 in W. H. van Dobben and R. H. Lowe-McConnell, editors. Unifying concepts in ecology. Junk B.V. Publishers, The Hague, The Netherlands.

Paine, R. T. 1980. Linkage, interaction strength and community infrastructure. Journal of Animal Ecology **49**:667–685.

Paine, R. T. 1974. Intertidal community structure: Experimental studies on the relationship between a dominant competitor and its principal predator. Oecologia **15**:93–120.

Perkins, J. H. 1982. Insects, experts and the insecticide crisis. Plenum Press, New York.

Perkins, P. D. 1980. North American insect status review. U.S. Fish and Wildlife Service, Washington, D.C.

Pimentel, D. 1991. The dimensions of the pesticide question. Pages 59–72 in F. H. Bormann and S. R. Kellert, editors. The broken circle. Yale University Press, New Haven, Connecticut.

Pimentel, D. 1980. Environmental risks associated with biological controls. Ecological Bulletin **23**:201–211.

Pimentel, D. 1975. Insects, science and society. Academic Press, New York.

Pimentel, D. 1961. Species diversity and insect population outbreaks. Bulletin of the Entomological Society of America **54**:76–86.

Pimentel, D. et al. 1980. Environmental quality and natural biota. Bioscience **30**:750–755.

Prescott-Allen, C., and R. Prescott-Allen. 1986. The primary resource. Yale University Press, New Haven, Connecticut.

Pyle, R. 1976. Conservation of Lepidoptera in the United States. Biological Conservation **9**:55–75.

Pyle, R. et al. 1981. Insect conservation. Annual Review of Entomology **26**:233–258.

Quammen, D. 1984. Avatars of the soul: Moth and butterfly, fact and idea. Outside **March**:25–27.

Rolston, H., III. 1991. Environmental ethics: Values in and duties to the natural world. Pages 73–98 in F. H. Bormann and

S. R. Kellert, editors. Unifying concepts in ecology. Junk B.V. Publishers, The Hague, The Netherlands.

Samways, M. 1990. Insect conservation ethics. Environmental Conservation **17:**7–8.

Smith, K. G. V., editor. 1973. Insects and other arthropods of medical importance. British Museum, London, England.

Southwood, T. R. E. 1977. Entomology and mankind. American Scientist **65:**30–39.

Tangley, L. 1984. Protecting the insignificant. Bioscience **34:**406–409.

Ulrich, R. S. 1993. Biophilia, biophobia, and natural landscapes. Pages 73–137 in S. R. Kellert and E. O. Wilson, editors. The biophilia hypothesis. Island Press, Washington, D.C.

U.S. Fish and Wildlife Service. 1992. Official list of endangered and threatened species. Office of Endangered Species, U.S. Department of Interior, Washington, D.C.

Usher, M. B. 1986. Wildlife conservation evaluation. Chapman and Hall, London, England.

Wells, K. 1987. One company treats a butterfly gruffly, another is friendlier. Wall Street Journal **116:**1,15.

Wells, S. M. et al. 1983. The IUCN invertebrate red data book. International Union for the Conservation of Nature and Natural Resources Gland, Switzerland.

Wigglesworth, V. B. 1980. Do insects feel pain? Journal of the Royal Entomological Society **4:**8–9.

Wilson, E. O. 1992. The diversity of life. Harvard University Press, Cambridge, Massachusetts.

Wilson, E. O. 1991. Biodiversity, prosperity, and value. Pages 3–10 in F. H. Bormann and S. R. Kellert, editors. Unifying concepts in ecology. Junk B.V. Publishers, The Hague, The Netherlands.

Wilson, E. O. 1987. The little things that run the world. Conservation Biology **1:**344–346.

Wilson, E. O. 1975. Sociobiology: The new synthesis. Harvard University Press, Cambridge, Massachusetts.

Forest-Dwelling Native Amazonians and the Conservation of Biodiversity: Interests in Common or in Collision?

KENT H. REDFORD

Program for Studies in Tropical Conservation
and Center for Latin American Studies
University of Florida
Gainesville, FL 32611, U.S.A.

ALLYN MACLEAN STEARMAN

Department of Sociology and Anthropology
University of Central Florida
Orlando, FL 32816, U.S.A.

Abstract: *Although concern for biodiversity and its conservation originated in the biological sciences, with growing international interest an increasing number of interest groups are claiming standing in discussions of the best ways to conserve biodiversity. One of these groups, formed by various indigenous peoples and their advocates, has repeatedly defended its claim to standing by stating that indigenous peoples are well equipped to conserve biodiversity. These claims have had far-reaching consequences, as millions of hectares of Amazonian forest have been deeded to indigenous groups, at least partially on the reasoning that such actions would conserve biodiversity. In this paper, we bring to the attention of the community of conservation biologists a group representing 229 native Amazonian groups comprising 1.2 million people in Peru, Bolivia, Ecuador, Brazil, and Colombia. In a document entitled "To the Community of Concerned Environmentalists," this group of indigenous peoples proposes a broad template for cooperation between conservation biologists and the indigenous peoples of the Amazon Basin. Following reiteration of the statement, we discuss the fact that these two groups define biodiversity and its conservation in different ways, with indigenous peoples focusing more on preservation of general habitat characteristics and exclusion of extensive habitat alteration. We conclude that the interests of conservation biologists may not be*

Paper submitted May 6, 1992; revised manuscript accepted November 10, 1992.

Habitantes nativos de la selva Amazónica y la conservación de la Biodiversidad: Intereses en común o en conflicto?

Resumen: *Si bien el interés por la biodiversidad y su conservación se originó en las ciencias biológicas, con el crecimiento del interés internacional, un creciente número de grupos de interés está reclamando posiciones en discusiones respecto a los mejores caminos para conservar la biodiversidad. Uno de estos grupos, formado por varias poblaciones indígenas y sus defensores, han sostenido repetidamente sus reclamos con respecto a su posición de que la población indígena se encuentra bien equipada para la conservación de la biodiversidad. Estos reclamos han tenido consecuencias de largo alcance, como los millones de hectáreas de la selva Amazónica que han sido cedidas a grupos indígenas, al menos parcialmente bajo el razonamiento de que tales acciones conservarán la biodiversidad. En este trabajo, nosotros presentamos a la comunidad de biólogos conservacionistas un grupo representativo de 229 grupos de nativos del Amazonas, los que comprenden 1.2 millones de la población en Perú, Bolivia, Ecuador, Brasil y Colombia. En un documento titulado "Para la Comunidad de Ambientalistas Interesados" este grupo de indígenas propone un amplio espectro de cooperación entre biólogos conservacionistas y poblaciones indígenas de la cuenca del Amazonas. Siguiendo este planteamiento, nosotros discutimos el hecho de que estos dos grupos definen biodiversidad y*

completely compatible with the agenda of indigenous peoples and their advocates but that by cooperating, valuable time is being bought by both sides.

conservación en forma distinta, con las poblaciones indígenas enfatizando la preservación de las características generales del hábitat y la exclusión de la alteración extensa del hábitat. Nosotros concluimos que el interés de los biólogos conservacionistas puede no ser totalmente compatible con los intereses de los grupos indígenas y sus defensores, pero a través de la cooperación ambos lados están ganando un tiempo valioso.

Biodiversity as a term and as a concept has its roots in the field of conservation biology. The first and strongest advocates for biodiversity conservation were biologists; the published experts on biodiversity are mainly biologists; the organizations that sponsor publications on biodiversity have primarily a biological focus: Wildlife Conservation International, the World Wildlife Fund, the International Union for the Conservation of Nature, and the National Academy of Sciences; and the journals that publish articles on biodiversity are written primarily by and for biologists and environmentalists.

In recent years, however, as biodiversity has moved from an academic working concept to a global concern, many people other than biologists have claimed standing in the arena of biodiversity (Redford & Sanderson 1992). In this paper, we wish to bring to the attention of the community of conservation biologists one of the groups of people with intense interest in the issues of biodiversity conservation—a group for which the issues are far from academic—the indigenous peoples inhabiting the rain forests of the Amazon basin.

An organization representing 229 native Amazonian groups comprising 1.2 million people in Perú, Bolivia, Ecuador, Brazil and Colombia, the Coordinadora de las Organizaciones Indígenas de la Cuenca Amazónica (COICA) held a meeting in Iquitos, Perú, to discuss the relationship between indigenous peoples and the environment. This meeting resulted in a document, "Two Agendas on Amazon Development," which was published in the journal Cultural Survival Quarterly (COICA 1989) and contained two parts—the first entitled "For Bilateral and Multilateral Funders" and the second "To the Community of Concerned Environmentalists." We have reproduced the second part below, because even though it was published in Cultural Survival Quarterly, it has not been read by many of the people to whom it was directed. Following the statement, we provide a commentary on the difficulties and challenges that inevitably will be raised by this statement.

The Coordinadora de las Organizaciones Indígenas de la Cuenca Amazónica Declaration

We, the Indigenous Peoples, have been an integral part of the Amazonian Biosphere for millennia. We use and care for the resources of that biosphere with respect, because it is our home, and because we know that our survival and that of our future generations depend on it. Our accumulated knowledge about the ecology of our forest home, our models for living within the Amazonian Biosphere, our reverence and respect for the tropical forest and its other inhabitants, both plant and animal, are the keys to guaranteeing the future of the Amazon Basin. A guarantee not only for our peoples, but also for all of humanity. Our experience, especially during the past 100 years, has taught us that when politicians and developers take charge of our Amazon, they are capable of destroying it because of their short-sightedness, their ignorance and their greed.

We are pleased and encouraged to see the interest and concern expressed by the environmentalist community for the future of our homeland. We are gratified by the efforts you have made in your country to educate your peoples about our homeland and the threat it now faces as well as the efforts you have made in South America to defend the Amazonian rain forests and to encourage proper management of their resources. We greatly appreciate and fully support the efforts some of you are making to lobby the U.S. Congress, the World Bank, USAID [U.S. Agency for International Development], and the InterAmerican Development Bank on behalf of the Amazonian Biosphere and its inhabitants. We recognize that through these efforts, the community of environmentalists has become an important political actor in determining the future of the Amazon Basin.

We are keenly aware that you share with us a common perception of the dangers which face our homeland. While we may differ about the methods to be used, we do share a fundamental concern for encouraging the long-term conservation and the intelligent use of the Amazonian rain forest. We have the same conservation goals.

Our concerns: We are concerned that you have left us, the Indigenous Peoples, out of your vision of the Amazonian Biosphere. The focus of concern of the environmental community has typically been the preservation of the tropical forests and its plant and animal inhabitants. You have shown little interest in its human inhabitants who are also part of that biosphere.

We are concerned about the "Debt for Nature Swaps" which put your organizations in a position of negotiating with our governments for the future of our homelands. We know of specific examples of such swaps which have shown the most brazen disregard for the rights of the indigenous inhabitants and which are resulting in the ultimate destruction of the very forests which they were meant to preserve.

We are concerned that you have left us Indigenous Peoples and our organizations out of the political process which is determining the future of our homeland. While we appreciate your efforts on our behalf, we want to make it clear that we never delegated any power of representation to the environmentalist community nor to any individual or organization within that community.

We are concerned about the violence and ecological destruction of our homeland caused by the increasing production and trafficking of cocaine, most of which is consumed here in the United States.

What We Want: We want you, the environmental community, to recognize that the most effective defense of the Amazonian biosphere is the recognition of our ownership rights over our territories and the promotion of our models for living within that biosphere.

We want you, the environmental community, to recognize that we Indigenous Peoples are an important and integral part of the Amazonian Biosphere.

We want you, the environmental community, to recognize and promote our rights as Indigenous Peoples as we have been defining those rights within the UN [United Nations] Working Group for Indigenous Peoples.

We want to represent ourselves and our interests directly in all negotiations concerning the future of our Amazonian homeland.

What We Propose: We propose that you work directly with our organizations on all your programs and campaigns which affect our homelands.

We propose that you swap "debt for indigenous stewardship" which would allow your organizations to help return areas of the Amazonian rain forest to our care and control.

We propose establishing a permanent dialogue with you to develop and implement new models for using the rain forest based on the list of alternatives presented with this document.

We propose joining hands with those members of the worldwide environmentalist community who:

- recognize our historical role as caretakers of the Amazon Basin.
- support our efforts to reclaim and defend our traditional territories.
- accept our organizations as legitimate and equal partners.

We propose reaching out to other Amazonian peoples such as the rubber tappers, the Brazil-nut gatherers, and others whose livelihood depends on the non-destructive extractive activities, many of [which] are of indigenous origin.

We propose that you consider allying yourselves with us, the Indigenous Peoples of the Amazon, in defense of our Amazonian homeland. (COICA 1989:77–78)

Indigenous Peoples as Natural Conservationists

It is important to note that the reasoning found in COICA's statement is not confined to that organization; rather it reflects a widespread belief that support for indigenous peoples is equivalent to, and even a prerequisite for, conservation of nature (Fig. 1). This is clearly revealed in a statement from a September 1989 meeting in New England that brought together indigenous representatives from around the world: "The message of the conference is that there is no option ... that tribal land rights and sovereignty must be supported in order to save both indigenous peoples *and the world's remaining natural areas.*" (van Lennep 1990:46; our emphasis) The statement continues "As indigenous peoples, we have always lived in accordance with the

Figure 1. Kayapó child with basket of hatchling parrots. Gorotire, Pará, Brazil. Photograph by Kent H. Redford.

sacred and natural laws, in balance with the natural world.... As indigenous peoples, we stand on the front lines of the struggle to defend the natural world.... We delegates to this conference recognize that indigenous people harvest resources in a sustainable manner..." (page 47).

These powerful statements are challenges to conservation biologists and environmentalists to involve indigenous peoples in all efforts to save not only the Amazon but all natural areas inhabited by native residents. The challenge is compelling, as witnessed by the fact that at the Iquitos meeting a statement entitled "First Summit between Indigenous Peoples and Environmentalists Iquitos Declaration," which contained, *inter alia,* an abbreviated version of the agenda printed above, was signed by over 20 representatives of indigenous groups and representatives of the Bank Information Center, Conservation International, Friends of the Earth U.S.A., Greenpeace, the National Wildlife Federation, the Rainforest Action Network, The Rainforest Alliance, the Sierra Club Legal Defense Fund, the World Resources Institute, the World Wildlife Fund, and Fundación Peruana para la Conservación de la Naturaleza (COICA n.d.).

But the world has moved well beyond simply signing statements concerning the compatibility of indigenous rights and biological conservation. In a dramatic move, Colombia has assigned land rights to 49,000 square miles of rain forest (half of its rain forest) to the local indigenous inhabitants, with the express reasoning that they are the people most likely to protect the biological diversity of the forest (Economist 1989). In 1992, in a similar action, the Brazilian government demarcated a 36,000-square-mile area for the Yanomami; as one journalist put it, "And because the nomadic Yanomami are de facto protectors of their habitat, the decision should also help conserve the Amazon rain forest" (Michaels 1991).

The question we wish to raise is whether the agendas of the indigenous peoples and their advocates are in fact as concordant with the interests of those primarily concerned with biological conservation as has been suggested. The statements quoted above, and many others (see Gray 1991) repeatedly claim this to be the case, but much evidence indicates that often the aims of these two groups are partially or completely in conflict. Our concern is that by failing to move beyond simple rhetoric, often for political expediency, conservationists and indigenous peoples alike will reap the whirlwind of often widely divergent expectations.

Indigenous Peoples, Conservation, and the Modern World

Let us begin by assessing what we believe to be the expectations of indigenous peoples in the Amazon. At present, native peoples are in a life and death struggle to hold onto the lands they have inhabited for hundreds if not thousands of years. In spite of a growing awareness of their plight among concerned members of the world community, indigenous peoples continue to be victimized at the hands of national governments. If some indigenous peoples have presented themselves uncritically as "natural conservationists," it is only because they recognize the power of this concept in rallying support for their struggle for land rights, particularly from important international conservation organizations such as those listed earlier. In the current world climate, the conservation community has significantly more resources, and hence political influence, than the indigenous movement does. Attempts by indigenous rights organizations to meld agendas with conservationists is an astute and understandable response.

Nonetheless, this well-meaning but perhaps overzealous attempt by numerous individuals and interest groups to generalize indigenous peoples as natural conservationists places a dangerous burden on many Native Americans (Redford 1991). Indigenous peoples are not a monolithic entity (Stearman 1992). Many, but not all, are acutely aware of their close cultural ties to the tropical forest and the necessity of conserving this resource for their continued well-being. For example, the Coconucos and Yanaconas of Colombia hold strong religious views about the preservation of an area that includes the Purace National Park. According to their belief system, this park is the dominion of the spirit being, Jucas, who is the source of all natural resources necessary for life. The Coconucos and Yanaconas work both formally and informally to protect the park, as park guards and by providing general oversight to the area (Faust 1991).

Other indigenous groups may not show such awareness or concern for the natural environment. Small groups in particular seem especially vulnerable to internal as well as external pressures to exploit their resources in ways that may not promote conservation (Stearman 1990). In most cases, these numerous small groups of indigenous peoples are but remnants of once thriving communities. Their population decline can usually be attributed to disease, genocide, and the brutalities suffered during the rubber boom of the early Twentieth Century, when entire regions of people were enslaved, tortured, and murdered (Taussig 1987). The small groups that survive are the descendents of those who escaped by retreating farther into the forest, out of harm's way. Today, as they face the conflicting demands and expectations of outsiders such as settlers, traders, miners, loggers, missionaries, government officials, and development agents, these small ethnic groups frequently find it virtually impossible to hold onto traditional cultural values, including those that may have supported a conservation ethic.

Even within a large, politically sophisticated native group with a great deal of cultural integrity and concern for preserving their natural environment, forces are at work that may run counter to conservation interests, and differences often exist between competing elements within the same ethnic group. Perhaps one of the most frequently cited cases of indigenous peoples successfully managing and conserving large, relatively undisturbed areas of their territory is that of the Kuna of Panama. Even though the Kuna do not live in the Amazon Basin, their case is often used as a model for other neotropical forest areas. Recent events indicate that the status quo among the Kuna may not be maintained. According to Mac Chapin (1991), an anthropologist who has lengthy experience with the Kuna, younger Kuna are beginning to question many of the old beliefs that inform the Kuna conservation ethic, or are simply failing to learn them. In the early 1980s, the Kuna were the first indigenous people in Latin America to set aside a large area of intact rain forest as a nature reserve. They received funding from several international sources for the study and management of this protected area. Like the Coconucos and Yanaconas of Colombia, their concern for setting aside this area derives from their belief in "spirit sanctuaries." But young Kuna are receiving western education that does not teach the "way of the Great Father," and in fact teaches them that these beliefs are primitive superstitions and largely irrelevant. The result, according to Chapin, is that the world view of younger Kuna is subtly moving away from the old traditions. Chapin expresses the fear that, "if the traditional belief system disappears from their culture, will the Kuna continue to treat the Earth and all of its creatures with the same respect? If the Kuna take on board the new ecological ethic of the Western scientific tradition, will it be able to supplant the traditional beliefs and perform anything approaching and same function?" (Chapin 1991:44). Chapin's concerns are echoed by Alcorn (1991:324): "Changes in the culture of the world's younger generation do not bode well for the continuation of traditional conservation systems in areas where the transition out of subsistence economies is, and may continue to be, incomplete."

In the Amazon, the widely publicized Yanomamo case again demonstrates the differences that can occur within an indigenous group and among those who are its advocates when conservation issues are at stake. In both Brazil and Venezuela, Yanomamo have been granted large areas of rain forest after a long and vituperative struggle for land rights. In both countries, the conservation ethic of the Yanomamo has often been cited as another justification for granting them this territory: "... indigenous groups, with the detailed knowledge necessary for conservation, have carefully managed this environment for millenia." (Arvelo-Jimenez & Cousins 1992:10). Although well-intentioned, these advocates choose not to focus on the more complex indigenous valuation of natural resources on their rain forest territory and how these resources may best be used to meet their needs in a changing society. In the words of one Yanomamo: "... ecologists, missionaries, and the government continue to see us as forest-dwelling animals, without the right to decide anything, without the right to potable water, electricity, or television, without the right to exploit the mineral wealth of Yanomamo lands; mineral wealth that any civilized person would unhesitatingly exploit if it were in their backyard.... Indians no longer want to live as if they were captive in a large zoo to be photographed by tourists" (Yonomami 1990:94, translation ours).

We must remember that indigenous peoples are increasingly becoming members of the modern world. They want to be able to choose what they will keep and what they will discard of their traditional ways of life. Virtually all are now linked to the market economy through barter or actual cash exchange. Indigenous people commonly want and have a right to health care, education, and material conveniences that improve their quality of life. While traditional knowledge of resource use may provide for these necessities in ways that conservationists find admirable and that perhaps serve as models for other peoples, the fact is that traditional ways often do not meet growing needs. Thus, indigenous peoples are forced to engage in activities that differ in type and intensity from traditional patterns of resource harvesting. As Hames (1991:192) has stated: "entrance into a market economy is undoubtedly the most potentially devastating change in aboriginal environmental relations."

To expect indigenous people to retain traditional, low-impact patterns of resource use is to deny them the right to grow and change in ways compatible with the rest of humanity. "Traditional culture-bearers should not be viewed as useful libraries of traditional information, but rather as essential participants in biodiversity conservation *during the transition to a fully integrated global capitalist economy* (Alcorn 1991:323; our emphasis)."

Biological Diversity and the Conservation Biologist

Even in those cases where indigenous peoples overtly profess a concern for conserving biological diversity for political, economic, religious, aesthetic, or moral reasons, they almost certainly do not ascribe the same meaning to this term as do biologists. The biodiversity that conservationists are interested in conserving is defined in a variety of ways but usually includes the full set of species, genetic variation within these species, the variety of ecosystems that contain the species, and the natural abundance in which these items occur (Office of Technology Assessment 1987). To conservationists, a forested landscape in which all biodiversity is conserved should include healthy populations of jaguars, woolly monkeys, white-lipped peccaries, large curassow, Scarlet Macaws, mahogany trees, tropical cedar trees, Brazil nuts, and large fruit-eating fish. It should be an area in which gene frequencies, species, landscapes or process—be they ecological or evolutionary—are allowed to follow a course largely unaffected by human activity.

Although there has been much discussion suggesting that low-level economic activity would be compatible with such biodiversity conservation, it is clear that if the *full range* of genetic, species, and ecosystem diversity is to be maintained *in its natural abundance* on a given piece of land, then virtually any significant activity by humans must not be allowed (see Redford 1992). Many studies have illustrated the point that even low levels of indigenous activity alter biodiversity as defined above (Fig. 2). For example, Vickers (1991) showed that although the Siona-Secoya of the Ecuadorian Amazon were able to hunt most species of game animals at con-

Figure 2. Yuquí hunter plucking macaw (Ara). Chimoré, Bolivia. Photograph by Kent H. Redford.

stant levels over a 10-year period, populations of woolly monkeys (*Lagothrix lagothricha*), curassow (*Mitu salvini*), and the trumpeter (*Psophia crepitans*) had been severely depleted. It should be stressed here that the vast majority of plant and animal species in the areas used by the Siona-Secoya appear not to have been strongly negatively affected by this group's activities—but not all species were unaffected.

Though not very well understood by ecologists, indications are that the loss of certain species (such as keystone species) will result in major changes in the community or ecosystem (see Terborgh 1988 and review in Redford 1992). It is possible that the loss of seed-dispersal services performed by those three large, frugivorous species negatively affected by Siona-Secoya hunting will change the interaction between large and small-seeded tree species, and will ultimately result in a different forest than the one in which these species were present.

Increasing evidence indicates that there is virtually no area in the Amazon that has not been affected in one way or another by human activity during the last several millenia (Balée 1989; Roosevelt 1989). In this light, there may be no such thing as "virgin forest." Even so, anthropogenic effects to a large extent were limited in scale and intensity. Following World War II, however, factors such as rapid acculturation, increased market participation, and large-scale development projects have inexorably changed the relationship between indigenous people and the Amazon forest. People who once made their own decisions about how to exploit their natural resources are now experiencing intense external pressures to conform to and participate in a consumer economy. What used to be local effects have become regional and even global effects, and humans now posses the capability to destroy much of what remains of the Amazon forest.

Biological Diversity and the Native Amazonian

In our experience, when indigenous peoples refer to biodiversity, they use a definition substantially different than the one used by conservation biologists. In the indigenous view, preserving biodiversity means preventing large-scale destruction, such as cutting and burning of forest for cattle ranches; building dams that displace native inhabitants from their homelands or alter the landscape in ways that make traditional subsistence from local resources impossible; or oil exploration and gold mining that bring in throngs of outsiders that disrupt indigenous life ways, introduce disease, and generally wreak havoc on the environment. Indigenous peoples rely on high levels of certain types of biodiversity. Their food security and social reproduction typically depend on high levels of intraspecific diversity (such as in manioc), high species diversity (for gathered foods), and ecological processes (pollination, succession) (Alcorn 1989).

To indigenous peoples, conserving biodiversity may not preclude slash-and-burn farming for the market, small-scale cattle ranching, selective logging from which they gain a profit, subsistence or even commercial hunting, or other forms of extractive activities that leave large areas of the forest or other natural habitat altered but still standing. This is clear from the call by the Amazonian natives' organizations for solidarity between indigenous communities and groups such as rubber tappers. But it must be recognized that, like indigenous reserves, extractive reserves are being created largely in response to issues involving social equity and land rights (Allegretti 1990). While extractive reserves do conserve important elements of biodiversity, the activities of rubber tappers have clearly been shown to alter forest biodiversity (see Almeida 1992).

Although multiple use areas such as biosphere reserves are characterized as meeting conservation as well as indigenous needs, neither group may fulfill its expectations in the end, creating conflict and causing inevitable recriminations. If indigenous peoples are ceded land, then it is unjust and unrealistic to expect them to conform to some preconceived stereotype of the "ecologically noble savage." And it is certainly reprehensible simply to create conservation areas and then to inform the indigenous inhabitants that they are now going to be "managed" as part of that unit. In those settings where the same piece of land is targeted to satisfy the agendas of both environmentalists and indigenous peoples, both groups must explicitly address questions of trade-offs.

While conservationists may hold the unrealistic expectation that native Amazonians will preserve land ceded to them in the same state in which they received it, indigenous peoples expect to be able to use those lands to assure their physical and cultural survival. Some indigenous groups may meet the expectations of conservationists, at least for the near future, by maintaining traditional ways of life that conserve "acceptable" levels of biodiversity in their territories. Others will not. In either case, indigenous peoples have the right to determine their own future.

At the same time, conservationists cannot demand that, in order to retain their land, native peoples must remain in isolated, stagnant communities using only limited forms of technology to exploit the environment in ways dictated by outsiders. As is often the case, unfortunately, this strategy denies the dynamic nature of culture and leads only to societies that inevitably suffer marginalization, poverty, and the destruction of those very traditions and values that were naively thought to be preserved by attempts to stop the forces of change. Rather, conservationists should work with native peoples as equal partners in developing alternative strategies to forest destruction, listening to their needs, and learning from indigenous experience. Native Amazonians know the forest better than anyone else. And while they many not be natural conservationists one and all, many respect the need to conserve the forest and the resources it holds, if for no other reason than to insure their own cultural survival. If indigenous people lose their land, they also risk losing their collective identity.

Much of this collective identity is imbedded in holding land communally, which gives indigenous patterns of land use much greater promise for conservation than western systems of individual property rights. Nonetheless, notwithstanding recent efforts, most Latin American nations with agrarian reform laws have difficulty dealing with the concepts of both communal land ownership and the granting of large tracts of land for purposes other than intensive agriculture. The philosophy, often codified in law, that the forest must be destroyed to qualify land as being used legitimately must be changed. As Gray (1991:18) states, "A people cannot live as a people unless they control their resources, their future, and their own development." And biologists must be prepared to recognize that an indigenous group has the right to decide the direction of its future, even if that future holds no place for the biodiversity conservationists so highly value.

Still, supporting indigenous land rights continues to offer the best hope for conserving and rationally using those tropical forests not contained in national parks. The communal territories of indigenous peoples maintain relatively intact large areas of land, buying time for the development of new ideas and for the creation of greater support for both biological conservation and the preservation of traditional cultural values. While the alliances forged between conservation groups and native Amazonians may be based on widely divergent agendas, they are not immutable. Through frank discussion and debate, the explicit recognition of different priorities and consequent trade-offs, and the understanding and compromise that this process engenders, indigenous peoples and conservation biologists can both work toward reaching the goals of common good that both are seeking.

Acknowledgments

We would like to thank Janis Alcorn, Mac Chapin, Jason Clay, and John Robinson for valuable comments. This is a contribution from the Program for Studies in Tropical Conservation, University of Florida, Gainesville, Florida.

Literature Cited

Alcorn, J. B. 1989. Process as resource: The traditional agricultural ideology of Bora and Huastec resource management and its implications for research. Pages 31–63 in D. A. Posey and W. Balée, editors. Resource management in Amazonia: Indigenous and folk strategies. Advances in economic botany, vol. 7.

Alcorn, J. B. 1991. Epilogue: Ethics, economics, and conservation. Pages 317–349 in M. L. Oldfield and J. B. Alcorn, editors. Biodiversity. Culture, conservation, and ecodevelopment. Westview Press, Boulder, Colorado.

Allegretti, M. H. 1990. Extractive reserves: An alternative for reconciling development and environmental conservation in Amazonia. Pages 252–264 in A. B. Anderson, editor. Alternatives to deforestation: Steps toward sustainable use of the Amazon rain forest. Columbia University Press, New York, New York.

Almeida, M. 1992. Reservas extrativistas como estrategia de conservação de fauna. pages 7–14 in Manejo da Vida Silvestre para a conservação na America Latina. Relatorio Tecnico. Belém, Pará, Brazil.

Arvelo-Jimenez, N., and A. L. Cousins. 1992. False promises. Cultural Survival Quarterly 16(1):10–13.

Balée, W. 1989. The culture of Amazonian forests. Pages 1–21 in D. A. Posey and W. Balee, editors. Resource management in

Amazonia: Indigenous and folk strategies. Advances in economic botany, vol. 7.

Chapin, M. 1991. Losing the way of the Great Father. New Scientist 10:40–44.

COICA (Coordinadora de las Organizaciones Indígenas de la Cuenca Amazónica). n.d. What is COICA?

COICA (Coordinadora de las Organizaciones Indígenas de la Cuenca Amazónica). 1989. Two agendas on Amazon development. Cultural Survival Quarterly 13(4):75–87.

Economist. 1989. This jungle's yours. The Economist, November 25, 313:46.

Faust, F. 1991. La cultura de los indígenas del macizo Colombiano y la protección de la naturaleza en el parque nacional de Purace. Novedades Colombianas. Museo de historia natural de la Universidad del Cauca. Nueva Epoca 3:54–62.

Gray, A. 1991. Between the spice of life and the melting pot: Biodiversity conservation and its impact on indigenous peoples. Document 70. International working group of indigenous affairs, Copenhagen, Denmark.

Hames, R. 1991. Wildlife conservation in tribal societies. Pages 172–199 in M. L. Oldfield and J. B. Alcorn, editors. Biodiversity: Culture, conservation, and ecodevelopment. Westview Press, Boulder, Colorado.

Michaels, J. 1991. Brazil creates homeland for Yanomamis. The Christian Science Monitor, November 19, 1991.

Office of Technology Assessment. 1987. Technologies to maintain biological diversity. U.S. Government Printing Office, Washington D.C.

Redford, K. H. 1991. The ecologically noble savage. Cultural Survival Quarterly 15(1):46–48.

Redford, K. H. 1992. The empty forest. BioScience 42(6):412–422.

Redford, K. H., and S. Sanderson. 1992. The brief, barren marriage of biodiversity and sustainability. Bulletin of the Ecological Society of America 73(1):36–39.

Roosevelt, A. 1989. Resource management in Amazonia before the conquest: Beyond ethnographic projection. Pages 30–62 in D. A. Posey and W. Balée, editors. Resource management in Amazonia: Indigenous and folk strategies. Advances in economic botany, vol. 7.

Stearman, A. M. 1990. The effects of settler incursion on fish and game resources of the Yuqui, a native Amazonian society of eastern Bolivia. Human Organization 49(4):373–385.

Stearman, A. M. 1992. Neotropical hunters and their neighbors: Effects of non-indigenous settlement patterns on three native Bolivian societies. Pages 108–128 in K. H. Redford and C. Padoch, editors. Conservation of neotropical forests: Building on traditional resource use. Columbia University Press, New York, New York.

Taussig, M. 1987. Shamanism, colonialism, and the wild man: A study in terror and healing. University of Chicago Press, Chicago, Illinois.

Terborgh, J. 1988. The big things that run the world—a sequel to E. O. Wilson. Conservation Biology 2:402–403.

van Lennep, E. 1990. Arctic to Amazonia: An alliance for the earth. Cultural Survival Quarterly 14(1):46–47.

Vickers, W. T. 1991. Hunting yields and game composition over ten years in an Amazon Indian territory. Pages 53–81 in J. G. Robinson and K. H. Redford, editors. Neotropical wildlife use and conservation. University of Chicago Press, Chicago, Illinois.

Yonomami, M. 1990. Que fiquem of garimpeiros. Veja 23(3):94.

Comments

Indigenous Peoples and Conservation

JANIS B. ALCORN

Biodiversity Support Program
℅ World Wildlife Fund
1250 24th Street NW
Washington, D.C. 20037, U.S.A.

Forest-dwelling peoples' organizations continue to express concern about destruction of their forests. The International Alliance of the Indigenous-Tribal Peoples of the Tropical Forests issued The Forest Peoples Charter in February 1992 (available from the World Rainforest Movement in the U.K. and Cultural Survival in the U.S.A.). The Charter sets out a conservation policy based on recognition of indigenous peoples' rights to conserve their forests and to regulate development activities currently imposed upon them without their consent. It is the first such statement from a global network of forest-dwelling peoples' organizations.

At the local level, forest-dwellers around the world have repeatedly made declarations deploring outsiders' destruction of forests for at least 500 years. The Forest Peoples Charter adds its weight to other international forest-dwellers' declarations, including the 1988 statement of the Coordinating Body for the Indigenous Peoples' Organizations of the Amazon Basin (COICA) highlighted in the essay by Redford and Stearman (1993). Many conservation groups are now supporting forest peoples' struggles for recognition of their rights. The Global Biodiversity Strategy (World Resources Institute et al. 1992) supports recognition of ancestral domains. Redford and Stearman also conclude that it is wise for conservationists to work with indigenous peoples. But like Clad (1984), they argue that the interests and agendas of the two groups are "partially or completely in conflict."

Conservation is a social and political process. Conservationists interested in achieving on-the-ground conservation of biodiversity have to choose among real options, not idealized academic options. Conservationists in an increasing number of countries are choosing the option of working with local peoples' organizations. Redford and Stearman are correct in stating that compromises are being made, but decisions about goals and compromises deserve further thought.

A decade ago, U.S. and European-based conservationists focused on supporting protected-areas strategies implemented through state governments. Park departments, with staff trained by academic centers that teach strategies based on protected areas, still espouse the conservationists' goals and agenda described by Redford and Stearman. But, despite the confluence of those two agendas, park departments and other state agencies have failed miserably at conserving biodiversity, globally and in Amazonia. Instead, paper parks abound, and deforestation rates have increased. While states have pleased conservationists by announcing the creation of parks, a careful look at state performance shows a general pattern whereby state-linked elites are continuing to log and mine in protected and reserved areas.

Not only have park strategies failed, but they have undermined forest-dwelling communities' ability to protect forests. They have been implemented at costs to local people in order to achieve global benefits (Wells 1992). The expected conservation benefits have not accrued, but local costs have been considerable (see Ghimire 1991). Conservationists' recognition of their "myth of the noble savage" (Redford 1990) is coupled with their recognition of what I call their "myth of the noble state." Conservationists are becoming more aware of real-world options and their costs and track records. And they are seeking to create positive partnerships with real indigenous peoples and real states.

Redford and Stearman pursue a question that merits further discussion in Conservation Biology: What interests do indigenous people and conservationists have in common? Redford and Stearman identify major differences in the two groups' conservation goals, based on their understanding of those goals. I would like to pursue their question further.

First, I will focus on their definitions of goals. The indigenous definition of conservation, based on Redford and Stearman's experiences, should be expanded. Based upon my own experiences in the Amazon, Central America, Asia, and the Pacific, and corroborated by other ethnobiologists and the ethnobiological literature, Redford and Stearman's definition is inadequate and misleading. Redford and Stearman accurately note that there is great heterogeneity within and between indigenous groups. Nonetheless, there are general patterns that provide a definition beyond the one they provide: "In the indigenous view, preserving biodiversity means preventing large-scale destruction." To my knowledge, there is no direct translation for the word "conservation" in any non-European language. It is generally translated as "respecting Nature," "taking care of things," or "doing things right." Indigenous peoples often find the Western idea of "conservation" as something to be separated from the rest of their activities as strange. A Karen man recently asked me why we always "put things in boxes." It makes things difficult, he said. To him, and to many others I've met in other countries, conservation is just part of making a living. Indigenous goals are different from the conservationists' goals characterized by Redford and Stearman. But the goal expressed in IUCN's updated World Conservation Strategy, "Caring for the Earth" (IUCN et al. 1991) is a close match for indigenous ideas of conservation.

Indigenous people demonstrate a concern for maintaining the ecological processes and the species that mediate those processes (Alcorn 1989a, 1989b). They often demonstrate a keen interest in the locations of rare plant species. Within any given community, there are usually several people who bring rare plants into cultivation in order to maintain them. There are fewer published examples of indigenous peoples' active efforts to maintain mammals. Many indigenous groups in Africa and Asia have a tradition of maintaining sacred forest areas where animals and plants are not disturbed. More common globally are community-enforced rules of forest and game use. In traditional societies, nature is viewed as part of human society, and proper relations with nature are necessary in order to have proper relations between people, including past and present generations. The commitment of indigenous peoples to conservation is complex and very old.

I strongly disagree with Redford and Stearman's statement that indigenous people have presented themselves as conservationists "*only* because they recognize the power of this concept in rallying support in their struggle for land rights" [emphasis added]. When indigenous people enter into discussions with powerful outsiders, they must meet on outsiders' terms and use their vocabulary. New use of the outsiders' concept of conservation coincides with the rise of international conservationists as a new player among powerful outsiders; this does not mean that conservation is new to indigenous peoples.

Most conservationists have broader goals than those defined by Redford and Stearman. Most U.S.-based conservation biologists do seem to share the narrower goals, although this continues to be debated within this journal. Conservation did not originate among biologists (as stated by Redford and Stearman), unless one accepts the narrow goals defined by Redford and Stearman that conservationists are to maintain ecosystems isolated from human beings (except the biologists who want to study them). Indigenous peoples' goals, as I have described them, don't completely match those narrow conservationists' goals. They more closely match the broader goals espoused by many conservationists who recognize that most of the world's biodiversity is found, and will continue to be found, in landscapes occupied by people.

Regarding Redford and Stearman's concern for the loss of the traditional conservation ethic, I would like to return to Chapin's comment quoted in their essay. Chapin notes that the modern conservation ethic may not be adequate to maintain biodiversity, compared to the traditional conservation ethic. I agree. I have argued elsewhere that the modern approach is too narrow and that conservationists have two goals: to stabilize the traditional conservation ethic wherever it still exists, and to improve the modern conservation ethic (Alcorn 1991).

Redford and Stearman characterize development as a threat to biodiversity and warn that indigenous peoples will cease to conserve biodiversity as they pursue development. There is evidence both to support and to contradict their warning. Some indigenous peoples pursuing economic and social development are moving to adopt the modern reserve concept to protect biodiversity from threats by commercialization. They are seeking the state's assistance to defend biodiversity. For example, in Mexico, rural communities sought and achieved establishment of "campesino ecological reserves" (Toledo 1992). Through the Union of Indian Nations, Brazilian Xavante sought assistance from the World Wildlife Fund to use scientific methods to monitor game populations in order to prevent poaching by outsiders and better regulate their own hunting (Butler 1992, personal communication). Likewise there is widespread evidence that, for centuries, traditional peoples around the world have intensified land use in certain areas of their territories in order to maintain forests in other areas.

Large-scale, outsider-driven development projects wreak devastating effects on levels of biodiversity. But examination of the range of development activities in the broader landscape shows that conventional wisdom about the impact of development is often wrong in areas where strong non-Western cultural roots are still intact. For hundreds of years, local communities have fought to keep commercial loggers out of their forests. Now they also fight "reforestation programs" that threaten to replace existing forests with plantations. The tradition of community forest defense continues around the world.

One of the main threats linked to development is commoditization of land and disruption of common property regimes. As Redford and Stearman note, indigenous patterns of communal land use offer "greater promise for conservation than Western systems of individual property rights." Indigenous peoples have held forest under complex, often-overlapping tenure rights that share benefits across their community and exclude noncommunity members. Overlapping rights protect the community from outside acquisition of their forests and from exclusive use by any one entity who might destroy it. Traditional systems are in effect partnerships between individuals and their community.

Partnerships with indigenous peoples offer the best option for achieving on-ground conservation both inside and outside parks. An internal World Bank evaluation of Latin American efforts found that even when indigenous lands have been demarcated and recognized by governments, they are still being exploited by settlers and logging operations. The state is not defending indigenous peoples' property rights, despite the facts that the state has recognized those rights and that a primary function of the ideal state is to defend property. This is particularly a problem among the smaller Amazonian groups described by Redford and Stearman. Strong partnerships with the state will be necessary for continued conservation of indigenous groups' forests. Building appropriate partnerships between states and indigenous communities may require new legislation, policies, institutional linkages, and processes. It requires creating communication networks and research linkages. It also requires adequate monitoring of biodiversity and institutional processes, an area where the collaboration of nongovernmental organizations can be particularly helpful.

One barrier to partnerships is the attitude that conservationists are in a position of authority to "cede" land, to "grant" rights to others, to speak for others, or to define others' knowledge. The 1988 COICA statement was itself issued in response to this problem. The COICA declaration specifically states, "We are concerned that you have left us Indigenous Peoples and our organizations out of the political process which is determining the future of our homeland.... [W]e never delegated any power of representation to the environmentalist community.... We want to represent ourselves and our interests directly in all negotiations concerning the future of our Amazonian homeland." When Redford and Stearman write about indigenous people "claiming standing" to enter conservation discussions, their statement implicity acknowledges the problem noted by COICA: "conservationists" are acting as gatekeepers to a discussion table that does not have a place set for those whose homeland's future hangs in the balance.

Until we recognize the authority of indigenous peoples as equals at the discussion table, we cannot join in partnerships with them. Chhatrapati Singh (1986) has noted: "Amongst ... externalities, the most destructive [to nature] is injustice or *adharma* ... [T]he consequences of such *adharma* have been borne by the rural poor, the tribals, and the flora and fauna" (Singh 1986:1). "[T]he issues actually at stake in the forest question ... are three: (a) justice to the people, forest dwellers and nondwellers; (b) justice to nature (trees, wild life, etc.); and (c) justice to coming generations" (Singh 1986:7). In the real world, conservation of forests and justice for biodiversity cannot be achieved until conservationists incorporate other peoples into their own moral universe and share indigenous peoples' goals of justice and recognition of human rights.

Acknowledgments

I thank Kent Redford and Allyn Stearman for sharing their manuscripts with me. I thank John Butler, Mac Chapin, Alejandro de Avila, Owen J. Lynch, and Toby MacGrath for comments and suggestions. The views expressed in this paper are my own and should not be attributed to the World Wildlife Fund, the Biodiversity Support Program, or the Agency for International Development.

Literature Cited

Alcorn, J. B. 1989a. Process as resource. Advances in Economic Botany 7:31–63.

Alcorn, J. B. 1989b. An economic analysis of Huastec Mayan forest management. Pages 182–206 in J. O. Browder, editor. Fragile lands of Latin America. Westview Press, Boulder, Colorado.

Alcorn, J. B. 1991. Ethics, economies, and conservation. Pages 317–349 in M. L. Oldfield and J. B. Alcorn, editors. Biodiversity: culture, conservation, and ecodevelopment. Westview Press, Boulder, Colorado.

Clad, J. 1984. Conservation and indigenous peoples: A study of convergent interests. Cultural Survival Quarterly 8:68–73.

Ghimire, K. B. 1991. Parks and people. Discussion Paper No. 29. United Nations Research Institute for Social Development, Geneva, Switzerland.

IUCN, UNEP, Worldwide Fund for Nature. 1991. Caring for the earth. International Union for the Conservation of Nature, Gland, Switzerland.

Redford, K. H. 1990. The ecologically noble savage. Cultural Survival Quarterly 15(1):46–48.

Redford, K. H., and A. M. Stearman. (1993). Forest-dwelling native Amazonians and the conservation of biodiversity. Conservation Biology 7(2):248–255.

Singh, C. 1986. Common property and common poverty: India's forests, forest dwellers, and the law. Oxford University Press, Oxford, England.

Toledo, V. 1992. Biodiversidad y campesinado: La modernización en conflicto. La Jornada del Campo 9:1–3.

Wells, M. 1992. Biodiversity conservation, affluence and poverty: Mismatched costs and benefits and efforts to remedy them. Ambio 21(3):237–243.

World Resources Institute, IUCN, UNEP. 1992. Global Biodiversity Strategy. World Resources Institute, Washington, D.C.

Response to Alcorn

On Common Ground? Response to Alcorn

KENT H. REDFORD

Program for Studies in Tropical Conservation
Center for Latin American Studies
University of Florida
Gainesville, FL 32611

ALLYN MACLEAN STEARMAN

Department of Sociology and Anthropology
University of Central Florida
Orlando, FL 32816

We welcome Janis Alcorn's response to our article "Forest-Dwelling Native Amazonians and the Conservation of Biodiversity: Interests in Common or in Collision?" One of our major purposes in writing the piece was to place on the table for discussion a very difficult and complicated issue. As is evident in her response, Alcorn is not in full agreement with us—however, she does agree with many of our major conclusions.

Rather than belabor the individual points, we would like to offer new evidence of the complicated nature of the question of cooperation between conservationists/environmentalists and indigenous peoples. A recent interview with Nicanor González, a Kuna Indian who was the international coordinator for the Second Interamerican Indian Congress on Natural Resources and the Environment, held in December 1991 in Bolivia, was entitled "We Are Not Conservationists" (Chelala 1992). In his thoughtful comments González states:

> What I have understood in talking with the indigenous authorities, indigenous groups, and individuals is that they are familiar with the laws of nature. They aren't conservationists; rather, they know how to interrelate humans and nature.... In this sense, then, I don't believe that you can say that indigenous people are conservationists as defined by ecologists. We aren't nature lovers. At no time have indigenous groups included the concepts of conservation and ecology in their traditional vocabulary. We speak, rather, of Mother Nature. Other organizations need to be clear about this before jumping in to solve some problem with the indigenous population (p. 45).

Alcorn accuses us of generalizing but says "Indigenous people demonstrate a concern for maintaining the ecological processes and the species that mediate those processes." This remark illustrates two things. First, generalizing is a double-edge sword (see for example Johnson 1989; and AMS's recent experiences with the Yuquí of Bolivia). Second, even if Alcorn's claim is generally true, the maintenance of these processes and critical species is not equivalent to the maintenance of biodiversity as we have defined it (see also Redford & Robinson in press).

While writing our piece, we struggled with the definition of "conservation" and how to characterize the position of "conservation biologists." We were roundly criticized by editors and reviewers for having a rigid stance on these terms, a stance that did not reflect the existing diversity of definitions and perspectives. In her response to our paper, Janis Alcorn falls into the same trap that caught us—an unclear definition of "conservation." Whereas we lean much more to the "conservation" of John Muir (i.e., preservation), Alcorn clearly lines up with Gifford Pinchot's definition of conservation, "conservation through use" (see Norton 1991). This is clear when she states in regard to a Karen man, "... and many others I've met in other countries, conservation is just part of making a living. Indigenous goals are different from the conservationists' goals characterized by Redford and Stearman. But the goals expressed in IUCN's updated World Conservation Strategy "Caring for the Earth" ... is a close match for indigenous ideas of conservation."

In "Caring for the Earth" "conservation" is defined as "the management of human use of organisms or ecosystems to ensure such use is sustainable. Besides sustainable use, conservation includes protection, maintenance, rehabilitation, restoration, and enhancement of populations and ecosystems" (I.U.C.N., U.N.E.P., W.W.F. 1991: p. 210). This definition is clearly an anthropocentric, use-oriented one.

Therefore, Alcorn's views of how indigenous people define conservation are very much in line with the indigenous leader quoted above—only he would say that it is *not* conservation. His perspective is in agreement with ours: that it is important to distinguish a type of nonutilitarian conservation, one that does not demand that every organism and process be valued strictly in terms of its usefulness to humans. Alcorn terms this a narrow definition of conservation. Maybe so; but it is vital if we are not to lose much of the biodiversity we value to forces John Muir described as "temple destroyers," "... devotees of ravaging commercialism, [who] seem to have a perfect contempt for Nature, and instead of lifting their eyes to the God of the mountains, lift them to the Almighty Dollar." (p. 8 in Norton 1991).

In closing we would like to cite two examples from recent publications that continue uncritically to claim congruence between the interests of conservationists and traditional forest-dwelling peoples:

(1) A declaration by the International Working Group on Indigenous Affairs from a meeting held in Paris in 1991 contains the following statement: "We the Indigenous People of the World since ancestral times, have been building up a culture, civilization, history and a world vision which has allowed us to co-exist harmoniously with Nature. This natural process was interrupted in America with the European invasion of the continent. ..." (p. 157). (IWGIA 1992).

(2) Finally, a publication entitled "Amazonia without Myths" commissioned by the Inter-American Development Bank, the United Nations Development Programme and the Amazon Cooperation Treaty contains the following statements: "Amazonian people possess a distinctive characteristic—their commitment to sustain nature. Indians on the one hand and rubber tappers (or seringueiros) and riverbank people, on the other, have lived for millennia and centuries, respectively, in intimate association with nature." (p. xiii) And continuing: "A fundamental trait of the Amazon peoples is their identification with nature. Anyone who endangers the Amazon ceases to be a part of the place and its culture. Certain conditions that will enable the Amazon peoples to act in the best interests of their future have not yet been met, however ..." (p. 68).

These claims obfuscate a very complicated terrain where common ground is not clearly defined. We are convinced, as is Alcorn, that the world's conservationists and the world's indigenous and traditional peoples have much in common. However, durable solutions will not come as a product of inflated rhetoric and wishful thinking but must be hammered out of precise definitions and acknowledged tradeoffs.

Acknowledgments

We would like to thank Dr. Steve Sanderson and Dr. Marianne Schmink for comments.

Literature Cited

Alcorn, J. 1993. Indigenous peoples and conservation. Conservation Biology 7:000–000.

Chelala, C. 1992. We are not conservationists. Cultural Survival Quarterly **Fall 1992**:43–45.

Commission on Development and Environment for Amazonia. 1992. Amazonia Without Myths. Inter-American Development Bank, United Nations Development Programme and Amazon Cooperation Treaty.

I.U.C.N., W.W.F., U.N.E.P. 1991. Caring for the earth: a strategy for sustainable development. Gland, Switzerland.

IWGIA [International Working Group on Indigenous Affairs]. 1992. Declaration by the Indigenous Peoples. pp. 157–163. IWGIA Yearbook 1991. Copenhagen, Denmark.

Johnson, A. 1989. How the Machiguenga manage resources: conservation or exploitation of nature? Pages 213–222 in D. A. Posey and W. Balée, editors. Resource Management in Amazonia: Indigenous and Folk Strategies. The New York Botanical Garden, New York.

Norton, B. 1991. Toward Unity among Environmentalists. Yale University Press, New Haven.

Redford, K. H., and A. M. Stearman. 1993. Forest-dwelling native Amazonians and the conservation of biodiversity: interests in common or in collision? Conservation Biology 7:248–255.

Redford, K. H., and J. G. Robinson. In press. The sustainability of wildlife and natural areas. Proceedings of the International Conference on the Definition and Measurement of Sustainability. United Nations University, New York.

Comments

Indigenous Reserves and Nature Conservation in Amazonian Forests

CARLOS A. PERES

Center for Tropical Conservation
Duke University
3705-C Erwin Road
Durham, NC 27705, U.S.A.

The growing duality among scientists between nature- and people–oriented conservation has been explicitly illustrated once again in a dispute between Redford and Stearman (1933a; 1993b) and Alcorn (1993) in a recent issue of *Conservation Biology*. Many conservation biologists would agree that the long-term objectives of indigenous peoples enpowered to determine their own destinations are incompatible with the conservation of tropical forest biodiversity. Cultural anthropologists and ethnobiologists, on the other hand, are quick to argue—under the banner of "use it or lose it"—that communities of forest peoples are the most legitimate inhabitants and effective guardians of what would otherwise be a doomed habitat. What at a glance might appear to be a small ideological clash, however, actually reflects a fundamental philosophical divergence that has been extended to small and corporate-sized nongovernment organizations, central-government decision makers, and the overall agenda of global environmental blueprints (Robinson 1993). Amidst this unresolved controversy, uninformed persons may be left wondering whether native peoples are an asset or a handicap to nature conservation.

Unfortunately, this otherwise healthy academic debate dramatically obscures common interests of all conservationists and undermines a potentially rational consensus on how to best protect tropical forests. Here I wish to refocus the recent discord over the definition of conservation by indigenous peoples on what may be considered a more practical problem.

Because most Amazonian countries recognize indigenous land rights and designate—even if on paper only—substantial chunks of their largely undisturbed forests to this category of conservation unit, the role of indigenous parks and reserves in Amazonian nature conservation is a substantial one. These countries thus provide a good number of (largely undecreed) indigenous and anthropological reserves, *resguardos,* and Amerindian lands where native peoples are allowed to continue their traditional livelihoods, often within large, legally notified land blocks. The process of setting aside Indian lands does not please both regional or central governments. For example, the recent demarcation of large tracts of Yanomami Indian lands in Roraima was met by harsh criticisms from Brazilian politicians across the country.

Mirroring the disagreements that have sparked these comments, Amazonian Indian reserves are usually administered by independent government agencies (such as the National Indian Foundation—FUNAI—in Brazil) whose goals and priorities do not necessarily coincide with those of natural resource and forestry institutes. This represents a profound obstacle to forest conservation policy and enforcement. Although logging permits may be required in some rapidly developing areas, indigenous groups are in practice given a blank check to exploit and manage their resources according to rapidly evolving "traditional practices." This has been widely witnessed on a vast development frontier in the southern and eastern Amazon, where several Indian tribes have begun a process of outright liquidation of their

---
* Current address: Departamento de Ecologia, Universidade de São Paulo, Caixa Postal 11.461, São Paulo, S.P. 05422-970, Brazil.
Paper submitted June 22, 1993; revised manuscript accepted July 16, 1993.

land resource capital in the form of western-style land concessions granted to logging companies and gold miners. For instance, having found their way to lucrative markets, the Kayapo of eastern Amazonia have logged $33 million in profits from mahogony extraction alone in 1988 (The Economist 1993). Under the current system their trees will continue to fall, because the Brazilian government will never be able to meet the Kayapo's demand for $50,000 per village per month for timber sales forgone. Other Brazilian Indian groups helping to shatter the myth of the noble savage include the Guajajara of the northeast, the Kaxarari of Acre, and the Nambikwara of Mato Grosso, all of which are involved in prime hardwood business. These market-oriented practices are clearly not what is generally considered genuine nature conservation, which raises serious questions over the role of indigenous reserves as conservation units. Other smaller-scale forms of resource depletion—including overhunting, overfishing, and overharvesting of timber and nontimber products—may also be incompatible with biodiversity conservation, yet they are under heavy subsistence and monetary pressure from increasingly consumptive societies. Current demands for material commodities range from pots and pans and shotguns to chain-saws and small aircrafts. Double standards in land- and resource-use policy directed to tribal and nontribal Amazonians is also a pervasive phenomenon. One therefore might wish to see greater uniformity between these policies—traditional wisdom alone will not prevent Indians from overharvesting resources—if Indian reserves are to fulfill a more useful conservation role in the long run.

The enormous magnitude and future consequences of this issue entitle it to be at the forefront of any Amazonian conservation discussion. For instance, Indian reserves account for 54% of all 459 Amazonian reserves with known areas in nine South American countries (including all recognized categories of "strictly protected" nature reserves, production and extractive forests and Amerindian lands). These include anthropological reserves and all forms of indigenous reserves, colonies, and parks. In terms of combined acreage, they represent 100.2 million ha in 371 reservations in Brazilian Amazonia alone (Fundção Nacional do Indio 1989), or 52% of the entire area receiving some form of nonprivate protection within the entire Amazonian region (adapted from World Conservation Monitoring Centre 1992a, 1992b). In addition, indigenous conservation units in Amazonia tend to comprise sizeable areas (mean = 400,256 ± 987,044 ha, n = 248) and outnumber all other forms of reserves, particularly in the range of 1000–10,000 ha (Fig. 1). This accounts for a substantial portion of yet relatively undisturbed Amazonian forests. Approximately 20% of Brazilian Amazonia, for instance, is represented by indigenous parks and reserves under the jurisdiction of Fundção Nacional do Indio FUNAI;

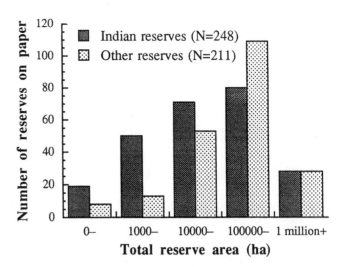

Figure 1. Updated number and size of existing Amazonian indigenous reserves "on paper" compared to those of all other categories of conservation units, including national parks, biological reserves, ecological stations, and other recognized forms of extractive and production forest (adapted from World Conservation Monitoring Centre 1992a, 1992b).

Rylands 1991). Moreover, the number of very large Amazonian Indian reserves (>1 million ha), which are likely to be the most viable from the viewpoint of long-term biodiversity maintenance, rival that of all other reserves combined. Finally, compared to all Amazonian nature reserves combined, the geographic distribution of Indian reserves is far more widespread across major river barriers, which greatly increases their complementary value to species conservation (Pressey et al. 1993).

Clearly, if Indian areas are to remain relatively stable and serve as strongholds of full complements of biological diversity, a substantial legislative revision of the Indian Code will need to be incorporated into national constitutions. This should be particularly the case where Indian reserves are being encroached by settlers or overrun by development projects. Subsequently, rational quotas on indigenous peoples' rights to manage, harvest, and convert their forest resources will have to be enforced. Negotiated commercial contracts with private companies dealing with live or mineral resource extraction shoud be terminally prohibited, except in a few cases where approval of the relevant authorities and consulting experts would become mandatory. This could be achieved initially by an increased dialogue between Indian and natural resource agencies, in Brazil represented by FUNAI and IBAMA (Brazilian Institute for the Environment and Renewable Natural Resources). The cost-efficiency of these agencies in a wide range of field operations would increase if they joined efforts, because many Amazonian Indian and nature reserves are located near or border one another. Joint operations could include the deployment of surveillance and patrol personnel, revision of logging and clear-cut permits, and inspection of boats and trucks from strategic guardposts intercepting important access routes. Forest-dwelling peoples' organizations could also be instrumental in this process, should they realize that the cause they are fighting for will gain fewer adepts as indigenous groups increasingly fail to serve the interests of conservation.

Indigenous lands, are, however, an extemely important reservoir for Amazonian biodiversity, whether in the form of isolated parks, adjacent buffer zones, or connecting corridors in a wider landscape of reserve networks. Every effort should thus be made to reform Indian land-use policy and secure Amazonian Indian lands from an increasingly broader development frontier. Perhaps then Amazonian Indians will be able to redeem some of their credit as conservationists—and satisfy skeptics such as Redford and Stearman (1993a)—promoting a badly needed consensus between biological and cultural conservation.

Acknowledgments

I would like to thank Nick Salafsky for comments on the manuscript.

Literature Cited

Alcorn, J. B. 1993. Indigenous peoples and conservation. Conservation Biology 7:424–426.

The Economist. 1993. The savage can also be ignoble **327** (June 12):54.

Fundação Nacional do Indio. 1989. Terra indígenas: Legislação e situação latifundiaria. Ministério do Interior, Superintendência de Assuntos Fundiários, e Documentação Interna Indígena, Brasília, D.F.

Pressey, R. L., C. J. Humphries, C. R. Margules, R. I. Vane-Wright, and P. H. Williams. 1993. Beyond opportunism: Key principles for systematic reserve selection. Trends in Ecology and Evolution **8**:124–128.

Redford, K. H., and A. M. Stearman. 1993a. Forest-dwelling native Amazonians and the conservation of biodiversity. Conservation Biology **7**:248–255.

Redford, K. H., and A. M. Stearman. 1993b. On common ground? Response to Alcorn. Conservation Biology **7**:427–428.

Robinson, J. G. 1993. The limits to caring: Sustainable living and the loss of biodiversity. Conservation Biology **7**:20–28.

Rylands, A. B. 1991. The status of conservation areas in the Brazilian Amazon. World Wildlife Fund, Washington, D.C.

World Conservation Monitoring Centre. 1992a. Protected areas of the world: A review of national systems, vol. 4. Neartic and neotropics. Cambridge, England.

World Conservation Monitoring Centre. 1992b. Mapping tropical managed areas. Cambridge, England.

Scientists as Advocates: The Point Reyes Bird Observatory and Gill Netting in Central California

JAMES E. SALZMAN

68 Francis Avenue
Cambridge, MA 02138, U.S.A.

Abstract: *Scientists and scientific organizations can play singularly important roles as advocates in the public resolution of conservation disputes. One scientific organization, the Point Reyes Bird Observatory, was instrumental in resolving an environmental controversy in the early 1980s over the use of gill nets off central California. The gill-net fishery was killing thousands of birds annually. The Point Reyes Bird Observatory's use of its scientific data facilitated negotiation and formed the basis for the threat of litigation. These efforts spurred enforcement of federal wildlife legislation and helped forge a permanent solution acceptable to state and federal environmental agencies, conservation groups, and local fishermen. But there are constraints on the active involvement of scientists in public disputes. The contentious nature of the decision-making forum, the exigencies and uncertainty inherent in policy decisions, and the bias popularly associated with lobbying efforts all tend to dissuade scientists from participation as advocates in public controversies. The gill-net case suggests that when coupled with supportive legislation, focused advocacy—the presentation of relevant data and insistence that it be interpreted accurately and acted upon—is an effective method of achieving biologically sound policies.*

Resumen: *Los científicos y las organizaciones científicas pueden jugar papeles singularmente importantes como defensores en la resolución pública de desacuerdos en conservación. Un organismo científico, el "Point Reyes Bird Observatory," fue crucial en la resolución de la controversia ambiental en los comienzos de los ochenta sobre el uso de redes de acarreo en las costas centrales de California. La pesca por redes de acarreo estaba matando miles de aves anualmente. El uso de datos científicos específicos por parte de "Point Reyes Bird Observatory" facilitó las negociaciones y formó la base para la amenaza de litigio. Estos esfuerzos estimularon la aplicación de la legislación federal en vida silvestre y ayudó a forjar una solución permanente y aceptable tanto para las agencias medio ambientales federales y estatales, como para grupos conservacionistas y pescadores locales. Pero, existen reatricciones en la participación activa de los científicos en disputas públicas. La naturaleza contenciosa del foro para la toma de decisiones; las exigencias y la inseguridad inherente en decisiones políticas y los prejuicios asociados con los esfuerzos de influenciar legislación, tienden a persuadir a los científicos de no participar en controversias públicas. Este caso de las redes de acarreo sugiere que el apoyo focalizado; la presentación de datos revelantes y la insistencia de que sean interpretados con precisión y ejecutados; al emparejarse con legislación de apoyo, es un método efectivo de lograr políticas biologicamente aceptables.*

Introduction

The last two decades have witnessed a tremendous growth of federal environmental legislation. Laws such as the National Environmental Policy Act of 1969, the Clean Air Act of 1970, the Marine Mammal Protection Act of 1972, and the Endangered Species Act of 1973 have come of age, and administrative and statutory frameworks now exist specifically to address environmental policies and disputes. These laws have furthered conservation goals not only because they direct the gov-

Paper submitted June 2, 1988; revised manuscript accepted November 14, 1988.

ernmental mandate to protect the environment, but because they provide private groups the means to shape this process through public comment and litigation. Scientists and scientific organizations can play an influential role in conservation disputes through both research and advocacy. In fact, 21 years ago a group of scientists concerned over the effects of DDT formed the Environmental Defense Fund, now an influential conservation organization. This paper presents a case study of the successful advocacy efforts of the Point Reyes Bird Observatory (PRBO), a research, conservation, and educational organization in northern California, in a recent controversy involving the use of gill nets off central California. The article explores this organization's effective use of scientific data, negotiation, and legislation to forge a permanent solution to a complicated environmental dispute. The concluding section examines the constraints on scientists' participation in public disputes and presents the perspectives of several influential environmental advocates.

A Brief History of the Point Reyes Bird Observatory

In March, 1965, the Point Reyes Bird Observatory was established in the hills of Point Reyes National Seashore just north of San Francisco. The founders, a group of scientists and bird watchers, envisioned a small banding station that would be used to study coastal birds and monitor land-bird migration through the Point Reyes headlands. With sufficient funds, they hoped to support at least one full-time biologist who would work with volunteers. PRBO was in full-time operation by the end of 1965, its residents and volunteers busy trapping, banding, and censusing local avian populations (McCrimmon 1985). PRBO grew rapidly, bolstered by the emergence of the environmental movement.

Neither purely an academic institution nor a research organization, PRBO from its early days recognized its ability to conduct long-term research studies while involving the public through volunteer field work. Its educational activities over the years have included a training program for environmental education teachers, a vigorous volunteer program, and publication of a newsletter. These efforts are consonant with the organization's commitment to raising the public's awareness of environmental issues. In conservation issues, PRBO has exercised influence through its basic and applied research and advocacy of its findings. PRBO's shorebird, seabird, and wetlands research and other activities have given it a respected voice in policy debates. These controversies have ranged from outer continental shelf oil leases, Army Corps of Engineers' dredging near snowy Plover (*Charadrius alexandrinus*) nesting sites, and damage assessments of the biological impact of oil spills, to the impact on bird populations of krill harvesting in the Antarctic. In fact, many of PRBO's long-term studies are directed toward obtaining scientific data to aid conservation and management efforts.

As PRBO has grown, it has retained its character as a singular research establishment. The 1986 budget stood at $860,000, 17 times greater than the budget of 1975. The sources of these funds include: gifts and dues, 30%; contracts, 43%; and grants, 14% (PRBO 1986). PRBO's present staff has 16 full-time biologists, 5 of whom hold doctoral degrees. Including the research associates and volunteers, however, there are more than 50 employees. Throughout its growth, PRBO has remained a scientist's cooperative, promoting conservation when its research efforts can play a constructive role. In the early 1980s, PRBO became directly involved in just such a controversy.

The Emergence of Gill Netting in Central California

Dorothy Hunt lives in the town of Aptos, on Monterey Bay. Dorothy often walks the same stretch of beach near her home to enjoy the fresh air, and to look for dead birds. As a volunteer for PRBO's Beached Bird Project, Dorothy's observations are useful in establishing a data baseline for how many seabirds "normally" die and wash ashore on beaches. On Dorothy's stretch of beach, five or six dead seabirds are a lot to find in one walk. One foggy morning, seven years ago in July, Dorothy found 108 birds washed up among the kelp and seaweed. Frances Bidstrup, another PRBO beach walker working a few miles south of Dorothy, found 185 beached birds on just a half-mile stretch of beach. Frances's and Dorothy's findings were not unique. On that 3-mile stretch of Manresa Beach, over 1,500 birds washed up in one day (Heneman 1981). Gill-net fishing had come to Monterey Bay.

The brightly-colored yellow and green nylon gill nets seem innocuous enough, but they are terribly effective at catching fish and other life that happens to become ensnared in their mesh. Fishermen generally use two types of gill net depending on the kinds of fish they wish to catch. In deep water they use 6-cm-mesh nets, while in shallower water they use a wider 20-cm mesh. Like giant curtains, the monofilament nets are lowered with lead weights to the ocean floor. The float line keeps the 3- to 6-meter-high nets vertical in the water. When large-enough fish hit these invisible threads, only their heads fit through the mesh. Struggling to get free, they tangle their gill covers, are unable to breathe, and suffocate. Two to 24 hours later, the nets are reeled onto large drums and the day's catch is removed. The 20-cm mesh nets work especially well for bottom fish such as Pacific Halibut (*Hippoglossus stenolepsis*), but the nets are indiscriminate, tangling and drowning diving seabirds and marine mammals as well.

Gill nets had actually been in use in Monterey Bay before 1981. In fact, there had been five or six gill-net boats on Monterey Bay throughout the 1970s. In 1981, however, there numbered 25 to 30 gill-net boats, most of them manned by refugees from South Vietnam. While the previous fishermen had used 0.4- to 0.8-km nets, the Vietnamese were setting nets up to 2.4 km long and fishing the same areas much more intensively. Coupled with this dramatic increase in gill-netting activity was an unusually large number of anchovies, which brought the seabirds into shallow waters, now filled with nets. This massive increase in dead seabirds did not go unnoticed. As many as 10,000 seabirds died in July alone, equaling the mortality caused by the major 1971 San Francisco oil spill (Heneman 1981). Spurred on by the Friends of the Sea Otter and other environmental groups, angered beach-goers, local newspapers, and residents of this popular tourist area, meetings began between the California Department of Fish and Game (CDFG), the U.S. Fish and Wildlife Service (USFWS), and private organizations, including PRBO. The parties hoped to find a solution that would maintain the fishery while reducing seabird mortality. As the use of gill nets was unregulated by state law, without enabling legislation CDFG had no power to restrict their use further.

The Search for a Legislative Solution

PRBO's participation in these initial discussions was mainly advisory, using its Beached Bird Project data to show that up to 90% of the seabird mortality was directly attributable to the use of gill nets (Heneman 1981). The main push for legislation, however, came from the Friends of the Sea Otter. In February, State Senator Henry Mello followed the recommendations of CDFG and sponsored bill SB 1475, which restricted gill netting in Monterey Bay to waters deeper than 10 fathoms (18.3 m). This bill also gave CDFG the authority to impose further restrictions if the initial actions were insufficient to reduce seabird mortality to "acceptable levels." As a result of budgetary constraints, however, the bill did not provide funds for CDFG to monitor the effectiveness of the regulations.

The omission of monitoring funds posed a significant problem. Senator Mello's bill presumed that placing the nets outside the shallow waters of Monterey Bay would reduce the drastic seabird mortality, but this was only an assumption. No one knew what level of mortality would occur in waters deeper than 10 fathoms. The Moss Landing Marine Lab had started monitoring the incidental catch by gill-net boats in 1981 to determine mortality (since many dead seabirds discarded from boats do not wash ashore) (Carter, personal communication). Unless the monitoring was continued, an unacceptable number of seabirds might still be killed, but a decline in seabirds found along the beaches would lead to the opposite, and incorrect, conclusion. In recognition of this danger, PRBO, along with the Monterey Peninsula Audubon Society, pledged to raise the $5,000 needed to continue the monitoring program through the summer (Heneman 1982). PRBO also solicited the aid of volunteers to help CDFG in their own beached bird censuses. Up to now, PRBO had played in important advisory and supporting role, but Monterey Bay was 100 miles south of PRBO, and the local organizations had taken the lead. This all changed in the summer of 1982.

The 17 June 1982 edition of the *Point Reyes Light*, a local Marin County newspaper, prominently featured an article titled, "Birds Face Net Peril" (PRL 1982*a*) (see Fig. 1). The paper reported that "in the past 2 years thousands of shore birds have died in gill nets used by fishermen in Monterey Bay. The same fishermen are now leaving that area and coming north seeking new territory.... Unfortunately, the fishermen are bringing the identical problems with them to our Bay" (PRL 1982*a*). In early June two Vietnamese boats, the *Cole* and the *Rhonda Lea*, mindful of the pending legislation, had moved up the coast from Monterey Bay and set 3.2

Figure 1. The Central California coast. Monterey Bay is just south of Devil's Slide. Source: Map by Jay Penniman in Point Reyes Bird Observatory Newsletter, 1986, 75:3.

km-long nets only 400 meters off Stinson Beach in Marin County (MIJ 1982*a*). The next day, two dead seals, an injured baby harbor porpoise (*Phocoena phocoena*), and 200 dead seabirds were found washed ashore (CP 1982*a*). PRBO's headquarters was only a few miles from that beach, and it quickly became involved.

Burr Heneman, then executive director of PRBO, had previously been involved with the gill-net issue in Monterey. He spoke to the press, was quoted in all the local papers, and wrote an article in the *PRBO Newsletter.* Within a week, Heneman traveled to Sacramento to obtain the consent of Senator Mello to amend his bill to include Marin and Sonoma counties and San Francisco Bay. Heneman met with E. C. Fullerton (director of CDFG), state legislators, and Zeke Grader (president of the Pacific Coast Federation of Fishermen's Associations). Neither Grader nor Fullerton supported a legislative ban on the use of gill nets along the entire coast. Thus a compromise was reached that granted CDFG the authority to prohibit or restrict the use of gill nets temporarily following a public hearing, if it was determined "that mortality in any local population of any species of seabird or marine mammal is occurring at a rate that threatens the viability of the local population, as a direct result of the use of gill nets" (Atkins & Heneman 1987). Governor Brown signed SB 1475 into law on 28 June, 1982.

The next day, a hearing was called by CDFG in Point Reyes to discuss the gill-net issue. By this time, the issue was extremely charged. Local residents had gathered over 700 signatures on petitions calling for the banning of gill nets off Marin County (MIJ 1982*b*). The *Cole* and the *Rhonda Lea* were still working up and down Stinson Beach, but now surrounded by government officials. As the *Coastal Post* reported: "Park lifeguards in their Zodiac vessel went out to the fishing boat and watched them pull in their nets. Although park employees estimate 200 dead California Murres [Common Murre = *Uria aalge*] and 10 or 15 Common Cormorants [Brandt's Cormorant = *Phalacrocorax penicillatus*] were visible in the nets, no mammals were seen. During the net pulling a news helicopter hung over the fishing operation for two hours. Federal and state officials waited on the beach to see if there was any evidence that the nets had caused the deaths and injury" (CP 1982*a*). Although none of the marine mammal deaths could be conclusively linked to gill-net fishing, the public regarded the connection as more than coincidental. The sudden appearance of Great White Sharks (*Carcharodon carcharias*), which led to the closing of the beach for a week in July, was also blamed on the dead fish and seals caused by gill netting (Heneman 1983). Indeed, by the time of the 29 June hearing, at least 13 Harbor Porpoises, fourteen Harbor Seals (*Phoca vitulina*), and a six-foot Blue Shark (*Prionace glauca*) had washed up on the Marin County coast (PRL 1982*b*).

At the hearing, PRBO testified along with other citizen groups, local government officials, and fishermen. But PRBO was the only group with actual data on what was happening. PRBO had observed six net pulls, about 10% of those that actually had occurred, and counted 400 dead seabirds (PRL 1982*b*). This testimony, as in Monterey, effectively dispelled the notion that "there wasn't a problem." As Heneman explained, "PRBO was not perceived as a biased advocate. We were an advocate for the right thing happening to wildlife and basing whatever happens on solid data. No other groups had that" (Heneman, personal communication). Zeke Grader, the fishermen's lobbyist, held a different perspective: "PRBO certainly played to a degree the role of advocates. But we kept their bias apart from their scientific information; and their information played a vital role. The fishermen felt the information was valid because it fit in with what they were seeing out there" (Grader, personal communication). In spite of PRBO's data, local fishermen, who for years had used hand-pulled gill nets (100 to 250 meters), were vocal about their concerns. Stated Josh Churchman, who fishes out of a 17-foot motorboat, "if you force us to move our nets to deeper water, I'm out of business—I'll lose my livelihood" (MIJ 1982*c*). While Churchman and others favored a ban on certain types of net, PRBO favored a ban on all gill netting under ten fathoms, along with continued monitoring. As Heneman stated in a *Coastal Post* editorial, "if you set any kind of nets where there are birds, you will catch birds" (CP 1982*b*). CDFG concurred with PRBO's proposal, adding that "we will not take a broad axe approach and just wipe out gill-net fishing. However, we may have to take a scalpel and go deeper until we find the source of the problem" (MIJ 1982*c*).

At the time of the hearing, seven gill-net boats were operating off Marin County, three of those Vietnamese. Within days of the regulations taking effect, PRBO learned of heavy seabird mortality down south along the San Mateo County coast. CDFG quickly responded with another hearing and new regulations. As Heneman explained, "the gill netters were finding new areas to fish faster than we were getting the locations covered by legislation. Now we were beginning to get reports of seabird and marine mammal kills following the familiar pattern along the Big Sur coast, near Morro Bay, and off Santa Barbara County. What was once a local Monterey problem was now regional in scope" (Heneman 1982). Still visiting Sacramento, Heneman helped draft a bill that would have given CDFG broad authority to regulate certain commercial fishing gear, such as nets that extensively damaged marine life. The bill was killed in the legislature by heavy lobbying from the southern California seafood industry (Heneman 1982). And the temporary ban on gill netting off Marin County expired in September. The monitoring, however, continued. CDFG assigned two department marine biologists to study further the impact of gill netting while PRBO and the International Council for Bird Preservation pledged to obtain the $5,000 necessary to continue the monitoring program (Heneman 1982).

More Temporary Restrictions

By the summer of 1983, many were dismayed by the situation. CDFG's temporary bans had expired the previous September, and in June, 1983, thousands of seabirds were once again washing up on the shores of San Francisco and Marin counties (PRL 1983a; CP 1983a). At the fourth CDFG hearing on gill netting held in early June, however, CDFG Director Fullerton concluded that the evidence presented did not warrant imposing temporary restrictions (PRL 1983b). The CDFG study started the previous year had not yielded any conclusive data.

CDFG's inaction may have seemed arbitrary, but it was consistent with the agency's philosophy of minimizing the impact to the gill-net fishery while protecting populations of species that were caught incidentally. There were still too many unanswered questions. As Heneman explained, "the phrase 'acceptable incidental kill' is the crux of the problem. Obviously, as long as netting continues you can't cut the losses to zero. But to tailor regulations that will reduce bird kills without bankrupting fishermen, you need a lot of hard data. Right now Fish and Game lacks the manpower and funding to do a thorough job of gathering all the information" (NF 1983). What species were being caught in nets made of different materials, with different mesh, set at different depths, at various times of year? A flat ban on all gill netting would make these questions moot, but Fullerton refused to consider such an approach. These questions were still unanswered when, in early August, hundreds more seabirds washed ashore along the San Francisco, San Mateo, and Marin county coasts (SFE 1983). The new head of CDFG, Howard Carper, called for a hearing which proved to be the most emotional gathering yet.

The August hearing began dramatically when the president of the San Francisco Society for the Prevention of Cruelty to Animals proclaimed, "we're talking about thousands and thousands and thousands of birds being slaughtered," while associates dumped a garbage can full of 72 foul-smelling individually wrapped seabirds onto the carpeted floor (MIJ 1983). On the opposing side, Dave Kruegel, the sponsor of a Vietnamese boat, claimed, to a chorus of boos, that a ban on gill nets would be a form of racism against Vietnamese fishermen (MIJ 1983). PRBO and CDFG biologists presented data that showed the continuing effect of gill netting on local seabird and marine mammal populations. The number of gill-net boats had increased from 34 boats in 1981 to 54 in 1982. At the time of the hearing, 74 gill-net boats were operating off central California (NF 1983). Part of this dramatic increase was due to an extremely poor salmon season, which forced salmon boats to fish with gill nets to make ends meet. PRBO concluded that this year's mortality of common murres alone could total 17,000 birds, on the order of 10% of the breeding population (Heneman 1983; MIJ 1983). Other environmental, civic, and sport-fishing groups spoke, and all but two of the gill-net fishermen present acknowledged the need for some form of regulation. Carper responded with temporary gill-net restrictions along the coast effective until 16 October, 1983 (Heneman 1983).

The Effect of Gill Netting on the Common Murre

The PRBO data proved especially useful at the hearing for it both reinforced and supplemented CDFG's findings. It also provided firsthand accounts of fishermen's attempts to conceal the incidental catch by weighting garbage bags full of dead birds or casting them out along rocky, inaccessible shores (Heneman 1983). Most important, though, was PRBO's ability to project the effects of mortality on the viability of the population. Since 1969, PRBO has served as the caretaker of the Farallon Islands, a National Wildlife Refuge that hosts the largest Common Murre breeding colony on the central California coast. In the mid-1800s, the murre colony on the Farallones may have numbered as high as 400,000 (Carter 1984). But by the end of the century the murre population on the Farallones had almost been wiped out by organized companies of egg gatherers, who satisfied the forty-niners' desire for "sunny side up," at a price. Over the years, the egg companies took as many as 12 million murre eggs (TW 1986). In later years, oil spills and bilge flushing outside the Golden Gate kept the population at low numbers. In 1972, three years after PRBO started managing the Farallones Islands, the common murre population stood at just 22,000. By 1982, the population had grown to 88,000 birds (Carter 1984). But over 90% of the seabird tangled in gill nets were common murres (NF 1983). Drawing on their data, PRBO scientists pointed out that such high levels of gill-net mortality would result in a rapid decline in the breeding population.

All the environmental groups could point out that birds were dying, and PRBO and CDFG could state how many, but only PRBO could show the long-term consequences on a specific population. As the *National Fisherman* reported, "Heneman's testimony provided the only concrete appraisal of the impact of net fishing on the overall bird population" (NF 1983). Don Schultz, a CDFG biologist, concurred: "PRBO was the public group with the most pertinent data regarding the status of the seabirds. In fact, they had more detailed information than the agency. When an outside entity provides data for far-reaching decisions, they are open to criticism; but the validity of PRBO's data was never challenged. They were recognized as experts in their field and the data was collected in a scientific manner. They were an integral part of the discussions" (Schultz, personal communication).

PRBO was convinced that the murre population could not sustain such a high level of mortality. Yet the pattern of a massive kill followed by a CDFG hearing and a temporary closure seemed to have become an endless loop. Five separate hearings had not stopped the mortality, and, on the fishermen's side, the stopgap measures simply pushed the incidental-kill problem around to other areas, where more boats competed for the same limited resource. Through the winter of 1983–1984, Heneman and Grader drafted a bill that provided for year-round and seasonal closures in particularly sensitive areas, such as that around the Farallon Islands. The bill also restricted the permissible length and height of nets and placed a one-year moratorium on the issuance of gill net permits. It did not, however, address marine mammal mortalities. The bill, SB 2266, was supported by all of the interested parties, including CDFG, California Gillnetters Association, National Audubon Society, and Marin County, and was signed into law on 20 June, 1984, by Governor Deukmajian (Heneman 1984).

While Heneman and Grader both told the press that "this is not a total solution" (MIJ 1984), the passage of the 1984 bill seemed, for the moment at least, to be an adequate resolution of the gill-net controversy. CDFG and PRBO monitoring studies continued, and, indeed, fewer numbers of murres were dying than in previous years. The question, though, was whether this represented a real decrease in mortality, or whether simply fewer birds were left to be killed. By early 1985, PRBO was convinced that the gill-netting situation was critical. In 1983, a cyclic global climatic event, called El Nino, had caused breeding failures and reduced seabird populations, so gill-net deaths had to be calculated from an already depleted population (PRL 1983c). Even taking the effects of El Nino on the murre colony into account, however, PRBO biologists found that gill netting still annually killed in the order of 10% of the murre population (Carter & Ainley 1986). Disseminating this data to other environmental groups, PRBO convinced them of the need for still stricter gill-net controls. But the commercial fishermen were unwilling to negotiate, believing the problem had been adequately resolved, and CDFG wanted to wait another season or two before evaluating the closures' adequacy. Ironically, while the closures had not halted the offshore mortality, they had been effective in eliminating the massive numbers of dead birds on beaches, thus removing the issue from the public's eye and denying the environmental groups their most effective tools: an outraged media and public. Frustrated at the state level, environmental organizations looked for the first time to the federal government.

The Initiative in Washington

Up until now, there had been little or no federal involvement in the gill-net affair beyond funding some of the monitoring programs. This seems odd, because the incidental kill of seabirds and marine mammals suggests conflict with the Migratory Bird Treaty Act, the Marine Mammal Protection Act, and the Endangered Species Act, all federal laws. The cause for the federal inaction lay in an informal agreement between USFWS, NMFS, and CDFG. As Heneman explained later, "unofficially, the position of the management staff at these [federal] agencies was that since the fisheries were state-managed and conducted in state waters, the solution was up to the state" (Atkins & Heneman 1987). The enforcement staff of the federal agencies did not want to get involved, either, because of the difficulty in obtaining incidental take convictions. At the same time, while California state law does incorporate the federal wildlife laws and CDFG did have the authority to enforce the laws, CDFG was understaffed and hesitant to step in and enforce provisions that the federal agencies were clearly avoiding and that contradicted their own stated policy of maintaining a viable gill-net fishery (Atkins & Heneman 1987). From the beginning, environmental groups had been content with federal agencies standing on the sidelines while CDFG tried various regulatory solutions. And the regulations had been effective to some extent, for the incidental take had been reduced from 1982 levels. But it was clear to PRBO that the problem had not gone away, and it convinced a number of national groups to take the struggle to Washington.

In the spring of 1985, Congresswoman Barbara Boxer and four other California legislators officially requested the USFWS and NMFS to carry out their obligations under federal law and act against illegal incidental take of seabirds and marine mammals by gill netting (CP 1985). They demanded a reply within sixty days. Under the Migratory Bird Treaty Act, 16 U.S.C. §703 et seq., it is illegal to kill birds included in the treaty unless the Secretary of the Interior adopts excepting legislation. The Common Murre is listed in the treaty. The Marine Mammal Protection Act, 16 U.S.C. §1531 et seq., similarly prohibits the killing of marine mammals subject to certain exceptions provided by the Secretary of Commerce. The Endangered Species Act, 16 U.S.C. §1531, prohibits the killing of species listed as "endangered" or "threatened," except pursuant to a research permit. The Southern Sea Otter (*Enhydra lutris nereis*) is listed as a threatened species. Thus, it appeared that the gill-net incidental kills were in blatant violation of federal and state law. USFWS's response acknowledged that the deaths had occurred and that proper preliminary steps might not have been taken. NMFS simply stated that they were presently working on the issue (Heneman, personal communication). Thus both agencies admitted that a problem existed, but neither committed itself to act. The initiative in Washington had apparently failed.

Up to now, PRBO had based its population data on the Farallon Island colonies. While USFWS's response in Washington had been noncommittal, the agency recognized that a problem existed and approached PRBO about the situation. Two PRBO biologists, David Ainley and Harry Carter, had previously studied the impact of gill netting in Alaska and British Columbia (Ainley et al. 1981; Carter & Sealy 1984). PRBO proposed that aerial censuses of murre colonies up and down the coast would provide exact and irrefutable evidence of the population's decline (Carter, personal communication). USFWS concurred, and beginning in the summer of 1985, USFWS and PRBO biologists took aerial photographs and conducted ground counts of murre colonies along the California coast. Meanwhile, Ainley continued an extensive study of the distribution of murres in the Gulf of the Farallones. This data would be helpful in identifying locations for gill netting where few seabirds would be affected. Heneman, now a consultant for the Marine Mammal Commission, was still working on forging a political solution. A workshop in January of 1986 succeeded in bringing together all the interested parties, including the federal agencies and the Vietnamese Fishermen's Association, but little was accomplished beyond heated debate. To many parties, gill netting was a resolved issue.

New Data and Threat of Litigation Break the Stalemate

In the summer of 1986, however, the aerial census data provided a shocking rebuttal. Flying over Devil's Slide Rock, USFWS and PRBO biologists expected to see the 2,300 murres that had bred there as recently as 1982. But there were no murres. The entire colony had been wiped out in only four years (Carter & Ainley 1986). As PRBO and USFWS biologists pored over the 1985 and 1986 aerial photographs, counting actual numbers of murres in each colony, they found that while murre colonies in northern California had remained stable since 1982, many colonies in central California had declined by up to 80%, even faster than the baseline colony monitored on the Farallones. Most of this decline was directly attributable to gill-netting (Takekawa et al. 1988). CDFG data also showed a harbor porpoise mortality rate of 200 to 300 per year out of a population as low as 3,000 porpoises (Heneman, personal communication). PRBO passed this new information to all the gill-net parties, issued a press release, and published a comprehensive article in its newsletter (Carter & Ainley 1986). The pace of events quickened.

The Washington-based Center for Environmental Education and the National Audubon Society, both of whom had previously watched from the sidelines, agreed to lobby USFWS and NMFS. Realizing that the threat of litigation was now the most effective alternative to break the political stalemate, Heneman encouraged a number of organizations, such as the United Anglers of California and Marin Peninsula Audubon Society, to raise funds to evaluate the strength of the gill-net case. The 17-page legal opinion by a Sacramento law firm concluded that the incidental gill-net mortalities of murres, Harbor Porpoises, Harbor Seals, and Southern Sea Otters were all violations of federal and state law. The attorneys declared that "the current problem does not seem to stem from any failure in law, but rather stems from a failure by those charged with the implementation of the law to fully carry out its mandate." In other words, the environmental groups had a very strong case and, thus, a new lever to force the government into action.

Encouraged by the strength of its legal position, later in the spring of 1986, the Center for Environmental Education and the National Audubon Society petitioned the Secretaries of Interior and Commerce (who oversee USFWS and NMFS, respectively) to enforce the Marine Mammal Protection Act and the Migratory Bird Treaty Act (Heneman 1986). In California, the Marin Audubon Society, the Whale Center, and the United Anglers of California petitioned the state attorney general to provide CDFG with a legal opinion of its responsibilities toward gill-net incidental mortality under state law. The attorney general's office issued its opinion to CDFG, stating that there was an "unequivocal prohibition against taking [of seabirds and harbor porpoises] under State and Federal law." Spurred by this strong opinion and concern from Washington and Sacramento, representatives from CDFG, USFWS, and NMFS met with the Center for Environmental Education lobbyists, Heneman, and PRBO biologists Ainley and Carter to discuss possible remedies. For the first time, the federal agencies conceded that serious action needed to be taken in order to protect the wildlife (Carter, personal communication). By September 1986, these meetings had expanded to include interested environmental groups and sport and commercial fishing interests: the same groups, in fact, who had reached an impasse nine months earlier.

No agreements were forthcoming. While the environmental groups had an undeniably strong legal case to force greater governmental enforcement efforts, the length and costs of litigation were daunting, as were the problems of enforcing a court order. For a long-term solution, the environmental groups needed final and effective legislation. But the backing of the fishermen was necessary for passage of any legislation, or their lobby would oppose it. In November, CDFG revealed its proposed regulation: closure of all fishing areas inside twenty fathoms. Although more comprehensive than the 1984 bill, the CDFG proposal angered both sides. The environmental organizations withheld support because PRBO's data on the distribution of murres indicated that gill-net fishing beyond twenty fathoms might actually increase mortality. The regulations were no benefit to the fishermen, either, for they removed the best gill-net fishing areas and made the White Croaker (*Gentonemus lineatus*) fishery, dominated by the Vietnamese, economically impractical. CDFG stood firm, however, and refused to consider lesser restrictions while likewise refusing to shut down the gill-net fishery (Carter, personal communication).

As Heneman, Grader, and others negotiated behind closed doors, another approach emerged: shifting some gill-net fishermen into a new fishery with alternative fishing methods (such as seine or traps). This proposal allowed CDFG to impose strict restrictions and permanent closures in the Gulf of the Farallones and other areas of high mortality. In exchange for the support of the Pacific Coast Federation of Fishermen's Associations and the Vietnamese Fishermen's Association of America, the compromise provided low-interest loans to fishermen who switched from gill nets to other equipment. In fact, the Vietnamese Fisherman's Association began to explore switching gear at this time with seed funding provided by fishing and environmental groups, including PRBO. The state, eager to bring the conflict to an end, did not oppose the funding provisions of the bill. With a long-term resolution of the problem in hand, the environmental groups agreed to a one-year period for introduction of the new regulations. Governor Deukmajian signed the bill into law in September 1987.

Today, the incidental take of seabirds and marine mammals from gill nets in central California is a resolved issue. A Nearshore Fisheries Research Advisory Committee has been formed from the various groups involved in the issue, including PRBO scientists. This committee is charged to evaluate the economic and biological consequences of alternative gear and methods, such as nighttime fishing, trolling, and trawling.

PRBO's Influence

Throughout this controversy, PRBO played a crucial role. And, surprisingly, neither government, fishermen, nor conservation groups viewed PRBO as a biased antagonist. Yet, as Heneman described later, at various times in the dispute, PRBO "provided its own gill-net impact data, helped collect and interpret agency data, pushed CDFG to conduct an adequate monitoring program, raised private funds to supplement the CDFG monitoring budget, negotiated regulations that were incorporated into several bills (all of which were lobbied through the legislature and across the governor's desk), pressured the federal agencies to become involved, assisted fishermen in switching to alternative gear, and kept PRBO members and the public informed through newsletters and the news media" (Atkins & Heneman 1987; Heneman, personal communication). In fact, PRBO and Heneman were mentioned by name in over 70 articles in newspapers and magazines. Central to PRBO's ability to maintain an objective position in the eyes of the participants was its role as a scientific research organization.

From the beginning, PRBO chose to restrict its contributions to those based on scientific data. As Heneman puts it, "we went down a road, right or wrong, of trying to nail down the fact that gill netting was having a biological impact." While this approach may seem indirect, it suited the situation well because CDFG had competing obligations: protection of wildlife and preservation of the fishing industry. The approach of environmental groups, translating support from the public into political action, was effective only to a point, Carter explained, because "the fishing industry also has strong emotional and political support and CDFG did not want to shut down the gill-net fishery." If CDFG was going to act incrementally, on what basis could its actions be challenged?

PRBO's use of its data in both public and private contexts was critical in preventing a stalemate and in insuring the adequacy of CDFG regulations. As Heneman reflected: "Information is power. You're in a terrible situation going into a room facing a bunch of fishermen with wives and children. 'You're killing birds.' 'No we're not.' Then what? You can't just develop the information, put it on paper and launch it into the blue. Nothing will happen. You must take the information, pursue it, and put the pressure on. You have a responsibility to find someone who can do something with it." Zeke Grader concurs: "The key to PRBO's effectiveness was the presentation of their information. They were very careful, very objective. They didn't say, 'it's your fault, fishermen.' They laid it out and said, 'here's the effect from oil spills, from gill netting, etc.' It came from the perspective of scientists who were actually out there, not groups who heard about it and came in later to say it was wrong. In all my dealings, that's pretty rare."

Basic research, such as investigating species' behavior and sources of mortality, is essential to improve our understanding of how to manage most effectively for biological diversity and viable populations. This research also aids management agencies and advocacy groups in identifying areas where their intervention will be biologically most significant. Yet, as the gill-net case study illustrates, research alone (i.e., simply reporting data) is often too faint a voice to be heard amid the din of controversial policy debates. PRBO did not simply report its findings and disappear, leaving the "policy makers" to resolve the issue. Because PRBO knew more about the biological impact of gill nets than anyone else, it remained an active participant to insure that informed judgments were made. Ainley's and Carter's previous experiences with gill netting proved important in providing irrefutable evidence of a population decline and in constructing scientifically sound solutions to reduce mortality. The volunteers at PRBO also were indispensable in collecting the sheer volume of data needed to make reliable conclusions. Heneman served a number of roles. He not only acted as the voice for PRBO biologists, but he also served as a fulcrum, facilitating movement and compromise from both environmentalists and fishermen.

In many ways, PRBO was an adjunct of the more vocal and visible environmental groups. While dumping a trash barrel of dead birds may have impressed the media, providing indisputable evidence of mortality at the population level was far more useful to CDFG. Both actions were important to the final outcome. But PRBO's data and scientific expertise might not have been accepted without question had it been viewed as part and parcel of advocacy. As Heneman and Carter contend, "being evenhanded and fair, being seen as scientists, is the main, precious thing that an organization like PRBO has to protect in a public policy issue. That's why it's inappropriate to go first to lawyers. That may be the right thing to do in that situation, and you can privately urge other groups. But PRBO can't do it." In fact, when, in 1986, the environmental groups were considering litigation, PRBO chose not to be included as a party, even though its data were the basis of the suit (Carter, personal communication).

Scientists and Focused Advocacy

There exists an ambiguity over the role of scientists as advocates. Judging from the gill-net case study and the nature of public decision making, advocacy seems far more effective in influencing policy than education or research alone. And surely, PRBO, despite its protestations, was an advocate at least on behalf of viable species populations. Yet it is at the mere recognition of advocacy that scientists and scientific organizations often balk. Dr. David Wilcove, senior ecologist with the Wilderness Society, acknowledges that "there exists the potential for far more cooperation between the advocacy groups and the scientific community than we enjoy at the present. Environmental disputes are becoming increasingly more technical in nature, and the input of the scientific community is definitely needed. But the nature of advocacy battles does not appeal to a lot of scientists" (Wilcove, personal communication). The contentious forum for resolving conservation disputes does seem to discourage many scientists from becoming involved. As Harvard Law School's Professor Richard Stewart, a prominent commentator on environmental law, explains, "scientists don't like the framework for resolving issues. They feel that the law oversimplifies and distorts science, and that the decisions are often made by scientific ignoramuses" (Stewart, personal communication). On the other side, advocates are often frustrated by the complexity of scientific issues. They want the scientists to provide strong, unambiguous statements that will sway a jury or policy maker (Black 1988).

Michael Bean, an attorney with the Environmental Defense Fund (EDF) and an authority on wildlife law, acknowledges the tensions inherent in scientists' involvement with advocacy groups. "Policy-makers and advocacy groups work in a milieu which requires making decisions every day, often without necessary knowledge. They must draw the best inferences they can with the available information. But some scientists require a higher degree of certainty about cause and effect" (Bean, personal communication). Doug Foy, director of the Boston-based Conservation Law Foundation, has reached the same conclusion: "Many scientists would rather be silent than give their best informed opinion, even if it is scientifically valid. The problem is not the absence of time; it's the refusal to be heard. Advocacy groups need scientists to stand up and be counted. We need their representation to offset the flaks" (Foy, personal communication). Scientists may well feel uncomfortable with the exigencies of decision making, and the uncertainties inherent in the process. As Wilcove describes, "when the expert witness begins to qualify everything he or she says, the advocate loses patience. Scientists must learn to convey their knowledge with sufficient clarity and persuasiveness to win in the advocacy arena."

To some extent, it appears that scientists view advocacy as antithetical to objectivity. Bean relates that "a number of scientists interested in joining EDF have expressed concern that they are burning their bridges and can't go back into academia and pure science, that they are tainted by joining a pure advocacy group. They believe it will interfere with their future opportunities." Scientists at PRBO also voiced concerns that an organization's participation in public disputes, if not carefully thought out, might have repercussions both in funding and in the esteem of other scientists. As Heneman explained, "PRBO wants to be the kind of organization that isn't perceived as a Sierra Club; PRBO is a science organization with deeply held convictions about a few issues it knows a lot about" (Heneman, personal communication). Heneman's distinction is critical, for the gill-net case study indicates that PRBO exercised influence through the parties' complete acceptance of its data, not through the mere fact that biologists were advocating a position.

More precisely, the mandate of government environmental agencies stems from legislation, and laws such as the Endangered Species Act and the Marine Mammal Protection Act are founded upon scientific definitions such as critical habitat and viable population. In any specific dispute, giving meaning to these legal terms requires data. Thus the value of PRBO's research prior to the advent of the gill-net controversy cannot be overestimated. The beached-bird counts, the murre colony data, and the harbor seal studies were all invaluable for understanding and quantifying the mortality caused by gill nets. Moreover, this data was the driving force behind the negotiated settlements. Every time legislation was passed, the bills had either the support or the neutrality of the fishing lobby. Each bill was written with fishing-group lobbyists, and each went a step farther toward a final solution. But negotiation alone proved insufficient to halt the gill-net mortality. The threat of litigation was crucial to resolving the conflict. Without the power of federal wildlife legislation, the 1984 bill would probably still be the law, coupled with new cycles of seasonal closures. Only when the federal agencies faced the likelihood of fighting, and losing, a lawsuit, did they pressure CDFG to reach a permanent solution.

PRBO's data were thus crucial, but advocacy organizations can rarely afford to pay researchers to count dead birds on the off chance that the data might prove useful at some future date. As Stewart notes, "it is fairly rare to have a scientific institution focused on a particular set of issues congruent with a conservation dispute." Yet this fortunate congruence assured PRBO access to the decision makers.

PRBO's involvement and influence in the gill-net controversy shows how scientists can be extremely effective in policy disputes and maintain their objectivity if they employ a focused advocacy: reporting data and pressing to insure that the information is interpreted correctly and acted upon. Central to focused advocacy is basic conservation-oriented research. In retrospect, PRBO was both smart and lucky. By intentionally directing its research efforts toward biologically sensitive areas and species, PRBO was able to produce data that addressed biological questions posed by gill netting. Litigation can take years, but basic research often takes more years. Moreover, there is no way prior to a conservation crisis to insure that scientists will have performed relevant basic research. The probability of confluence can be increased, however, by supporting research efforts directed toward specific environmentally sensitive areas.

It remains an open question whether a lawsuit would have resulted in a preferable outcome for the fishermen and environmentalists. In retrospect, negotiation did

Figure 2. Muir beach, summer 1984.

forge a long-term solution acceptable to all parties, a result that a court-ordered settlement probably would not have achieved. Negotiation may also have been preferable for the wildlife, for gill-net mortality is not restricted to California. Each year hundreds of thousands of seabirds are killed in the North Pacific and North Atlantic fisheries (Ainley et al. 1981; Piatt et al. 1984; Piatt & Reddin 1984). Because there are a number of nations involved, such as Japan, Greenland, and Denmark, legal obligations are indistinct and difficult to enforce over such large ocean areas. These conservation problems differ substantially from those present in California, where a small number of fishermen worked in a discrete area and were subject solely to American law. Nonetheless, the resolution of the gill-net controversy off central California does offer a model and hope for future multinational agreement.

Acknowledgments

This article was written under the Winter Term Writing Program at the Harvard Law School. The author is espe-

Figure 3. Murres on the beach.

cially grateful to Harry Carter for his extensive editorial and scientific suggestions. Burr Heneman, David Wilcove, Martha Minow, and Eric Fajer also critically reviewed the manuscript and provided many helpful comments. The author thanks Richard Stewart, William Bossert, Michael Bean, Doug Foy, and Ira Rubinoff for sharing their thoughts on science and advocacy. The Point Reyes Bird Observatory was most generous in providing the use of its facilities, resources, and photographs.

Literature Cited

Ainley, D. G., A. R. DeGange, L. L. Jones, and R. J. Beach. 1981. Mortality of seabirds in high-seas salmon gill nets. Fisheries Bulletin **79**:800–806.

Atkins, N., and B. Heneman. 1987. The dangers of gill netting to seabirds. American Birds **41**:1395–1403.

Black, B. 1988. Evolving legal standards for the admissibility of scientific evidence. Science **239**:1508–1512.

Carter, H. R. 1984. Rise and fall of the Farallon common murre. PRBO Newsletter **72**:1–3, 11.

Carter, H. R., and D. G. Ainley. 1986. Disappearing murres. PRBO Newsletter **75**:4.

Carter, H. R., and S. G. Sealy. 1984. Marbled murrelet mortality due to gill-net fishing in Barkley Sound, British Columbia. Pages 212–220 in D. N. Nettleship, G. A. Sanger, and P. F. Springer, editors. Marine birds: their feeding ecology and commercial fisheries relationships. Canadian Wildlife Service Special Publication.

Coastal Post. 1982a. Vietnamese fishing nets may be responsible for two dead seals and injured porpoise pup. 14 June, 1982. Newspaper article, Marin County, California.

Coastal Post. 1982b. 6 July, 1982. Editorial, Marin County, California.

Coastal Post. 1983a. Off-limits gill netting area asked. 6 June, 1983. Newspaper article, Marin County, California.

Coastal Post. 1985. Feds asked to help with illegal gillnetting. 4 March, 1985. Newspaper article, Marin County, California.

Heneman, B. 1981. Why count dead seabirds? PRBO Newsletter **55**:2.

Heneman, B. 1982. Monterey Bay gill net legislation — help needed. PRBO Newsletter **58**:7.

Heneman, B. 1983. Gill nets and seabirds 1983. PRBO Newsletter **63**:1, 15, 10.

Heneman, B. 1984. Gill net news. PRBO Newsletter **65**:7.

Heneman, B. 1986. Gill net news. PRBO Newsletter **75**:1–3, 10.

Marin Independent Journal. 1982a. New law cracks down on gill net fishing. 1 July, 1982. Newspaper article, Marin County, California.

Marin Independent Journal. 1982b. Bill limiting gill net fishing faces Point Reyes hearing. 23 June, 1982. Newspaper article, Marin County, California.

Marin Independent Journal. 1982c. Fishermen fear new gill net law. 29 June, 1982. Newspaper article, Marin County, California.

Marin Independent Journal. 1983. Gill-netting debate renewed. 1 August, 1983. Newspaper article, Marin County, California.

Marin Independent Journal. 1984. Fishermen, environmentalists hail gill-net law. 22 June, 1984. Newspaper article, Marin County, California.

McCrimmon, D. A. Quo vadis? 1985. PRBO begins its second 20 years. PRBO Newsletter **69**:1, 6.

National Fisherman. 1983. Fish and Game orders more closures for California gillnetters. November.

Piatt, J. F., D. N. Nettleship, and W. Threlfall. 1984. Net mortality of common murres and Atlantic puffins in Newfoundland, 1951–81. Pages 196–206 in D. N. Nettleship, G. A. Sanger, and P. F. Springer, editors. Marine birds: their feeding ecology and commercial fisheries relationships. Canadian Wildlife Service Special Publication.

Piatt, J. F., and D. G. Reddin. 1984. Recent trends in the west Greenland salmon fishery, and implications for thick-billed murres. Pages 208–210 in D. N. Nettleship, G. A. Sanger, and P. F. Springer, editors. Marine birds: their feeding ecology and commercial fisheries relationships. Canadian Wildlife Service Special Publication.

Point Reyes Bird Observatory. 1986. Annual report, 1986. Point Reyes Bird Observatory, Point Reyes, California.

Point Reyes Light. 1982a. Birds face net peril. 17 June, 1982. Newspaper article, Point Reyes, California.

Point Reyes Light. 1982b. State can't control gill nets. 1 July, 1982. Newspaper article, Point Reyes, California.

Point Reyes Light. 1983a. More birds killed by gill nets. 2 June, 1983. Newspaper article, Point Reyes, California.

Point Reyes Light. 1983b. Fish and Game refuses to regulate gill nets. 30 June, 1983. Newspaper article, Point Reyes, California.

Point Reyes Light. 1983c. Gill nets. 4 August, 1983. Newspaper article, Point Reyes, California.

San Francisco Examiner. 1984. Gill nets kill hundreds of birds. 1 August, 1983. Newspaper article, San Francisco, California.

Takekawa, J. E., H. R. Carter, and T. E. Harvey. 1988. Decline of the common murre in central California. San Francisco National Wildlife Refuge. Unpublished manuscript.

This World. 1986. Gill nets. 14 September, 1986. Newspaper article, San Francisco, California.

The Sweetwater Rattlesnake Round-Up: A Case Study in Environmental Ethics

JACK WEIR

Professor of Philosophy
Morehead State University
UPO 0662
Morehead, KY 40351, U.S.A.

Abstract: *This study is an ethical analysis and evaluation of an ecologically and environmentally intrusive social phenomenon, the Sweetwater (Texas) Jaycees Rattlesnake Round-Up, an annual event that results in the killing of thousands of rattlesnakes (*Crotalus atrox*). The largest and most publicized of several such festivals in the United States, the Sweetwater Round-Up has developed into a huge community extravaganza with lucrative benefits to the local economy. After overviews of the Round-Up and of recent biological research, the main reasons justifying the practice are evaluated from both a local and environmental perspective. Because the annual take has remained high, the proponents claim that the Round-Ups are not reducing the rattlesnake population but are in fact "helping the balance of nature." Although probably false due to lost reproductive potential, this claim cannot be decisively refuted due to a lack of reliable scientific data. Therefore, recommendations are made for restructuring the rattlesnake hunt to obtain scientifically credible data. Paramount is that the time span and territory be limited, because otherwise conclusions based on annual take cannot be made. Recommendations are also presented for conducting additional research and for limiting the ecological impact of the festival while at the same time preserving its economic and social benefits.*

Resumen: *Este estudio es un análisis ético y una evaluación de un fenómeno social intrusivo tanto en sentido ecológico como ambiental: el rodeo de la serpiente de cascabel "Jaycees" en Sweetwater (Texas); un evento anual que tiene como resultado la matanza de miles de serpientes de cascabel (*Crotalus atrox*). El rodeo de Sweetwater, el festival más grande y más publicitado en su tipo, se ha convertido en un evento espectacular con beneficios lucrativos para la economia local. Las principales razones de esta práctica son evaluadas desde una perspectiva tanto local como ambiental. Dado que la captura anual ha permanecido alta, los adherentes afirman que los rodeos no están reduciendo la población de serpientes de cascabel, sino que por el contrario están contribuyendo al equilibrio de la naturaleza. A pesar de que muy probablemente esta afirmación sea falsa, debido a la pérdida del potencial reproductivo, no es posible refutarla por la falta de datos científicos confiables. Por consiguiente, se hacen recomendaciones para reestructurar la caza de la serpiente de cascabel a los efectos de obtener datos científicos confiables. Resulta importante limitar el tiempo y el territorio ya que en caso contrario no se pueden obtener conclusiones basadas en la captura anual. También se presentan recomendaciones para llevar a cabo investigaciones adicionales y para limitar el impacto ecológico del festival preservando al mismo tiempo sus beneficios económicos y sociales.*

"There are numerous areas so overpopulated that we'll never be able to hunt the rattlesnake out. Ever. Not ever."
—J. Shaddix, Snakeskin Seller, at the 1989 Sweetwater Rattlesnake Round-Up (*Sweetwater Reporter* 1989)

Paper submitted October 22, 1990; revised manuscript accepted July 25, 1991.

Promoted as "the World's Largest," the Sweetwater Jaycees Rattlesnake Round-Up draws up to 35,000 spectators, including celebrities, and has resulted in the capture and killing of up to 18,000 snakes at a single weekend event. The Sweetwater Round-Up is unquestionably the biggest and most spectacular of several similar events occurring elsewhere in the United States.

Now in its thirty-fourth year, the annual festival regularly receives national and international media attention. Remarkably, only recently has the festival been studied by environmental and ecological scientists. It has never been studied by social scientists, and not until 1989 was it picketed by environmental activists.

In addition to the Sweetwater Round-Up, less spectacular events are held in Pennsylvania, Georgia, Alabama, Florida, Oklahoma, and several other Texas cities. Reportedly the oldest dates to 1939 in Okeene, Oklahoma. According to Klauber (1972), "These drives, ostensibly to rid the surrounding country of dangerous snakes, and to profit from their sale, are really publicity projects evolved by Junior Chambers of Commerce." Communitywide rattlesnake hunts date back to colonial times when they were sometimes called "rattlesnake bees." Snake killing around dens was a regular practice. In Massachusetts, as early as 1680 hunters were employed for two shillings a day to kill rattlesnakes, and in 1740 a particular day, evidently at fall hibernating time, was decreed for a general snake hunt. These practices spread into the Midwest. Bounties have been offered, traps set, dens poisoned, and statewide control officers employed. Iowa had a rattlesnake hunt in 1849 in which hunters were divided into two contestant groups and 3750 snakes were killed (Klauber 1972).

Admittedly evaluative, this paper is a critical examination, from an ethical perspective, of this bizarre social phenomenon; I analyze and evaluate the arguments given by the Sweetwater Jaycees to justify the Round-Up. These arguments are important to more than Texans, and for more than rattlesnake round-ups, because similar reasons are routinely given throughout the United States to defend a wide variety of environmentally intrusive actions and policies. Killing so-called "pests" is made into sport and festival, such as the recent nationally publicized contests involving shooting prairie dogs in Colorado and pigeons in Pennsylvania. These cases can be especially emotive and dramatic because they involve species that can threaten human health, individual livelihoods, and community economies. These various cases also present similar difficulties in finding solutions satisfactory to environmental scientists, environmental activists, and local residents, who inevitably have deeply vested personal and economic interests in the festivals.

Historical Overview

The Sweetwater Jaycees Rattlesnake Round-Up is held annually on the second weekend of March (Thursday through Sunday) at and around the Nolan County Coliseum in Sweetwater, a West Texas county-seat town of 12,000 that is largely dependent upon ranching and oil for its economy. The nearest city is Abilene, 45 miles (72 km) to the east, and the nearest metroplex is Dallas–Fort Worth, 150 miles (240 km) beyond Abilene.

The Sweetwater Rattlesnake Round-Up is one of the oldest and is undisputed as "the world's largest." It is of particular interest to biologists and social scientists because statistical records have been kept (see Table 1). The Round-Up began in 1958 when some ranchers proposed a snake hunt "to rid the area of rattlesnakes" because "the snakes were damaging their livestock" (Sweetwater Jaycees 1990a). The 1958 "Rattlesnake Hunt" was promoted by the city's Board of Development and was loosely organized with no admission fees. After unsuccessfully trying to kill them with exhaust from a pickup truck, the "thousands" of captured snakes were decapitated and deposited in the city dump. Overwhelmed by the crowds, the Board of Development decided to turn the project over to the Jaycees, who then sent a delegation to the round-up in Okeene, Oklahoma, to gather ideas. In the early years and until B. Ransberger, a local Jaycee, was sufficiently trained, professional snake handlers B. Jenni from Oklahoma City, S. Downs from Florida, and P. Burchfield from Columbus, Ohio, were hired to assist and give snake-handling demonstrations. Today, however, the "gala affair" is "strictly a hometown project" (Shelton 1981). Initially the Round-Up lost money, but its income now enables the Jaycees to contribute to numerous community benevolences (Sweetwater Jaycees 1990a). For instance, in 1989 the Round-Up's gross income was approximately $55,000. About $30,000 was profit, of which 90% went to charitable projects (Cox 1991).

Over the decades, the event has evolved into a spring community extravaganza with huge crowds and numerous collateral activities. Recent attendance has been

Table 1. Sweetwater Rattlesnake Round-Up statistics.

Year	Pounds (kg)	Snakes	Year	Pounds (kg)	Snakes
1958	Undetermined		1975	5730 (2601)	
1959	3128 (1420)	3000	1976	2397 (1088)	
1960	8989 (4081)	6881	1977	6348 (2882)	
1961	4584 (2081)	4584	1978	3343 (1518)	
1962	2392 (1086)	2392	1979	5839 (2651)	
1963	4500 (2043)	5000	1980	4470 (2029)	
1964	3762 (1708)	3900	1981	5155 (2340)	
1965	2340 (1062)	1900	1982	17,986 (8166)	
1966	3400 (1544)	4021	1983	15,053 (6834)	
1967	4000 (1816)	4300	1984	6281 (2852)	
1968	Undetermined	3400	1985	12,797 (5810)	
1969	2474 (1123)	3500	1986	16,086 (7303)	
1970	8886 (3768)	9017	1987	11,359 (5157)	
1971	3700 (1680)	4246	1988	11,709 (5316)	
1972	7274 (3302)	6379	1989	3620 (1643)	
1973	3584 (1627)	4103[a]	1990	3129 (1421)	
1974	2456 (1115)		1991	4474 (2031)	

[a] Statistics on the total number of snakes captured have not been kept since 1973.
Sources: Pounds (Sweetwater Jaycees 1990g, 1991; Shelton 1981), Snakes (Shelton 1981).

30,000–35,000 ($5.00 admission for adults and $2.00 for children). Current functions directly associated with the rattlesnake hunt itself include hunting competition by individuals and teams for trophies and prize money ($15.00 per person registration fee), with a list of lands opened for hunting provided to the contestants (weigh-in of snakes lasting from Friday at 8:00 A.M. until Sunday 12:00 noon); "educational" and "safety" demonstrations of snake handling and striking by snake handlers, notably Ransberger, a Sweetwater Jaycee who has developed a national and international reputation; guided hunts ($50 per person for the first day, $35 for the second day); guided photographic excursions; bus tours ($3.00 per person); "milking" demonstrations to obtain venom to be sold for research and pharmaceutical purposes; killing, butchering, and skinning of snakes, with spectators voluntarily participating in the skinning; cooking and selling of "chicken-fried" snake meat; a rattler meat–eating contest; selling of snake-handling and hunting paraphernalia, such as hooks, tongs, mirrors, and so on; and selling of curios and souvenirs, such as snake-rattle keyrings and official T-shirts. Shelton (1981) declares that "every part of the rattlesnake is used, including the venom for research." Prize money and/or trophies are awarded for the most pounds collected (from $700 for first place to $300 for fifth place), the longest snake ($300), the smallest snake (no prize money), and the record longest snake if captured ($500). Moreover, the snakes captured by the registered hunters (and any other rattlesnakes brought in) are purchased by the Jaycees ($8.50 per pound [0.454 kg] in 1989; $5.50 in 1991) and then resold at the current market price (usually to exotic skin dealers and curio manufacturers) ($12.00 per pound [0.454 kg] in 1989). As an indication of the amount of money involved, the first-place winner in 1989 collected 738 pounds (335 kg) of rattlesnakes (roughly 642 snakes) and received $6973 in prize money, and the second-place winner in 1989 collected 492 pounds (223 kg) of rattlesnakes (roughly 428 snakes) and received $4782 in prize money (Sweetwater Jaycees 1989, 1990*a,b,c,d,e,f*; Zelisko 1990*b*).

In addition to the activities and events directly linked to snake hunting, recent Round-Ups have included the following collateral festivities: Rattlesnake Review Parade: Miss Snake Charmer Queen Contest; Men's Beard Growing Contest; Rattlesnake Dances with country and western bands; Rattlesnake Run (2-mile [3.2 km] walk/run and 6.2-mile [10 km] run); Curio Show for displaying and selling of arts and crafts; Gun, Knife, and Coin Show for displaying and selling these and other items; Open Cook-Off contest of domestic meat or wild game; Chili/Brisket Cook-Off contest; archery contest; and an antique car show called Snake, Rattle, and Roll.

The hunters capture the snakes and bring them to the Nolan County Coliseum, headquarters for the festival, where the snakes are weighed and deposited in a snake pit for live demonstrations and eventually either butchering or sale to dealers. The recommended equipment for the hunt is: "pressurized spray can (approximately three (3) gallon [11.4 liters] size with 10′ × ¼″ [3.05 m × 6.4 mm] copper tubing extension on hose), container for snakes (20 gallon [75.8 liters] trash can or large burlap bags), high-top boots, snake-bite kit, and snake hooks" (Shelton 1981; *cf.* Sweetwater Jaycees 1990*d*).

To force the snakes out of their dens, rattlesnake hunters throughout the United States use a gasoline sprayer technique that was perfected in Sweetwater. The method "was accidentally discovered by J. S. 'Slim' Staton and Jim Dulaney well before the Rattlesnake Round-Up started in 1958" (Shelton 1981). Finding a large den under huge boulders, Staton and Dulaney decided to use a garden-type sprayer to inject gasoline into the den and then burn the snakes. But when they sprayed gasoline into the den and before they were able to ignite it, "rattlers began pouring out" (Shelton 1981). Perfecting the procedure over the years, hunters attach a long copper tube to the sprayer and try to spray the back of the den. Shelton (1981) explains: "Experienced hunters take great pains to get their copper tubing to [the] back side of the den before spraying fumes. If spray is in front of the rattleers [sic], they will move back instead of making exit. If too much gasoline is sprayed into the den and on the snakes, they do not emerge but die inside the den. Patience and experience develops this type of hunting to a fine degree."

Often other species of snakes are gassed and captured by this method, including many harmless snakes. Although no quantitative data are available, biologists at the 1986 Round-Up observed that, in addition to western diamondbacks (*Crotalus atrox*), the following species were brought in by hunters: western rattlesnakes (*Crotalus veridis*), massasauga rattlesnakes (*Crotalus catenatus*), racers (*Coluber constrictor*), coachwhips (*Masticophis flagellum*), glossy snakes (*Arizona elegans*), and western bullsnakes (*Pituophis melanoleucus*) (Campbell et al. 1989). The rattlesnake control officer in South Dakato, A. M. Jackley, reported in 1946 that for every 100 rattlers killed by poisoning dens, about 40 harmless snakes are also killed (Klauber 1972).

Based on calculations from the data listed in Table 1, the amount of rattlesnakes captured per Round-Up has averaged 6289 pounds (2855 kg) (or 5469 snakes), calculated at 1.15 pounds [0.52 kg] per rattlesnake, the estimate reached by Campbell et al. (1989). Campbell et al.'s estimate, however, may be high; my own calculations, based on the data from 1959 to 1973, omitting 1968, when weight was undetermined, suggest 0.996 pounds [0.452 kg] per snake—63,013 pounds [28,608 kg] and 63,223 snakes. Nonetheless, I have used their higher figure because they had data unavailable to me. (For some years [1963, 1967, 1968] the data in Table 1 appear to be inaccurate because the figures are too imprecise.) Including 1991, a total of 201,245 pounds (91,365 kg) (174,996 snakes) have been captured. The number of rattlesnakes captured has varied from a low in 1965 of 2340 pounds (1062 kg) (2035 snakes) to a high in 1982 of 17,986 pounds (8166 kg) (15,640 snakes). Generally peak years have been followed by low years, which have been followed by high years. Perhaps most significant are the ten-year totals:

1959–69	39,569 lbs. (17,964 kg)
1970–79	49,557 lbs. (22,499 kg)
1980–89	104,516 lbs. (47,450 kg)

Factors influencing the number of snakes captured include the number of hunters, their skill and aggressiveness, and—most importantly—the weather, since early warm temperatures and mild winters can cause the snakes to vacate their dens before the Round-Up. In addition, the Jaycees pay below market price for the snakes ($8.50 per pound [0.454 kg] as opposed to $12.00 in 1989), and consequently some hunters do not sell their snakes to the Jaycees. For instance, in 1989 the low snake poundage (3692 lbs. [1,676 kg]) was attributed to a warm winter and high prices elsewhere for snakes (Sweetwater Reporter 1989). In addition, many hunters probably keep a specimen or two for their own collections. Hence, the totals reported by the Jaycees are probably low.

Ecological Impact

The ecological impact of rattlesnake round-ups has been examined by Campbell et al. (1989) by using on-site observations, statistical analyses, and laboratory tests. To date, this has been the only scientific study, and additional research is greatly needed. (For standard information on *Crotalus atrox* biology and ecology, see Gloyd 1978; Goin et al. 1978; Klauber 1972; and Tennant 1985.) The available statistics are difficult to interpret due to the absence of data on undisturbed populations. Campbell et al. scrutinized two aspects of the round-ups: (1) the impact of the hunts upon population distribution, and (2) the effect of the gasoline-spraying technique upon the habitat and larger community. Regarding the former, rattlesnake hunters and round-up organizers have contended that the snake population has not been significantly affected by the massive collecting efforts because large numbers of snakes continue to be found during the hunts. After analyzing the available statistics for 1978–1986 regarding the number of snakes captured and the number of registered snake hunters, Campbell et al. concluded that the data neither support nor decisively refute the claims of the proponents of the rattlesnake round-ups. The increases in the number of snakes collected per year were due primarily to increases in the number of hunters since the number of snakes captured per hunter per year did not change significantly. Nevertheless, estimates based on the number of females captured and the resultant lost reproductive potential indicate a significant positive correlation and probable long-term consequences. Because the snakes live and reproduce for about ten years, it is possible that the hunters are capturing yearling adults that have not and will not be able to reproduce.

Importantly, Campbell et al. (1989) recognize that the stability in the number of snakes captured per hunter may have resulted from broadening the territories from which the snakes have been collected, from more intensive hunting efforts among decreased snake populations, or both. During interviews conducted with hunters at Freer, Texas, several hunters "indicated that many snakes were brought in from long distances."

Campbell et al. (1989) comment that, to obtain conclusive data, a control group of snakes would need to be monitored to gauge the relative fluctuations in population density due to such natural events as drought, disease, and predation. In addition, some estimate of the size (weight) of the snakes collected would be helpful for determining maturity. Currently unavailable, such data are expensive and difficult to obtain.

Regarding the second aspect of the research, spraying gasoline into the dens has numerous side effects. Other species living in the rattlesnake community are affected, including harmless species of colubrid snakes, lizards, toads, insects, and so on. Gasoline undoubtedly directly contacts some of the species and kills some. Laboratory tests were conducted on the effects of gasoline fumes upon a variety of species (western diamondback rattlesnake [*Crotalus atrox*], rat snake [*Elaphe obsoleta*], common kingsnake [*Lampropeltus getulus*], prairie kingsnake [*Lampropeltus calligaster*], racer [*Coluber constrictor*], collared lizard [*Crotaphytus collaris*], Woodhouse's Toad [*Bufo woodhousei*], and house cricket [*Acheta domesticus*]). Campbell et al. (1989) found that exposure to fumes (30 and 60 minutes for the snakes; 10 and 30 minutes for the lizards, toads, and crickets) significantly affects the species' abilities to right themselves and forage for food. Only one snake and one lizard did not recover from the 30-minute exposure, but none of the toads and none of the crickets recovered from the 30-minute exposure. Moreover, some species' abilities to forage was affected even 7 days after exposure. Long exposures (60 minutes) killed more of the specimens (1 out of 10 rattlesnakes, 4 out of 9 colubrid snakes). Campbell et al. (1989) concluded that "spraying of gasoline into dens may have direct effects on the health of individuals of other species in the habitat and indirect effects through decreases in quality (decreased prey availability, decreased availability of suitable dens or burrows) of those habitats for the resident species."

Proponents' Arguments

The Sweetwater Jaycees and other proponents give four explicit reasons to justify the Rattlesnake Round-Up. To these can be added the social benefits of the festival, the impact of the festival on the city's economy, and the sporting values derived by the hunters and spectators. Because of space limitations, only the four explicit claims can be considered. The first is examined in depth and the others briefly.

Protecting Livestock by Helping the Balance of Nature

In response to accusations by environmentalists that the Sweetwater Round-Ups are exterminating the snake, the Jaycees and other proponents have argued that the annual hunts are helping the balance of nature and not significantly affecting snake population densities as evidenced by the large numbers of snakes that continue to be brought in each year by hunters (Shelton 1981; Campbell et al. 1989; Sweetwater Reporter 1989). This claim is probably false due to the destruction of reproductive potential, the expansion of the hunt to include ever-larger geographical territory, more aggressive hunting, and other factors. Inconsistently with this defensive claim, Jaycees's publications explicitly state that the Round-Up was initially conceived as a means of reducing—if not exterminating—the rattlesnake population: "In 1958, a group of ranchers got together to rid the area of rattlesnakes. The snakes were damaging their livestock. Thus was born the Rattlesnake Round-Up" (Sweetwater Jaycees 1990a). Nevertheless, when defending the Round-Up against environmentalists, the sponsors are usually careful to claim that they do not want to exterminate the snake, as when Shelton (1981) states: "Purpose of the Sweetwater Jaycees Rattlesnake Round-Up has never been to exterminate the specie [sic] but to aid nature in controlling over population of rattlers in the area. Western Diamondbacks are very prolific and in earlier years became a real hazard at times. Much livestock was lost in addition to numerous citizens being struck."

Amazingly, by citing hunting as a means of increasing deer populations, Shelton (1981) next implies that the snake hunts may actually be *increasing* the number of rattlesnakes, a speculation surely of interest to the putatively infested ranchers. Again inconsistently, in a less defensive context, Shelton (1981) willingly concedes that the population has been noticeably reduced. Regarding the 18 Ranch north of Sweetwater, he states: "Although less amount of these critters infest this country today, there remains ample supply and there is no danger of extermination." Another proponent, rare skin dealer J. Shaddix, asserts that more than 100 million rattlesnakes inhabit Texas and that the snakes are so prolific that they will never be hunted to extinction (Sweetwater Reporter 1989). Shaddix's estimate is perhaps high; approximately two-thirds of Texas is suitable habitat (about 111 million acres), and rattlesnakes in other regions have been known to populate their habitat at about one snake per acre (see Klauber 1972). The Texas habitat, however, has been reduced significantly by farming, cleared land, and urban areas.

Rather than complicity or self-deception, the inconsistency among the Jaycees and other proponents may indicate no more than differing opinions among a diverse and independent club membership. Nonetheless, the comments show that the original purpose "to rid the area of rattlesnakes" has been abandoned and that the festival now has taken on a life of its own. On the one hand, if the purpose originally was to reduce the snake population and the hunt is failing to do so, as some proponents recently have claimed, then why should the futile effort be continued—unless the social and civic purposes are now paramount? But, on the other hand, if the Round-Ups are significantly reducing the snake population and the intent is not to exterminate the species from the region, then the festival should either be discontinued or appropriately curtailed to permit more snakes to survive. I suspect that many West Texans, including both townspeople and ranchers, would like the snake totally exterminated from all except remote wilderness regions, especially if the rattlesnake's niche in the ecosystem could be replaced by species harmless to humans and livestock.

Significantly, an economic factor that has developed since the Round-Up was started in 1958 should not be overlooked. Rattlesnake skins now have high market value for making exotic boots, belts, purses, and so on. This market has resulted from the sensationalism attached to rattlesnakes, not from the intrinsic utility of the products. So lucrative is the market that snake hunters and skin dealers can earn livings as professionals capturing and selling their quarry. These professionals are dependent upon rattlesnake round-ups since during these events landowners open their properties to hunting, and large numbers of snakes are captured. With vested interest, hunters, skin dealers, and manufacturers want the festivals to continue. So lucrative is the hunt that it is surprising that landowners have not started posting their property to hunters and harvesting the snakes themselves. Like other types of exotic animal farms, perhaps even rattlesnake ranches could be profitable. So far, however, rattlesnake farming has been unsuccessful. In general, rattlesnakes in captivity are not cared for easily, require expensive food, catch diseases easily, and do not reproduce (see Klauber 1972).

Some of the empirical claims in the proponents' argument are unsubstantiated by the evidence. Rattlesnakes seldom bite livestock and, when they do, the bite is seldom fatal. Even bites of humans are seldom fatal. Regrettably, no quantitative data are available on the densities or dynamics of diamondback populations in Texas. Nevertheless, Klauber (1972) reports two studies that, despite being completed several years ago, are still valuable because they were completed when snake populations were probably larger than now. The first was from 1936–1948 at the San Joaquin Experimental Range in Madera County, California, when the northern Pacific rattlesnake (*Crotalus viridis oreganus*) population was estimated at about 1.2 snakes per acre. The annual snakebite frequency in the herd of cattle (about 190 cattle on 4600 acres) was 1.4% of the herd with a mortality rate of 0.22% of the herd (16% of those bitten). Of the bites, 80% were on the head and neck, probably because, based on observation, the cattle tried to sniff the snakes when they rattled. Usually the bites caused swelling and sickness for a few days and some weight loss, but recovery was rapid and complete in most cases. Even calves recovered. Obviously the frequency of bites would depend upon the density of both the livestock and the snakes. Second, first reported in 1956, Klauber's poll of 134 Texas County Agricultural Agents found that none of the agents considered damage to livestock from snakebites "serious," 77 considered damage "negligible," 31 considered it "unimportant," and 26 considered it "moderate" in effect. Not insignificantly, the damage of bites to horses, sheep, goats, and hogs is even less than to cattle, perhaps because of cattle's inquisitiveness about the snakes.

Evidently overlooked by the Jaycees is the economic value of rattlesnakes to ranching and farming ecology. Rattlesnakes consume small mammals, such as rats, mice, ground squirrels, and gophers, that otherwise would cause great damage to crops, pastures, and grazing ranges. Rodents often carry disease and their burrowing habits can cause erosion, although burrowing may be beneficial to the soil in some situations. As long ago as 1951 and before recent inflation, Webber (1951) claimed that every bullsnake was worth $50 per year to a Texas farmer as a rodent destroyer. As reported by Klauber (1972), when den cohabitants are killed along with rattlesnakes, as many as 40 harmless snakes may be killed for every 100 rattlers. Even if these figures are exaggerated, the long-term ecological and environmental consequences of the rattlesnake round-ups upon the ranch lands are severe and probably counterproductive to the economic interests of ranchers.

A helpful distinction made by Rolston (1986, 1988), an environmental ethicist and biologist, is between three environments: urban, rural, and wilderness. Each environment has its own values—intrinsic, instrumental, and systemic. What is justifiable for one environment may not be justifiable for another, although the overarching obligation, according to Rolston, is homologously to follow nature since we all live within one interdependent planetary ecosystem. Following Leopold (1949), Rolston generally favors environmental holism and the preservation and maximization of the values of beauty, stability, and integrity. Rural and urban usage should conform broadly to the homologous principle of following sustainable natural patterns and to the value-added principle of increasing, not decreasing, the values of beauty, stability, and integrity.

Utilizing Rolston's distinctions, several inferences can be made. Except perhaps on rare occasions to control overpopulation resulting from human intervention, no hunting—including rattlesnake round-ups—should be permitted in such wilderness areas as game preserves, national parks and forests, and state parks. As much as possible, such areas should be allowed to function in "natural" ecological balance. In some cases, recreational use and the risk of snakebite might necessitate removing venomous snakes, but in such cases the sites in question obviously would no longer be undisturbed wilderness areas. Moreover, the uncontrolled exploitation of private property should not be permitted, whether for agricultural, industrial, or personal use. The goal for using rural lands should be long-term sustainable agriculture and the preservation of the beauty, diversity, and stability of the land.

Since rattlesnakes on grazing range present no actual threat to ranch profits—indeed, they may enhance profits—rattlesnake round-ups should be curtailed in such areas, unless it can be established that the snakes are overpopulated. Possibly in Texas in the late nineteenth and early twentieth centuries as a result of agricultural development, the oil industry, and urban growth, snake population densities may have increased on the remaining areas of suitable habitat due to relocation and migration of the snakes and their prey. Although unlikely, this perhaps may be inferred from the ranchers' pleas in the 1950s for help in ridding their lands of rattlesnakes. Even without intervention, competition for food would probably have soon lowered the snake populations.

Civic Benevolences: Does the End Justify the Means?

Always mentioned as a reason for the Round-Up are the civic projects that its revenue makes possible. The Jaycees (1990a) list the following as recipients of Round-Up money: Boy Scouts of America (troop sponsor and sustaining member), Hospice Program, Sweetwater High School Blow Out Party (drug and alcohol free), Sweetwater Fire Department (over $6500 in equipment contributed), Rolling Plains Memorial Hospital (donations of equipment), AJRA Rodeo (sponsorship of annual local rodeo), Sweetwater Youth Baseball (3 leagues), Nolan County Girl's Softball Association, Tai Kwan Do Karate (state and nationwide travel assistance), National Jaycee Daisy BB Gun Training, Labor

Day Rest Stops (72 continual hours), Muscular Dystrophy (over $5600 in donations collected last 4 years), Ben Richie Boy's Ranch (ranch home for troubled children), Thanksgiving Day Feast for Underprivileged (over 450 served yearly), Sweetwater High School Power Lifters, Crimestoppers, American Cancer Society, Miss Snake Charmer Queen Pageant ($1900 scholarship money awarded), and Sunshine Inn Training Center (for the mentally handicapped). Perhaps with only a few questionable items, these are undeniably worthy community and humanitarian projects. At issue are whether the ends justify *this* means, whether other means to the same ends would be better or more justifiable, and whether the bad side effects outweigh this particular means.

Estimates of net utility need to include long-term negative side effects. The worst side effects of the Round-Ups are probably two: (1) long-term damage to the agricultural ecosystem, including probable economic losses to livestock owners and possible extermination of the rattlesnakes; and (2) the fostering of irresponsible exploitative attitudes and practices toward wild animals and the environment. Two other risks seldom taken seriously and never considered in calculating the worthiness or cost of the project are not insignificant: the risk of snakebite to hunters, handlers, and spectators, and the risk of range fires due to spraying gasoline into the dens during a dry, fire-prone time of year. Surely the snakebite risk should be taken more seriously, especially since Shelton (1981) records nine snakebite victims from 1959–1980, evidently all hunters or handlers. Clearly the hunters and handlers voluntarily assume the snakebite risk. Nevertheless, should communities encourage and reward their citizens for taking such risks and also provide them with opportunities to do so? In regard to fire, the "Round-Up Rules & Regulations" urge the hunters to "Use extreme caution in fire protection" (Shelton 1981). Huge range fires have occurred in recent years in early March, but snake hunters have never been suggested as causing the fires, although they now hunt great distances from Sweetwater.

Education and Safety

The Jaycees claim that their Round-Up emphasizes education and safety rather than the sensationalism that other roundups typically include. Their own "History of the Sweetwater Jaycees" (1990*a*) states: "Sweetwater's Rattlesnake Round-Up differs from most in that safety is the focus. We feature safety demonstrations at the Round-Up and travel throughout the year giving safety demonstrations to interested groups and organizations. We 'don't' have sacking contests, play with the snakes or do anything that might give the wrong impression to an onlooker, especially our younger guests." The Jaycees' hometown snake handler, Ransberger, asserts that he aims his demonstrations at education and safety. Shelton (1981) explains: "Ransberger's demonstration in the pit is strictly educational. He stresses safety in every respect. He explains his program to instill a high regard for the rattlers to youngsters.... He has been bitten several times and states that each time it was carelessness on his part. His only serious bite was at Big Spring where the pit was too small and snakes piled so high that one struck over his boot top. He was hospitalized for 10 days."

Several snakes are placed in a fenced pit with Ransberger, who in short sleeves and without gloves exposes the fangs and shows how to catch, grasp, and milk the snakes. Using balloons as a target, he provokes the snakes to strike and pop the balloons. A retired railroad engineer, Ransberger travels the United States, especially Texas, giving demonstrations. Although a self-professed teacher of safety, he has been bitten 42 times, most recently in February 1989 at Andrews, Texas—a bite that resulted in the surgical removal of 23 inches of muscle from his right arm (Shelton 1981; Sweetwater Reporter 1989; Zelisko 1990*a*).

Surely it is misleading to say that such activities are not sensationalism. The threat of bites is real, and the threat is used to exploit the crowds. Also contributing to the bizarre excitement, snakes are publicly milked, butchered, skinned (by spectators if they wish), deep-fried, and eaten. Regardless of the proponents' claims, such activities are undeniably far more sensational than educational. In fact, the "education" received by the spectators is one of bad environmental science and senseless risk-taking. Surely the risk-taking, drama, and heroism of Ransberger—confined in a small pit with killer snakes—does not go unnoticed by the enthralled crowds, especially the children. Rather than teaching safety, the Round-Ups currently teach senseless risk-taking for personal thrills, public adulation, and prize money. Rather than instilling respect for natural species and ecosystems, the snake hunts and pit demonstrations teach disrespect for nature and engender attitudes of cruelty to animals and harsh domination of the environment rather than attitudes of systemic harmony and sustainable cohabitation.

From their perspective—their "place," to use the term proposed by environmental ethicist Norton (1991)—West Texans do not see it this way. Although not specifically addressing rattlesnake round-ups, Norton has predicted—correctly in this case—that environmental problems will take the form of disputes about "place." According to Norton's insight, environmental problems have a similar structure in which a resource is exploited by individuals seeking to further their own self-interests. From their "place," these individuals tend to see the immediate, local context. However, due to the interconnected nature of ecosystemic hierarchies, environmental problems usually emerge at the higher contextual levels. Due to their local perspective, the individuals using the resource often are unable or reluctant to see the problem. Moreover, when the problem is finally acknowledged, the scale of resolution is also a function of the observer's place because one's sense of place determines (1) the object or unit selected for management, and (2) the conceptualization of the boundaries of the system needing management.

According to Norton, environmentalists see the bigger picture. From their place, they tend to see the impact of actions and policies on interdependent collective systems, ecologically bounded spaces, and generations of time. In Leopold's (1949) phrase, they "think like a mountain." Believing in the long-term resiliency of natural ecosystems and aware of the imprecision of their sciences, environmentalists generally tend to take a "hands-off" approach and then, when a crisis seems imminent, to become reactive prohibiters of the environmentally intrusive actions of others. Aware of hierarchical interdependencies and threshold limits, their greatest fear is a sudden and irreversible decline in ecosystems that will have pervasive and cataclysmic ramifications.

Due to the pressures of our competitive free-market economy, the impact of technology, and our own burgeoning human population, Norton explains, the exploitation of the land by individuals tends inevitably to become more and more intense. The logic of the "commons," to use Hardin's (1968) familiar term, makes such exploitation inevitable. Landowners are forced to think locally and take the short-term perspective because otherwise they face financial ruin and the loss of the land itself. After exploiting a natural resource for several years, especially one that has seemed inexhaustible, and when their economic welfare or capital investments have become tied to the resource, individuals in our culture usually feel that they have a "right" to the resource, even when the resource is actually public or wild. The availability of alternative explanations allows them, due to their biases and vested interests, to find "reasons" to continue the exploitation. One's place inevitably affects one's interpretation. Ironically, due to the multifaceted dynamics of ecosystems and the inexactness of the ecological sciences, the exploiter's interpretation sometimes is right.

From their way of seeing things, West Texas ranchers and farmers must exploit the land if they are to survive. Even when fully informed of the rattler's probable benefit to land ecology and overall profits, they stubbornly still hate the snakes. Who hasn't lost a prize bull, horse, colt, cow, or calf? And who wants to take a chance with children or grandchildren, exposing them to snakebite, regardless of how slim the risk? We've all heard tell of folks coming home and finding rattlers in their houses and bedrooms. "Bleeding-heart liberals" just don't see things right. What's a few more mice compared to rattlers? The land is not friendly; it breeds mesquite thorns, prickly pear cactus, and horned toads. Rugged, desert-hot, and dusty, it has to be fought, killed, conquered. Only the tough survive.

Enmeshed in folklore and the cowboy mythos, rattlesnake round-ups have become a contemporary rite of passage. Rattlesnakes are the enemy; they are part of the uncivilized wilderness to be overcome and conquered. Destroying the killer rattler has become a way of prolonging the conquest of the frontier, continuing the excitement and adventure of the cowboy saga, proving one's manhood, and identifying with the local ethos. Seldom today does one get such an opportunity to display courage and to join the ranks of our heroic forefathers.

Added to the myth and the machoism are the festival and its economic benefits. What has helped make the Round-Up thrive, and so resulted in its continuation, is the sense of community and the sensational excitement, money, and publicity that it brings. Apparently nothing unites a community better than a spirited festival and a fight against a common enemy. For four days each spring, the world notices Sweetwater; they exist, have an identity, and prosper economically. Every small town needs a festival; from their "place," the festival simply must continue.

Miscellaneus Goods: Exotics and Venom

Considered worthless when the Round-Up started in 1958, rattlesnakes are now a key part of the exotic skin industry. At issue with all exotic animal products is whether the exploitation of the animals and the extraction of the environmental resources are harmonious with natural sustainable patterns and proportional to the benefits obtained. Rolston (1989) explains:

> Following the natural, I accept fur coats on Eskimos, but not on fashion models. The fur on the Eskimo is doing what the fur on the seal is doing, protecting against the cold. The fur on the fashion model is flattering her vanity; seals are not vain, nor can they be flattered. I eat cows and make shoes of their hides but disapprove of anaconda boots worn as a status symbol by the coach at a professional football game. If substitutes are readily available, I will not raise and kill animals merely for the leather. That is pointless. But if I need a covering for my feet, like a covering for my back, this seems vital enough to warrant killing an animal. The leather protects my feet, the hide protects the cow. If I am going to eat the cow, why waste the hide? This seems pointless.

When adequate substitutes are readily available—or, as in this case, *better* substitutes—the environmental risks and costs associated with the exotics are frivolous and extravagant.

The proponents also claim that the snakes are needed for research and medicine (Shelton 1981; Sweetwater Jaycees 1990a). According to Shelton (1981), "consideration should be given to the scores of researchers [sic] and scientists that are making much effort with rattlesnake venom, striving to find treatment and elimination of several of our most dread diseases." Shelton states that venom is being studied as "a potential cure for cancer, heart conditions, tumors and multiple sclerosis," that rattlesnake blood is being used in enzyme research, and that rattlesnake fat and oil are being used to treat arthritis and rheumatism.

If true, these needs would justify capturing only those snakes actually required for research and medicine, not the mass removals of the Round-Ups. Due to the adverse effects of gasoline upon the snakes as shown by Campbell et al. (1989), it is doubtful whether the snakes captured during the Round-Ups are suitable subjects for scientific studies. Moreover, after being milked, the snakes could be released back into the wild—or perhaps kept on snake dairies.

Recommendations

Below are five recommendations. Obviously, resistance is likely to the more restrictive ones. If the Jaycees are sincere about their desire to "help the balance of nature," then they should at least be willing to adopt policies, especially ones without cost, designed to allow biologists and wildlife officials to monitor the rattlesnake population in a scientifically reliable manner.

(1) Obtain Reliable Data by Imposing Territorial and Time Limits

Because the ecosystemic factors influencing snake populations are so numerous, the statistics from past Round-Ups are not scientifically credible. These variables include such natural (rattlesnake) factors as disease, predation, prey (food) availability, and weather, and such human factors as the size of the geographical region, the length of the hunt, the number of hunters, their expertise, the intensity of their individual efforts, the type and intensity of agricultural and industrial usage of the land, the technology used during the hunt (such as motorcycles, land rovers, gasoline spraying, and nighttime spotlighting on roadways), and the amount of promotional and training efforts by the competing snake-hunting clubs.

The least expensive and most scientifically credible recourse would be to limit the hunts to a given period of time and to a specific geographical region with a boundary that would not be changed for several years. This can be done without any immediate or direct cost to the Jaycees. Since they already publish for hunters a list of private lands open for rattlesnake hunting, the Jaycees could easily adapt their regulations and materials to a restricted territory. Although the borders of this territory should perhaps be patrolled to discourage violations, even this would not be essential since the probable impact of violations would be minor. More scientifically reliable data would be obtained if the geographical region for the hunt could be mapped out according to such biological criteria as ecosystemic barriers to rattlesnake migration rather than by such political criteria as county lines, roadways, and private property. If the statistics remain stable or improve after several hunts in the specified territory, then the conclusion would be warranted that the population is replenishing itself; on the other hand, if the statistics worsen after several hunts in the specified territory, then the conclusion would be warranted that the population is not replenishing itself.

In addition to their current practice of weighing the snakes, it would be helpful if the Jaycees would resume their former practice of recording the number of individuals captured (see Table 1), thereby allowing biologists to estimate the average maturity (age) of the snakes. Furthermore, if funds or volunteers are available, more extensive scientific data could be gathered. Of great value would be the following: (1) the size and weight of each individual captured, which together would precisely determine the number and percentages of yearlings and mature adults being captured; (2) the number of males and females captured, which would allow estimates of lost reproductive capacity; and (3) the number and species of cohabitants captured, which would help gauge the broader, ecosystemic impact of the hunts.

(2) Monitor a Control Territory

Although costly, an undisturbed control territory and snake population could be intensively monitored by biologists to gauge the impact of numerous natural factors (disease, predation, food, weather, and so on). The population in the control zone could then be compared with the population in hunted areas, and conclusions could be made regarding the relative impact of the rattlesnake hunts. Perhaps state, federal, or private agencies would be willing to fund the research and the training of volunteers.

An unacceptable proposal would be to count and weigh all of the snakes captured each year in a designated test area while still permitting unrestricted regular hunting outside the test area. In other words, rather than restricting the territory of the entire hunt (the first recommendation above), the snakes captured in a smaller area would be carefully counted. This would not work because the data would be unreliable. Aggressive hunters would probably avoid the test site because they would not want to be monitored by biologists and because taking their catch to a special checkpoint would be confusing and time-consuming, especially with wide-open hunting nearby. More importantly, the hunters would be biased in favor of allowing the rattlesnakes to repopulate the test site, thereby silencing the environmentalists and preserving the hunters' unrestricted freedom. This bias probably would not be overcome by having a secret test area or by special training or instructions for the hunters. Only by restricting the range of the entire hunt will hunters be sufficiently encouraged to hunt on the monitored lands. Reliable scientific data cannot be obtained unless the range of the entire hunt is restricted.

(3) Prohibit Gasoline Spraying

The gasoline spraying technique should be prohibited for four reasons: (1) to prevent overkilling and eventually exterminating rattlesnakes; (2) to prevent injuring and killing cohabitants, especially beneficial harmless snakes; (3) to prevent the environmental pollution caused by the gasoline spray and fumes; and (4) to eliminate the danger of fire. Banning gasoline would probably result in the development of an environmentally safer alternative spray. By taking advantage of normal hibernation, reproductive, migration, and sunning patterns, herpetologists and wildlife authorities have collected hundreds of specimens from a few dens, both in spring and fall, without using chemicals to compel the snakes out of the dens (Klauber 1972). Nevertheless, as long as capturing snakes remains the central activity of the Round-Ups, total prohibition of chemical spraying is probably unrealistic. The actual capturing of snakes, however, should become less and less a feature of the festival, as recommended below.

Consistent with the hunter's ethic, another reason to prohibit spraying is to give the snakes a "sporting chance." Admittedly, this would increase the risk of snakebite. Spraying snakes while they are sleeping in their dens is not sport; it is slaughter by chemical extermination. An analogous form of deer hunting would be to fence the woods, leaving only a narrow opening at one end, and then burn the forest, shooting the deer and other animals as they run through the gate. Such methods are hardly sport.

(4) Voluntary Self-Regulation

Voluntary self-regulation is the surest way for the Jaycees to preserve their festival and guarantee a continuous supply of rattlesnakes. Regardless of whether they consciously acknowledge it, Round-Up leaders are primarily conducting a community festival and not aiding agriculture and the balance of nature. Given current trends and attitudes, the festival will probably continue until the snake has largely been exterminated or until the market price has made private harvesting more attractive than the hundreds of enthusiastic hunters spraying volatile gasoline on dry, tinderbox range. If these unlikely events occur, the festival may still continue with a symbol or token replacing the rattlesnake, as with many other religio-political festivals, including some rattlesnake festivals. One such festival is held annually in San Antonia, Florida, on the third Saturday of October. The festival has an educational rattlesnake lecture, music, arts, crafts, antiques, wildlife exhibits, food, curios, and a 5-mile (8 km) "Rattlesnake Run," but no snake hunting. Some regular participants reportedly have never seen a snake at the festival. The profits are distributed in a community-chest type program by the sponsoring charitable organization, Rattlesnake and Gopher Enthusiasts (RAGE). Foresightful Round-Up leaders should begin now the shift from real hunts to symbolic ones, lest the folk fest and their civic livelihood be exterminated by nature or economics. Certainly such a shift is warranted by the principles of environmental holism and sustainable agriculture.

Ideally, the massive snake-hunting aspect of the Sweetwater Rattlesnake Round-Up should be discontinued. Realistically, however, it will be continued, at least for several years. Therefore, reasonable compromises and interim conservation measures need to be introduced. Round-Up officials should begin to phase in restrictions, beginning with rigid time and territory limitations. Over several years, the time and territory should be gradually and progressively limited. Eventually the hunting could be restricted to 2–3 hours each day. If these "minihunts" (or rattlesnake "rodeos") were highly publicized and sensationalized—with lucrative prize money—they could easily replace the existing methods without loss of fanfare or income. In fact, such highly competitive minihunts could be started immediately as an adjunct to the existing weekend-long hunt. Then, when the minihunts become popular, the long hunt could be discontinued. The contestants could all start and finish at a central arena where the fans could be entertained while waiting for the hunters to return. Enough snakes would still be captured to more than placate and enthrall the crowds, but environmental damage would be vastly reduced. And, as long as the prize money was kept high, the hunters would be happy.

Fire and snakebite risks would be minimized if the Jaycees more carefully monitored and regulated the Round-Up. Methods of safety and control could include restricting the hunt to a geographical area small enough to be carefully managed and requiring all hunters to wear protective gloves, puttees, and boots. As long as gasoline continues to be used, all hunters should be required to carry fire extinguishers.

Instead of sensationalism, the festival should use proven educational activities. As in classes at schools, zoos, and nature centers, prepared dead specimens and some live ones—securely confined—would instill respect for the actual dangers that rattlesnakes pose and provide better educational results. Instead of famous snake handlers, the Jaycees could obtain biologists, ecologists, and wildlife officials from regional universities and state agencies to give lectures and guide field trips. These officials would provide positive role models for children and would probably provide their services without cost.

(5) Legislation and Government Regulation

If efforts at voluntary self-regulation fail and if it can be established that the rattlesnake round-ups are decreasing the snake population, then snake hunting and rattlesnake round-ups should be regulated by law by the U.S. Fish and Wildlife Service and the Texas Parks and Wildlife Department, like other types of commercial and sport hunting and fishing. Extensive and credible scientific data, such as that proposed above, would enable legislators to formulate accurate and effective regulations. Such regulations could include limited geographic regions and times for hunting, restrictions on methods used for hunting (especially gasoline spraying), prohibitions against taking beneficial snakes, licensing of hunters and round-ups, and bag limits on the size and number of individual rattlesnakes that can be collected or killed. Landowners could be encouraged to post their property to prevent hunting, and wildlife and agricultural officials could monitor snake population densities.

Conclusion

The Sweetwater Rattlesnake Round-Up has become an end in itself, a social phenomenon with a life of its own. Even assuming that the ranchers' pleas are not pretentious or exaggerated, the actual value of the festival has probably never been the actual round-up of snakes, since the destruction of rattlesnakes has probably never benefited livestock or grazing lands or "the balance of nature." The festival has become a symbol of community identity, a publicity extravaganza, and a boon to the local economy.

The principles of environmental holism and sustainable agriculture encourage us to find ways to live harmoniously with nature. Rattlesnakes occupy an essential niche in rural ecosystems, especially semi-arid ranch ecosystems, and usually pose little actual threat to either humans or livestock. Moreover, long-term profitability, species diversity, and control of other agriculturally harmful species warrant allowing the rattlesnakes to reach population densities consistent with the availability of food supplies in the ranch ecosystem. The round-ups generate unecological attitudes of domination and destruction of nature and the wild. For these reasons, the competitive capture of large numbers of rattlesnakes ideally should be discontinued. With the rattlesnake hunts replaced by token hunts, field trips, and nonsensational educational programs and exhibits, the spring festivals could still continue. Nevertheless, given political realities and implacable attitudes toward snakes, rattlesnake round-ups are likely to continue for several years to come, perhaps even until the species population is no longer viable. Not encouraging are past experiences with such other species as passenger pigeons and bison, and rattlesnakes have already been eliminated from numerous highly developed regions where they once thrived. Perhaps someday a more ecologically holistic viewpoint will prevail.

Acknowledgments

For helpful suggestions and comments, I wish to thank Bryan G. Norton, Holmes Rolston, III, and the anonymous reviewers of *Conservation Biology*. An earlier version of this paper was presented in August 1990 at the Realia Annual Conference in Estes Park, Colorado, and comments by the participants were helpful. I would especially like to thank the Sweetwater Jaycees, the Sweetwater Chamber of Commerce, and the Sweetwater Public Library for their courteous and friendly assistance. The Sweetwater Jaycees have consistently shown their good will by making their records openly available to researchers. For keeping the records, and especially for their good will, the Jaycees should be commended.

Literature Cited

Campbell, J. A., D. R. Formanowicz, Jr., and E. D. Brodie, Jr. 1989. Potential impact of rattlesnake roundups on natural populations. The Texas Journal of Science **41**:301–317.

Cox, J. 1991. Reptilian rodeos. Texas Parks and Wildlife **49**(3):22–27.

Gloyd, H. K. 1978. The rattlesnakes: genera *Sistrurus* and *Crotolus*. Society for the Study of Amphibians and Reptiles, Milwaukee, Wisconsin.

Goin, C. J., O. B. Goin, and G. R. Zug. 1978. Introduction to herpetology. 3rd ed. W. H. Freeman, San Francisco, California.

Hardin, G. 1968. "The tragedy of the commons." Science **162**:1243–1248

Klauber, L. M. 1972. Rattlesnakes: their habits, life histories, and influence on mankind. 2 volumes. University of California Press, Berkeley, California.

Leopold, A. 1949. A Sand County almanac. Oxford University Press, Ballantine Books, New York.

Norton, B. G. 1991. Toward unity among environmentalists. Oxford University Press, New York.

Rolston, H. III. 1986. Philosophy gone wild: essays in environmental ethics. Prometheus Books, Buffalo, New York.

Rolston, H. III. 1988. Environmental ethics: duties to and values in the natural world. Temple University Press, Philadelphia, Pennsylvania.

Rolston, H. III. 1989. Treating animals naturally? Between the Species **5**:131–137.

Shelton, H. 1981. A history of the Sweetwater Jaycees Rattlesnake Roundup. Pages 95–234 in J. Kilmon and H. Shelton, editors. Rattlesnakes in America. Shelton Press, Sweetwater, Texas.

Sweetwater Jaycees. 1989. World's largest rattlesnake roundup. Promotional brochure for 1989–93 Round-Ups. Sweetwater Jaycees, Sweetwater, Texas.

Sweetwater Jaycees. 1990a. History of the Sweetwater Jaycees. Sweetwater Jaycees, Sweetwater, Texas.

Sweetwater Jaycees. 1990b. 1989 Sweetwater Jaycees Rattlesnake Round-Up statistics. Sweetwater Jaycees, Sweetwater, Texas.

Sweetwater Jaycees. 1990c. 1990 Sweetwater Jaycees Rattlesnake Round-Up statistics. Sweetwater Jaycees, Sweetwater, Texas.

Sweetwater Jaycees. 1990d. 1990 Sweetwater Rattlesnake Round-Up equipment list. Sweetwater Jaycees, Sweetwater, Texas.

Sweetwater Jaycees. 1990e. 1990 Sweetwater Rattlesnake Round-Up price list. Sweetwater Jaycees, Sweetwater, Texas.

Sweetwater Jaycees. 1990f. Questions answers. Sweetwater Jaycees, Sweetwater, Texas.

Sweetwater Jaycees. 1990g. Statistics from the 1st to the 31st annual Rattlesnake Round-Ups, 1958–90. Sweetwater Jaycees, Sweetwater, Texas.

Sweetwater Jaycees. 1991. 1991 Sweetwater Jaycees Rattlesnake Round-Up statistics. Sweetwater Jaycees, Sweetwater, Texas.

Sweetwater Reporter. 1989. Snake poundage totals 3,692 at 31st roundup. Sweetwater Reporter (**13 March**):1.

Tennant, A. 1985. A field guide to Texas snakes. Texas Monthly Press, Austin, Texas.

Webber, J. P. 1951. Snake facts and fiction. Texas Game and Fish **9**(6):12–13, 30.

Zelisko, L. 1990a. Ransberger heads for last roundup. Abilene Reporter-News (**4 March**):11A, 13A.

Zelisko, L. 1990b. 30,000 fans are expected. Abilene Reporter-News (**4 March**):11A, 13A.

The Olympic Goat Controversy: A Perspective

VICTOR B. SCHEFFER

14806 SE 54th Street
Bellevue, WA 98006, U.S.A.

Abstract: *Mountain goats (*Oreamnos americanus*) introduced into Olympic National Park are multiplying and causing soil erosion and changes in floral composition. Park managers want the goats removed or, if necessary, killed. But the Fund for Animals, a national humane society, argues that the present goat population should be left undisturbed as a replacement of a presumed indigenous stock that disappeared long ago. (I side with the park managers.) The debate underscores the value of both logic (or reason) and sentiment (or emotion) in making wildlife management decisions.*

La controversia de la Cabra Olímpica: Una perspectiva

Resumen: *Las cabras de montaña (*Oreamnos americanus*) introducidas en el Parque Nacional Olímpico se están multiplicando y causando ersión del suelo y cambios en la composición florística. Quienes manejan los Parques quieren que las cabras sean removidas y si es necesario eliminadas. Sin embargo la fundación para los animales, una sociedad nacional humanitaria, argumenta que la presente población de cabras debe ser dejada sin perturbar como reemplazodel supuesto stock indígena que desapareció hace tiempo. (yo estoy de parte de los que manejan el parque.). El debate toma en cuenta tanto el valor lógico (o de la razón) como el sentimental (o emotivo) para tomar decisiones de manejo de la fauna silvestre.*

Introduction

The 20-year experience of the National Park Service (NPS) in dealing with the mountain goats of Olympic National Park is a useful case history in wildlife management. It is well documented (NPS 1987, 1988; Carlquist 1990; Houston et al. 1991a, 1991b). The NPS has concluded that, if nonlethal means of preventing damage by goats should prove infeasible, goats must be shot. The Fund for Animals (1992) disagrees. In this paper I examine the arguments offered by both sides in the debate.

Background

Goats were translocated during the 1920s from Canada and Alaska to the Olympic Peninsula (Fig. 1). In 1937 they numbered about 25 (Scheffer 1949:237) and by 1983 about 1200 (Houston et al. 1986). Although 155 goats are known to have been removed between those years, the population grew at an average rate of about 9% a year.

But the soils and biotas of the park had evolved on a goat free "land-bridge island" (Newmark 1987). By the late 1980s, the park's drier regions were beginning to show changes as a result of goat grazing, wallowing, and trampling. Goats were even "mining" bare soil where hikers had urinated! Most conspicuous were changes in floral composition, such as the disappearance of lichen and moss cover, which stabilizes bare soil surfaces in the absence of vascular plants (NPS 1987:7–8). And the NPS perceived threats to certain unique endemic plants—nine species and varieties—growing in areas used by goats.

Between 1981 and 1989, humans removed 509 animals from the goat population (Houston et al. 1991b: 89). Of these, 360 were captured alive, 28 accidentally killed during capture, 19 shot for research, 99 killed by

Paper submitted April 29, 1992; revised manuscript accepted March 2, 1993.

Figure 1. Native range of the mountain goat in northwestern North America, 1988 (National Park Service map).

sport hunters outside the park, and 3 killed by poachers (Fig. 2). At the end of 1990 the estimated population on the peninsula was only 389 ± 106 (172 goats seen). That total showed clearly that removals by humans outnumbered natural recruitment.

The NPS also tested population control by contraception (NPS 1987:46–47). Later, an independent five-member panel comprised of veterinarians, wildlife biologists, and a reproductive physiologist evaluated the potential of goat control by contraception (Scientific Panel 1992). Panel members visited the goat range, studied past research by the NPS, and brought to bear their collective experience with the application of contraceptives to overabundant wild or feral animals. They concluded that "current contraceptive or sterilant technologies will not eliminate mountain goats from ONP."

Although the reestablishment of wolves (*Canis lupis*) in the park would impose a degree of control on the goat population, the NPS has never included this possibility in its management plans. (The last Olympic wolf was killed in the 1920s.) Students at Evergreen State College have suggested that the Peninsula could support at least 40–60 wolves (Students 1975:57).

In 1987 the NPS released an environmental assessment that gave preference to settling the goat contro-

Figure 2. Two mountain goats, tranquilized by aerial darting, are removed from Olympic National Park, 1988 (National Park Service photo by Richard W. Olson).

versy by removing all goats from the core of the park and thereafter removing—by capturing or killing—any that appeared along its borders (NPS 1987:52–54, 65–67). Later, the NPS announced that it would release in 1993 a Final Environmental Impact Statement (Interagency Goat Management Team 1992:6).

Conflict: Factual Considerations

The Fund for Animals (1992), a national society of 150,000 members, claims that goats occupied the Olympic Peninsula into the nineteenth century. If so, the present population is a replacement or "restoration" (my term) entitled to protection. The Fund builds its case partly on a model drawn by anthropologist R. Lee Lyman (1988) and partly on "documented and scientific historical evidence." Lyman examined the known distribution of goats in five northwestern states in relation to the postulated distribution of Pleistocene ice lobes. From a "dispersal model" of goat occurrences at various times and places, he concluded that by 10,000 years ago goats could have reached the Olympics. He suggested that, if goat remains dating from the recent thousand years ever *should* be found here, the NPS should rethink its policy, quit calling the planted animals exotics, and leave them undisturbed. The Fund for Animals also points to narratives published between 1844 and 1917 that mentioned the goat as a member of the Olympic fauna.

The NPS rests its case on the present distribution of mammals in western Washington and on the unreliability of reports of Olympic goats before 1925.

First, the goat is one of 11 species of mammals native to the Cascade Range of Washington that are *not* native in the Olympic Range only 120 km away (Dalquest 1948; Scheffer 1949). Among the missing are six species characteristic of alpine or subalpine habitats. Conversely, one mammal species (*Marmota olympus*) native to the Olympics is not recorded from the Cascades. Geologic clues indicate that continental ice in the Puget Sound Basin would have isolated the high Olympics from the high Cascades long before the first goats reached North America, perhaps 40,000 years ago (NPS 1987:7, 17; Kruckeberg 1991:2–33).

Second, early reports of Olympic goats cannot be taken seriously. For example, John Dunn visited the Indians living near Cape Flattery and reported that they "manufacture some of their blankets from the wool of the wild goat" (1844:231). But ethnologist Erna Gunther later learned from descendants of those Indians that "the mountain goat does not occur on the Olympic Peninsula.... Mountain-goat wool was bought in Victoria [British Columbia] through the Klallam" (1936:117). Albert B. Reagan, Indian Agent at Lapush in the early 1900s, excavated middens along the seacoast, where he found remains of bighorn sheep and mountain goat "usually only in the ladle form of the horns" (1917:16). These, again, would surely have been trade goods. Eight years earlier, Reagan (1909) had published a list of the animals of the Olympic Peninsula; it did not include the goat.

Two other narratives briefly mentioned Olympic goats (Seattle Press 1890:20; Gilman 1896:138). The first, composed after a five-month crossing of the Olympic Range in winter and spring (the first crossing ever) stated simply that "one goat was seen by the party." The second included "mountain goat" and "pelican," among other species, as "game animals" of the Olympics. These narratives can hardly be taken as zoological records.

The strongest evidence—albeit negative—that goats were not indigenous comes from the published accounts of the dozen or more zoologists who explored the Olympics between 1895 and 1921 on expeditions of the U.S. Biological Survey and the Field Museum of Natural History (Hall 1932:74). These explorers reported no goats.

Ethical Considerations

But the goat controversy is basically a clash of human values—the sort of controversy that is settled through agreement rather than discovery. Informed public opinion will ultimately determine whether Americans want a goat-free Olympic Park at the cost of routinely exiling or killing goats. The Fund for Animals has chosen unwisely to offer what it calls "historic and scientific evidence" (1992) in defending its case. Would not the Fund gain wider public support by relying purely on moral persuasion? Philosopher Mary Midgley has asked (1983: 33), "What does it mean to say that scruples on behalf of animals are merely emotional, or emotive or sentimental? What else ought they to be?"

Two lessons can be read in the Olympic Park experience with its unwanted goats.

First, national park managers will increasingly deal with exotic species as they deal with wildfires, hurricanes, and floods: with patience yet with steady resolve to maintain indigenous biosystems as nearly natural as possible. While "natural" as a state unperturbed by humans has long been an unreality—an abstraction—it is still useful as a goal. And all land managers need goals, however visionary or remote.

Second, animal welfare, an umbrella term for kindness to animals, humaneness, animal protection, anticruelty, and (lately) animal rights, will continue to grow in American thought. As a societal endeavor to win greater consideration for the interests of all living things, animal welfare began in the 1960s to draw energy from the "liberation" and "ecology" movements of that era (Scheffer 1991:29–30). The significance of the animal welfare ethic for national park managers is that they will increasingly become more sensitive to public opinion—a set of preferences compounded of sentiment (or emotion) and logic (or reason). Park managers will increasingly turn for advice to social scientists, who will sample public attitudes and preferences with respect to park uses; will develop new technologies for interpreting park values; will assist in the drafting of regulations; and will join in mediating disputes over the status of exotic species, such as goats.

Conclusions

The planting of foreign goats in the Olympics seemed a good idea at the time and even 10 years later (1935) when I first worked as a biologist in the Olympic National Forest. But today, public attitudes toward natural areas and their biota are changing. The more we humans shape and color the landforms around us according to the designs of each new generation, the more we treasure those fragments kept undesigned. Wild places. Places to which we respond with all our senses, places where we bond with the earthly systems that nourish our civilization and our species. If a personal thought may be injected here it is this: the humane removal of goats is a small price to pay for keeping the Olympics wild.

Acknowledgments

For study materials and ideas, I thank Roger Anunson of the Fund for Animals, Robert L. Wood of Olympic Park Associates, and the generous wildlife biologists of Olympic National Park.

Literature Cited

Carlquist, B. 1990. An effective management plan for the exotic mountain goats in Olympic National Park. Natural Areas Journal **(10)(1)**:12–18.

Dalquest, W. W. 1948. Mammals of Washington. University of Kansas Publications, Museum of Natural History **2**:1–144.

Dunn, J. 1844. History of the Oregon Territory and British North American fur trade. Edwards and Hughes, London, England.

Fund for Animals. 1992. Historic and scientific evidence that mountain goats are native to the Olympic Peninsula. Press release, Fund for Animals, Northwest Region, Salem, Oregon.

Gilman, C. 1896. The Olympic country. National Geographic Magazine **7(4)**:133–140, map.

Gunther, E. 1936. A preliminary report on the zoological knowledge of the Makah. Pages 105–118 in Essays in Honor of Alfred Louis Kroeber. University of California Press, Berkeley, California.

Hall, F. S. 1932. A historical resume of exploration and survey: Mammal types and their collectors in the State of Washington. Murrelet **13**:63–91.

Houston, D. B., B. B. Moorhead, and R. W. Olson. 1986. An aerial census of mountain goats in the Olympic Mountain Range, Washington. Northwest Science **60(2)**:131–136.

Houston, D. B., B. B. Moorhead, and R. W. Olson. 1991a. Mountain goat population trends in the Olympic Mountain Range, Washington. Northwest Science **65(5)**:212–216.

Houston, D. B., E. G. Schreiner, B. B. Moorhead, and R. W. Olson. 1991b. Mountain goat management in Olympic National Park: A progress report. Natural Areas Journal **11(2)**:87–92.

Interagency Goat Management Team (IGMT). 1992. Newsletter no. 1, January. IGMT, Olympic National Park, Port Angeles, Washington.

Kruckeberg, A. R. 1991. The natural history of Puget Sound Country. University of Washington Press, Seattle, Washington.

Lyman, R. L. 1988. Significance for wildlife management of the Late Quaternary biogeography of mountain goats (*Oreamnos americanus*) in the Pacific Northwest, U.S.A. Arctic and Alpine Research **20(1)**:13–23.

Midgley, M. 1983. Animals and why they matter. University of Georgia Press, Athens, Georgia.

National Park Service. 1987. Environmental assessment: Mountain goat management in Olympic National Park. September. Port Angeles, Washington.

National Park Service. 1988. Decision record: Mountain goat management in Olympic National Park. March. Port Angeles, Washington.

Newmark, W. D. 1987. A land-bridge island perspective on extinctions in western North American parks. Nature **325(6103)**:430–432.

Reagan, A. B. 1909. Animals of the Olympic Peninsula, Washington. Proceedings of the Indiana Academy of Sciences **1908**:193–199.

Reagan, A. B. 1917. Archaeological notes on western Washington and adjacent British Columbia. Proceedings of the California Academy of Sciences, 4th series **7(1)**:1–31.

Scheffer, V. B. 1949. Mammals of the Olympic National Park and vicinity. U.S. Fish and Wildlife Service, Seattle, Washington. Unpublished.

Scheffer, V. B. 1991. The shaping of environmentalism in America. University of Washington Press, Seattle, Washington.

Scientific Panel (5 authors). 1992. The applicability of contraceptives in the elimination or control of exotic mountain goats from Olympic National Park. Final report to the National Park Service. January. Unpublished.

Seattle Press. 1890. The Olympics: An account of the explorations made by the "Press" explorers. Seattle Press July 16, **18(10)**:1–11. (With an editorial and "resume of the natural resources of explored region" on p. 20.)

Students of Evergreen State College (8 authors). 1975. A case study for species reintroduction: The wolf in Olympic National Park. Olympia, Washington.

Response to Scheffer

CATHY SUE ANUNSEN
ROGER ANUNSEN
The Fund for Animals
5778 Commercial Street SE
Salem, OR 97306, U.S.A.

The Fund for Animals has had a longstanding and unwavering commitment to the protection of endangered species. The Fund demonstrated that commitment in December 1992 in a legal settlement with the Interior Department that "will expedite federal protection for hundreds of imperiled species," both plants and animals (Washington Post 1992). The Fund supports humane management when a need for control is documented, but it opposes the extermination of animals simply on the basis of their having been arbitrarily labeled an exotic species.

It is the position of the Fund that officials of Olympic National Park have not substantiated their claims that (1) the goat is an exotic species or that (2) the goat's impact on park flora warrants the radical solution they propose: total extermination. If a need for control is established and verified by disinterested scientists, the Fund recommends, as it has since its 1988 appointment to the park's Technical Advisory Committee, the use of nonlethal control methods such as contraception.

The park's 1978 Management Policies states that "control programs will most likely be taken against exotic species which have a high impact on protected park resources and where the program has a reasonable chance for successful control" (National Park Service 1987a:4). There are no park plants federally listed as threatened or endangered. The park has not been able to document that the goat is exotic or that it has a significant, let alone "high impact," on park plants. A panel of four scientists, including two who are colleagues of the author, conducted what the park purports to be a "peer review" of the vegetation impact data in April 1992. The author, Ed Schreiner, had not completed the manuscript at the time of the "peer review." This vegetation study was supposed to support the park's damage claims. It was reportedly in its "last phase" in July 1990 but now, three years later, it still has not been released to the public (Finnerty 1990).

In 1987, however, when the goat population was at its highest—1200—the park's Environmental Assessment stated that "there is no apparent danger that these (plant) species will be extirpated" (National Park Service 1987a:22). In an August 1992 interview with the National Geographic News Service, Schreiner conceded that "None of the estimated 60 alpine and subalpine plant species are in danger from extinction."

Finally, excerpts from the as yet unreleased vegetation report include the following admission: "Specific relationships between mountain goats and rare plants were difficult to assess because few rare taxa occurred in either recon plots or permanent plots at Klahhane Ridge, Mount Dana, and Avalanche Canyon" (Schreiner 1993:84).

It is the Fund's belief that park officials have grossly exaggerated the impact on park flora from the 386 goats that remain in the 900,000-acre park.

Significant omissions occur in Scheffer's materials, particularly in the areas of (1) soil erosion; (2) the viability of contraception as a nonlethal means to control the goat population; and (3) the "land-bridge island" analysis.

Scheffer reports that goats were "mining" the soil in a particular area. But the role park biologists played in creating the most severely damaged area was not discussed. In the late 1960s, park and college researchers placed artificial salt licks at the park's Klahhane Ridge to lure the goats to be counted, studied, and later trapped. The 14-year-long program resulted in an artificial concentration of goats, with one third of the entire park's population living on 0.5 percent of the park. The population there soared from 29 goats in 1971 to 229 goats in 1981. As the licks, which had been placed directly on the ground, dissolved, the salt leached deep into the soil and the mineral-starved goats dug into the ground, displacing the soil and trampling the surrounding vegetation. Park officials continue to characterize this area as a typical product of "destructive" goat behavior (Anunsen 1993).

In his discussion of contraception, Scheffer quotes the scientific panel as determining that contraception "will not eliminate mountain goats from the ONP." He omits the next sentence of the report: "Indeed, even with the use of lethal shooting, it will likely be very expensive and difficult to totally eliminate mountain goats from ONP" (Scientific Panel 1992:25). The proposal for total extermination becomes even more far-fetched in light of the position of Washington State's Department of Wildlife that the goats are native to the state and that they should stay—at least on Forest Service land surrounding the park.

Scheffer also fails to acknowledge a highly publicized letter to the park from a member of that contraception panel, Jay Kirkpatrick, senior research scientist at the Deaconess Research Institute and whom Olympic officials acknowledge as a nationally recognized authority on wildlife fertility. In a letter of February 8, 1993, Kirkpatrick "scolded" the park for misleading the public about the potential use of contraceptives for mountain goat control. In an interview with Seattle Times reporter Ron Judd, Kirkpatrick said he and others were taken aback by the park's seeming predisposition to shoot the goats, when logic suggested that simply limiting herds might suffice. Kirkpatrick added:

> We were given a very narrow charge: Can contraception be used to eliminate goats from the park? We had the answer in about five minutes—no. The thing that bothered us was, we were given no latitude to make other recommendations ... Can contraceptives be used to control wildlife populations? Clearly, yes. The technology has advanced to the point it could be tried on Olympic goats. The odds of it working on a mountain goat are very, very high. (Seattle Times 1993)

Dr Kirkpatrick said he wrote the park because Olympic officials repeatedly took his study panel's conclusion out of context, suggesting that contraception to control goats was not a viable alternative to shooting. That wasn't true then, and it's even less true now, he said.

Scheffer's conclusion that the peninsula evolved on a goat-free land-bridge island is convenient but suspect. As noted by R. Lee Lyman, much has been learned since Dalquist outlined his hypothesis in 1946 (Lyman 1988:16). The more recent information "suggests the Puget lowland served as a biogeographic filter rather than a barrier for mammalian taxa" (Lyman 1988:13).

Scheffer's admonition to the Fund is two-fold: (1) he asserts that the Fund should stay out of the scientific kitchen or, in this case the library, and not use the information we have discovered in our defense of the goats, and (2) he advises the Fund to limit its involvement in ecological issue to emotional pleas based on moral persuasion. Surely he doesn't mean to say that only an elite few should have access to archival records?

Scheffer argues that an "informed public opinion will ultimately determine whether Americans want a goat-free park." The public can only become informed if it has all of the facts, not just those spoon-fed to them by an anointed few who feel they have an exclusive right to find and interpret information relating to the public's wildlife. It was Scheffer's scientists and historians who failed to find the early references to the existence of mountain goats in the Olympics. If the public had to rely on those professionals, they would still believe there was no prior record of mountain goats in the Olympic range, not even a suggestion.

An examination of the record of those scientists, local authors, and park officials charged with the responsibility of informing the public about the history of the mountain goat on the Olympic peninsula provides the best argument for the inclusion of all ideas and information in controversies involving public resources.

Scheffer. In *Mammals of the Olympic Peninsula* (1946), Scheffer's discussion of the history of the mountain goat is predicated on the assumption that the first goats on the peninsula were those that were planted. He relied on "scattered notes" in manuscript files and on "a few published references" to the planting (Scheffer 1946:124). Scheffer dismisses the Fund's references to the accounts of early explorers because they can "hardly be taken as zoological records," but he accepted "scattered notes" for his own paper.

Robert L. Wood. The author of *Across the Olympic Mountains* (1967), a book about the Seattle Press expedition, failed to find the reference to a goat in the 1890 Seattle Press account by Captain Barnes. The article contains an account of the wildlife encountered during the famous expedition and includes a specific reference to an 1890 sighting of "one goat" by the entire party (Seattle Press 1890). In late 1991, Mr. Woods reiterated to Fund researchers the often-repeated conclusions of his research: "There is no record of anyone in the Press party ever seeing mountain goats" (personal communication). Fund researchers found the Seattle Press reference a few months later.

Park officials. In an effort to support the Park's contention that the goats were not native, park spokesperson Paul Crawford told the Bremerton Sun in January 1992 that "none of the early peninsula explorers reported seeing goats" (Bremerton Sun 1992). Mr Crawford made this statement on the same day that the paper reported that the Fund had discovered an 1896 National Geographic Magazine account of an 1889 exploration of the Olympics by the respected explorer Samuel C. Gilman. In the article, Gilman lists the wildlife of the Peninsula, including "mountain goat" (Gilman 1896).

The degree of credibility accorded the reports of the early explorers by the park and their allies has totally depended on the perceived content of those reports. When the park believed that none of the early explorers saw goats, they repeatedly cited this "fact" as a cornerstone of their proof that the goat was not native. After the Fund found that two of the first three major expeditions into the Olympics reported mountain goats, the park reversed its position and dismissed these early explorers' reports as unreliable.

Scheffer diminishes Reagan's credibility as a scientist when he identifies the Stanford Ph.D. as "Albert B. Reagan, Indian Agent." A fellow of the American Association for the Advancement of Science, a member of many learned societies, including the American Ethnologists Society, the American Anthropological Association, and the New York and California Academy of Sciences, Reagan's work included archaeological research for the Laboratory of Anthropology Institution at Santa Fe, New Mexico. He authored over 500 papers, which he contributed to various publications both in the U.S. and abroad.

Scheffer did not cite Reagan's report in his 1946 work (Scheffer 1946), and he now dismisses the mountain goat bones identified by Dr. Reagan as "surely" being the result of trade because the reference says "usually only in the ladle form" (Reagan 1917:16). Scheffer overlooks the significance of the word "usually." Clearly, some were not in ladle form or Reagan would not have made this distinction. The important point for this discussion, however, is that had the public record been limited to the conclusions of the park biologist, there would be no debate.

The primary author of the park's publications regarding the mountain goat, including the 1987 Environmental Assessment, is research biologist Bruce Moorhead. Moorhead claimed that his research was "exhaustive," but he had to confess at a public meeting in January 1992 that he did not know of the Reagan paper, found by Fund researchers. Moorhead maintained for over a decade that "there is no evidence to suggest that they [the goats] ever inhabited the Olympic Peninsula" (National Park Service 1987b:17). (For other statements that there was no evidence that goats were native, see National Park Service [1987b:5, 6] and Moorhead [1988].)

At the time the 1987 statement was made, Moorhead's files contained a computer list entitled, "A chronology of early mountain goat reports." This list included the Gilman reference as well as the quote "one goat was seen by the party." He did not cite these references and discuss their merit; he said they did not exist. Regarding the reliability of Gilman's work, the Seattle Post-Intelligencer wrote that "[n]o work of equal importance and value in a practical sense has been accomplished on the American continent in the past quarter of a century" (Seattle Post-Intelligencer 1890), and in a 1983 study the National Park Service proclaimed the Gilman accounts as "accurate and thorough" (Evans 1983:25). Neither goat report reference was pursued by park biologists or researchers.

Contrary to Scheffer's contention that the public should stay out of the scientific arena of ecological debates, the goat controversy and the Fund's discoveries show the importance of the checks and balances provided by the involvement of citizens and wildlife advocacy organizations who have no scientific turf to protect nor jobs to perpetuate. The Fund's role in the policy debate over the goat issue is not and should not be limited to mere sentimentalism, but to full exposure of all issues—scientific, historical, archaeological, and ethical.

Archaeological documents and historical accounts cast strong doubt on the contention that goats are exotic to the Olympic peninsula. An absence of any scientific evidence documenting ecological damage by goats stretches the faith of interested parties who are told that the animals imperil rare plants. And statements from respected scientists about a high probability of success in applying immunocontraceptive techniques to goats put the lie to the suggestion that killing the goats is the only feasible management option. Despite Dr. Scheffer's valiant, if misguided, defense of the Park Service, the agency has not met its burden of proof.

Acknowledgments

We thank the librarians for hours of anonymous assistance and for keeping the doors to historical documents open to the public.

Literature Cited

Anunsen, C. S. 1993. Saving the goats: A showdown in Washington State. *Animals Agenda* **Jan/Feb**:24.

Bremerton Sun, January 9, 1992:1.

Evans, G. 1983. Historic resources study, National Park Service, Pacific Northwest Region, Seattle, Washington.

Gilman, S. C. 1896. The Olympic County. National Geographic Magazine **7(4)133**:140.

National Park Service. 1987a Environmental impact statement.

National Park Service. 1987b. Environmental assessment: Mountain goat management in Olympic National Park.

Lyman, R. L. 1988. Significance for wildlife management of the Late Quaternary biolgeography of mountain goats (*Oreamnos americanus*) in the Pacific Northwest, U.S.A. Arctic and Alpine Research **20(1)**:13–23.

Moorhead, B. 1989. Non-native mountain goat management undertaken at Olympic National Park. Park Science (National Park Service publication) **9(3)**:10–11.

National Park Service. 1981. An environmental assessment on the management of introduced mountain goats in Olympic National Park.

1990 July 20. Letter from Superintendent Maureen Finnerty to the Fund for Animals Regional Coordinator Cathy Sue Anunsen.

Reagan, A. B. 1917. Archaeological notes on western Washington and adjacent British Columbia. Proceedings of the California Academy of Sciences. 4th Series. **7(1)**:16.

Scheffer, V. 1946. Mammals of the Olympic Peninsula, Washington. Fish and Wildlife Service, p. 129.

Schreiner, E., A. Woodward, and M. Gracz. 1993. Vegetation in relation to introduced mountain goats in Olympic National Park: A technical report. National Park Service,

Scientific Panel. 1992. The applicability of contraceptives in the elimination or control of exotic mountain goats from Olympic National Park. Unpublished.

Seattle Post-Intelligencer. 1890. June 5. P. 9.

Seattle Press (later known as the Seattle Times). 1890. Resume of the natural resources of explored region. July 16:20.

Seattle Times. 1993. April 8:1.

Washington Post. 1992. December 16:1.

Wood, R. L. 1967. Across the Olympic Mountains. University of Washington Press and the Mountaineers, Seattle, Washington.

Reply to the Anunsens

VICTOR B. SCHEFFER

14806 SE 54th Street
Bellevue, WA 98006, U.S.A.

I'm pleased that the Fund for Animals has commented on the Olympic goat controversy. The Fund is a New York–based anti-cruelty organization with 150,000 members. I respond to important points raised by the Anunsens.

"There are no park plants federally listed as threatened or endangered." True, though I do not use this argument. I emphasize that "changes in floral composition" have resulted from heavy grazing by goats.

"The park has not been able to document that the goat is exotic. . . ." I reiterate that the expeditions of the Biological Survey between 1897 and 1921, and the expedition of the Field Museum in 1898, found no evidence of goats. Published allusions to goats in the park before 1925 (Dunn 1844; Seattle Press 1890; Gilman 1896) are bare statements without elaboration. There exist no specimen records of endemic goats, with the improbable exception of "big horn" and "mountain goat" remains that Regan found in seacoastal middens, "usually only in the ladle form of the horns." (Regan 1917:16) That either species was endemic to the Olympic Mountains is equally unlikely. Ethnologist Erna Gunther's statement (1936:117) that "the mountain goat does not occur on the Olympic Peninsula" is important. Working in the Washington State Museum for over 30 years, she spent hundreds of hours interviewing the Clallam and Makah about their lifeways, traditions, and knowledge of the Olympic environment.

With respect to Lyman's (1988) theory, I agree that goats *could* have reached the Olympics in the late Quaternary. Lyman elaborates on the theory in an unpublished 1993 paper, "Indirect Evidence for the Pre-1925 Presence of Mountain Goats on the Olympic Mountains of Washington State."

"Scheffer also fails to acknowledge [the comments of Jay Kirkpatrick on the value of goat contraceptives]." I first learned of these comments in September 1993. I accept Kirkpatrick's opinion that a vaccine administered by dart might keep a female sterile for up to three years.

But Kirkpatrick's voice was only one in a five-man panel which concluded that "treating mountain goats with these agents [sterilants] would represent a very expensive, never-ending program that, at best, would only partially control the population." If goats are to be eliminated, not simply controlled, "lethal shooting appears to be the only feasible option." (Scientific Panel 1992:2)

The National Park Service's Final Environmental Impact Statement on goat management may appear while the present article is in press. I trust that it will recommend what is best for Olympic National Park and, in the long run, best for people.

Literature Cited

Dunn, J. 1844. History of the Oregon Territory and British North American fur trade. Edwards and Hughes, London.

Gilman, S. C. 1896. The Olympic country. National Geographic Magazine 7(4)April:133–140, map.

Gunther, E. 1936. A preliminary report on the zoological knowledge of the Makah. Pages 105–118 in Essays in Honor of Alfred Louis Kroeber. University of California Press, Berkeley, California.

Lyman, R. L. 1988. Significance for wildlife management of the Late Quaternary biogeography of mountain goats (*Oreamnos americanus*) in the Pacific Northwest, U.S.A. Arctic and Alpine Research 20(1):13–23.

Regan, A. B. 1917. Archaeological notes on western Washington and adjacent British Columbia. Proceedings of the California Academy of Sciences, 4th ser., 7(1):1–31.

Scientific panel (5 authors). 1992. The applicability of contraceptives in the elimination or control of exotic mountain goats from Olympic National Park. Final report to the National Park Service, January, unpublished.

Seattle Press. 1890. The Olympics: An account of the explorations made by the "Press" explorers. Seattle WA, July 16, 18(10):1–11 with an editorial and "resume of the natural resources of explored region" on p. 20.

Essay

Assessing Extinction Threats: Toward a Reevaluation of IUCN Threatened Species Categories

GEORGINA M. MACE
Institute of Zoology
Zoological Society of London
Regent's Park, London NW1 4RY, U.K.

RUSSELL LANDE
Department of Ecology and Evolution
University of Chicago
Chicago, Illinois 60637, U.S.A.

Abstract: *IUCN categories of threat (Endangered, Vulnerable, Rare, Indeterminate, and others) are widely used in 'Red lists' of endangered species and have become an important tool in conservation action at international, national, regional, and thematic levels. The existing definitions are largely subjective, and as a result, categorizations made by different authorities differ and may not accurately reflect actual extinction risks. We present proposals to redefine categories in terms of the probability of extinction within a specific time period, based on the theory of extinction times for single populations and on meaningful time scales for conservation action. Three categories are proposed (CRITICAL, ENDANGERED, VULNERABLE) with decreasing levels of threat over increasing time scales for species estimated to have at least a 10% probability of extinction within 100 years. The process of assigning species to categories may need to vary among different taxonomic groups, but we present some simple qualitative criteria based on population biology theory, which we suggest are appropriate at least for most large vertebrates. The process of assessing threat is clearly distinguished from that of setting priorities for conservation action, and only the former is discussed here.*

Resumen: *La categorización de la Unión Internacional para la Conservación de la Naturaleza (UICN) de las especies amenazadas (en peligro, vulnerables, raras, indeterminadas y otras) son ampliamente utilizadas en las Listas Rojas de especies en peligro y se han convertido en una herramienta importante para las acciones de conservación al nivel internacional, nacional, regional y temático. Las definiciones de las categorías existentes son muy subjetivas y, como resultado, las categorizaciones hechas por diferentes autores difieren y quizás no reflejen con certeza el riesgo real de extinción. Presentamos propuestas para re-definir las categorías en términos de la probabilidad de extinción dentro de un período de tiempo específico. Las propuestas están basadas en la teoría del tiempo de extinción para poblaciones individuales y en escalas de tiempo que tengan significado para las acciones de conservación. Se proponen tres categorías (CRITICA, EN PELIGRO, VULNERABLE) con niveles decrecientes de amenaza sobre escalas de tiempo en aumento para especies que se estima tengan cuando ménos un 10% de probabilidad de extinción en 100 años. El proceso de asignar especies a categorías puede que necesite variar dentro de los diferentes grupos taxonómicos pero nosotros presentamos algunos criterios cualitativos simples basados en la teoría de la biología de las poblaciones, las cuales sugerimos son apropiadas para cuando ménos la mayoría de los grandes vertebrados. El proceso de evaluar la amenaza se distingue claramente del de definir las prioridades para las acciones de conservación, sólamente el primero se discute aquí.*

Paper submitted February 12, 1990; revised manuscript accepted October 8, 1990.

Introduction

Background

The Steering Committee of the Species Survival Commission (SSC) of the IUCN has initiated a review of the overall functioning of the Red Data Books. The review will cover three elements: (1) the form, format, content, and publication of Red Data Books; (2) the categories of threat used in Red Data Books and the IUCN Red List (Extinct, Endangered, Vulnerable, Rare, and Indeterminate); and (3) the system for assigning species to categories. This paper is concerned with the second element and includes proposals to improve the objectivity and scientific basis for the threatened species categories currently used in Red Data Books (see IUCN 1988 for current definitions).

There are at least three reasons why a review of the categorization system is now appropriate: (1) the existing system is somewhat circular in nature and excessively subjective. When practiced by a few people who are experienced with its use in a variety of contexts it can be a robust and workable system, but increasingly, different groups with particular regional or taxonomic interests are using the Red Data Book format to develop local or specific publications. Although this is generally of great benefit, the interpretation and use of the present threatened species categories are now diverging widely. This leads to disputes and uncertainties over particular species that are not easily resolved and that ultimately may negatively affect species conservation. (2) Increasingly, the categories of threat are being used in setting priorities for action, for example, through specialist group action plans (e.g., Oates 1986; Eudey 1988; East 1988, 1989; Schreiber et al. 1989). If the categories are to be used for planning then it is essential that the system used to establish the level of threat be consistent and clearly understood, which at present it does not seem to be. (3) A variety of recent developments in the study of population viability have resulted in techniques that can be helpful in assessing extinction risks.

Assessing Threats Versus Setting Priorities

In the first place it is important to distinguish systems for assessing threats of extinction from systems designed to help set priorities for action. The categories of threat should simply provide an assessment of the likelihood that if current circumstances prevail the species will go extinct within a given period of time. This should be a scientific assessment, which ideally should be completely objective. In contrast, a system for setting priorities for action will include the likelihood of extinction, but will also embrace numerous other factors, such as the likelihood that restorative action will be successful; economic, political, and logistical considerations; and perhaps the taxonomic distinctiveness of the species under review. Various categorization systems used in the past, and proposed more recently, have confounded these two processes (see Fitter & Fitter 1987; Munton 1987). To devise a general system for setting priorities is not useful because different concerns predominate within different taxonomic, ecological, geographical, and political units. The process of setting priorities is therefore best left to specific plans developed by specialist bodies such as the national and international agencies, the specialist groups, and other regional bodies that can devise priority assessments in the appropriate regional or taxonomic context. An objective assessment of extinction risk may also then contribute to the decisions taken by governments on which among a variety of recommendations to implement. The present paper is therefore confined to a discussion of assessing threats.

Aims of the System of Categorization

For Whom?

Holt (1987) identifies three different groups whose needs from Red Data Books (and therefore categories of threat) may not be mutually compatible: the lay public, national and international legislators, and conservation professionals. In each case the purpose is to highlight taxa with a high extinction risk, but there are differences in the quality and quantity of information needed to support the assessment. Scott et al. (1987) make the point that in many cases simple inclusion in a Red Data Book has had as much effect on raising awareness as any of the supporting data (see also Fitter 1974). Legislators need a simple, but objective and soundly based system because this is most easily incorporated into legislation (Bean 1987). Legislators frequently require some statement about status for every case they consider, however weak the available information might be. Inevitably, therefore, there is a conflict between expediency and the desire for scientific credibility and objectivity. Conservationists generally require more precision, particularly if they are involved in planning conservation programs that aim to make maximal use of limited resources.

Characteristics of an Ideal System

With this multiplicity of purposes in mind it is appropriate to consider various characteristics of an ideal system:

(1) The system should be essentially simple, providing easily assimilated data on the risk of extinction. In terms of assessing risk, there seems to be little virtue in developing numerous categories, or in categorizing risk on the basis of a range of different parameters (e.g., abundance, nature of threat, likelihood of persistence of threat, etc.). The categories should be few in number,

should have a clear relationship to one another (Holt 1987; Munton 1987), and should be based around a probabilistic assessment of extinction risk.

(2) The system for categorization has to be flexible in terms of data required. The nature and amount of data available to assess extinction risks varies widely from almost none (in the vast majority of species) to highly detailed population data (in a very few cases). The categorization system should make maximum use of whatever data are available. One beneficial consequence of this process would be to identify key population data for field workers to collect that would be useful in assessing extinction risk.

(3) The categorization system also needs to be flexible in terms of the population unit to which it applies. Throughout this discussion, it is assumed that the system being developed will apply to any species, subspecies, or geographically separate population. The categorization system therefore needs to be equally applicable to limited lower taxonomic levels and to more limited geographical scope. Action planning will need to be focused on particular taxonomic groups or geographical areas, and can then incorporate an additional system for setting priorities that reflect taxonomic distinctiveness and extinction risks outside the local area (e.g., see East 1988, 1989; Schreiber et al. 1989).

(4) The terminology used in categorization should be appropriate, and the various terms used should have a clear relationship to each other. For example, among the current terms both 'endangered' and 'vulnerable' are readily comprehended, but 'rare' is confusing. It can be interpreted as a statement about distribution status, level of threat, or local population size, and the relationships between these factors are complex (Rabinowitz et al. 1986). Rare (i.e., low-density) species are not always at risk and many species at risk are not numerically rare (King 1987; Munton 1987; Heywood 1988). The relationship of 'rare' to 'endangered' and 'vulnerable' is also unclear.

(5) If the system is to be objectively based upon sound scientific principles, it should include some assessment of uncertainty. This might be in terms of confidence levels, sensitivity analyses, or, most simply, on an ordinal scale reflecting the adequacy of the data and models in any particular case.

(6) The categories should incorporate a time scale. On a geological time scale all species are doomed to extinction, so terms such as "in danger of extinction" are rather meaningless. The concern we are addressing here is the high background level of the current rates of extinction, and one aim is therefore preservation over the upcoming centuries (Soulé & Simberloff 1986). Therefore, the probability of extinction should be expressed in terms of a finite time scale, for example, 100 years. Munton (1987) suggests using a measure of number of years until extinction. However, since most models of population extinction times result in approximately exponential distributions, as in Goodman's (1987) model of density-dependent population growth in a fluctuating environment, mean extinction time may not accurately reflect the high probability that the species will go extinct within a time period considerably shorter than the mean (see Fig. 1). More useful are measures such as "95% likelihood of persistence for 100 years."

Population Viability Analysis and Extinction Factors

Various approaches to defining viable populations have been taken recently (Shaffer 1981, 1990; Gilpin & Soulé, 1986; Soulé 1987). These have emphasized that there is no simple solution to the question of what constitutes a viable population. Rather, through an analysis of extinction factors and their interactions it is possible to assess probabilities and time scales for population persistence for a particular taxon at a particular time and place. The development of population viability analyses has led to the definition of intrinsic and extrinsic factors that determine extinction risks (see Soulé 1983; Soulé 1987; Gilpin & Soulé 1986; see also King 1987). Briefly these can be summarized as population dynamics (number of individuals, life history and age or stage distribution, geographic structure, growth rate, variation in demographic parameters), population characteristics (morphology, physiology, genetic variation, behavior and dispersal patterns), and environmental effects (habitat quality and quantity, patterns and rates of environmental disturbance and change, interactions with other species including man).

Preliminary models are available to assess a population's expected persistence under various extinction pressures, for example, demographic variation (Goodman 1987a, b; Belovsky 1987; CBSG 1989), catastrophes (Shaffer 1987), inbreeding and loss of genetic diversity (Lande & Barrowclough 1987; Lacy 1987), metapopulation structure (Gilpin 1987; Quinn & Hastings 1987; Murphy et al. 1990). In addition, various approaches have been made to modeling extinction in populations threatened by habitat loss (e.g., Gutiérrez & Carey 1985; Maguire et al. 1987; Lande 1988), disease (e.g., Anderson & May 1979; Dobson & May 1986; Seal et al. 1989), parasites (e.g., May & Anderson 1979; May & Robinson 1985; Dobson & May 1986), competitors, poaching (e.g., Caughley 1988), and harvesting or hunting (e.g., Holt 1987).

So far, the development of these models has been rather limited, and in particular they often fail to successfully incorporate several different extinction factors and their interactions (Lande 1988). Nevertheless the approach has been applied in particular cases even with existing models (e.g., grizzly bear: Shaffer 1983; spotted owl: Gutiérrez & Carey 1985; Florida panther: CBSG 1989), and there is much potential for further development.

Although different extinction factors may be critical for different species, other, noncritical factors cannot be ignored. For example, it seems likely that for many species, habitat loss constitutes the most immediate threat. However, simply preserving habitats may not be sufficient to permit long term persistence if surviving populations are small and subdivided and therefore have a high probability of extinction from demographic or genetic causes. Extinction factors may also have cumulative or synergistic effects; for example, the hunting of a species may not have been a problem before the population was fragmented by habitat loss. In every case, therefore, all the various extinction factors and their interactions need to be considered. To this end more attention needs to be directed toward development of models that reflect the random influences that are significant to most populations, that incorporate the effects of many different factors, and that relate to the many plant, invertebrate, and lower vertebrate species whose population biology has only rarely been considered so far by these methods.

Viability analysis should suggest the appropriate kind of data for assigning extinction risks to species, though much additional effort will be needed to develop appropriate models and collect appropriate field data.

Proposal

Three Categories and Their Justification

We propose the recognition of three categories of threat (plus EXTINCT), defined as follows:

CRITICAL: 50% probability of extinction within 5 years or 2 generations, whichever is longer.
ENDANGERED: 20% probability of extinction within 20 years or 10 generations, whichever is longer.
VULNERABLE: 10% probability of extinction within 100 years.

These definitions are based on a consideration of the theory of extinction times for single populations as well as on meaningful time scales for conservation action. If biological diversity is to be maintained for the foreseeable future at anywhere near recent levels occurring in natural ecosystems, fairly stringent criteria must be adopted for the lowest level of extinction risk, which we call VULNERABLE. A 10% probability of extinction within 100 years has been suggested as the highest level of risk that is biologically acceptable (Shaffer 1981) and seems appropriate for this category. Furthermore, events more than about 100 years in the future are hard to foresee, and this may be the longest duration that legislative systems are capable of dealing with effectively.

It seems desirable to establish a CRITICAL category to emphasize that some species or populations have a very high risk of extinction in the immediate future. We propose that this category include species or populations with a 50% chance of extinction within 5 years or two generations, and which are clearly at very high risk.

An intermediate category, ENDANGERED, seems desirable to focus attention on species or populations that are in substantial danger of extinction within our lifetimes. A 20% chance of extinction within 20 years or 10 generations seems to be appropriate in this context.

For increasing levels of risk represented by the categories VULNERABLE, ENDANGERED, and CRITICAL, it is necessary to increase the probability of extinction or to decrease the time scale, or both. We have chosen to do both for the following reasons. First, as already mentioned, decreasing the time scale emphasizes the immediacy of the situation. Ideally, the time scale should be expressed in natural biological units of generation time of the species or population (Leslie 1966), but there is also a natural time scale for human activities such as conservation efforts, so we have given time scales in years and in generations for the CRITICAL and ENDANGERED categories.

Second, the uncertainty of estimates of extinction probabilities decreases with increasing risk levels. In population models incorporating fluctuating environments and catastrophes, the probability distribution of extinction times is approximately exponential (Nobile et al. 1985; Goodman 1987). In a fluctuating environment where a population can become extinct only through a series of unfavorable events, there is an initial, relatively brief period in which the chance of extinction is near zero, as in the inverse Gaussian distribution of extinction times for density-independent fluctuations (Ginzburg et al. 1982; Lande & Orzack 1988). If catastrophes that can extinguish the population occur with probability p per unit time, and are much more important than normal environmental fluctuations, the probability distribution of extinction times is approximately exponential, pe^{-pt}, and the cumulative probability of extinction up to time t is approximately $1 - e^{-pt}$. Thus, typical probability distributions of extinction times look like the curves in Figures 1A and 1B, and the cumulative probabilities of extinction up to any given time look like the curves in Figures 1C and 1D. Dashed curves represent different distributions of extinction times and cumulative extinction probabilities obtained by changing the model parameters in a formal population viability analysis (e.g., different amounts of environmental variation in demographic parameters). The uncertainty in an estimate of cumulative extinction probability up to a certain time can be measured by its coefficient of variation, that is, the standard deviation among different estimates of the cumulative extinction probability with respect to reasonable variation in model parameters, divided by the best estimate. It is apparent from Figures 1C and 1D that at least for small variations in the parameters (if the parameters are reasonably well known), the uncertainty of estimates of cumulative extinction probability at particular times decreases as the level of risk increases. Thus at times, t_1, t_2, and t_3 when the best estimates of the cumulative extinction probabilities are 10%, 20%, and 50% respectively, the corresponding ranges of extinction probabilities in Figure 1C are 6.5%–14.8%, 13.2%–28.6%, and 35.1%–65.0%, and in Figure 1D are 6.8%–13.1%, 13.9%–25.7%, and 37.2%–60.2%. Taking half the range as a rough approximation of the standard deviation in this simple illustration gives uncertainty measures of 0.41, 0.38, and 0.30 in Figure 1C, and 0.31, 0.29, and 0.23 in Figure 1D, corresponding to the three levels of risk. Given that for practical reasons we have chosen to shorten the time scales for the more threatened categories, these results suggest that to maintain low levels of uncertainty, we should also increase the probabilities of extinction in the definition of the ENDANGERED and CRITICAL categories.

These definitions are based on general principles of population biology with broad applicability, and we believe them to be appropriate across a wide range of life forms. Although we expect the process of assigning species to categories (see below) to be an evolving (though closely controlled and monitored) process, and one that might vary across broad taxonomic groups, we recommend that the definitions be constant both across taxonomic groups and over time.

Assigning Species or Populations to Categories

We recognize that in most cases, there are insufficient data and imperfect models on which to base a formal probabilistic analysis. Even when considerable information does exist there may be substantial uncertainties in the extinction risks obtained from population models containing many parameters that are difficult to estimate accurately. Parameters such as environmental stochasticity (temporal fluctuations in demographic parameters such as age- or developmental stage–specific mortality and fertility rates), rare catastrophic events, as well as inbreeding depression and genetic variability in particular characters required for adaptation are all difficult to estimate accurately. Therefore it may not be possible to do an accurate probabilistic viability analysis even for some very well studied species. We suggest that the categorization of many species should be based on more qualitative criteria derived from the same body of theory as the definitions above, which will broaden the scope and applicability of the categorization system. In these more qualitative criteria we use measures of effective population size (N_e) and give approximate equivalents in actual population size (N). It is important to recognize that the relationship between N_e and N depends upon a variety of interacting factors. Estimating N_e for a particular population will require quite extensive information on breeding structure and life history characteristics of the population and may then produce only an approximate figure (Lande & Barrowclough 1987). In addition, different methods of estimating N_e will give variable results (Harris & Allendorf 1989). N_e/N ratios vary widely across species, but are typically in the range 0.2 to 0.5. In the criteria below we give a value for N_e as well as an approximate value of N assuming that the N_e/N ratio is 0.2.

We suggest the following criteria for the three categories:

CRITICAL: 50% probability of extinction within 5 years or 2 generations, whichever is longer, or
(1) Any **two** of the following criteria:
(a) Total population $N_e < 50$ (corresponding to actual $N < 250$).
(b) Population fragmented: ≤2 subpopulations with $N_e > 25$ ($N > 125$) with immigration rates <1 per generation.
(c) Census data of >20% annual decline in numbers over the past 2 years, or >50% decline in the last generation, or equivalent projected declines based on demographic projections after allowing for known cycles.
(d) Population subject to catastrophic crashes (>50% reduction) per 5 to 10 years, or 2 to 4 generations, with subpopulations highly correlated in their fluctuations.
or (2) Observed, inferred, or projected habitat alteration (i.e., degradation, loss, or fragmentation) resulting in characteristics of (1).
or (3) Observed, inferred, or projected commercial exploitation or ecological interactions with introduced species (predators, competitors, pathogens, or parasites) resulting in characteristics of (1).

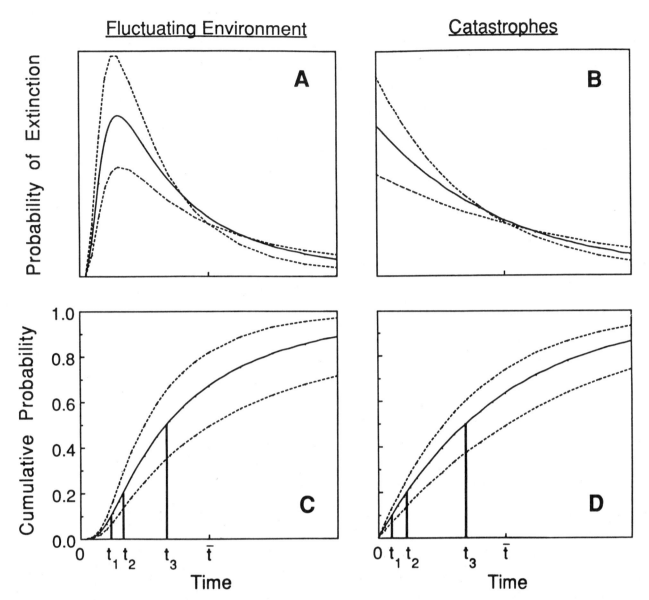

Figure 1. Probability distributions of time to extinction in a fluctuating environment, inverse Gaussian distributions (A), or with catastrophes, exponential distributions (B). Corresponding cumulative extinction probabilities of extinction up to any given time are shown below (C and D). Solid curves represent the best estimates from available data and dashed curves represent different estimates based upon the likely range of variation in the parameters. t_1, t_2, and t_3 are times at which the best estimates of cumulative extinction probabilities are 10%, 20%, and 50%. $\bar{t}$ is the expected time to extinction in the solid curves.

ENDANGERED:

20% probability of extinction within 20 years or 10 generations, whichever is longer, or

(1) Any **two** of the following or any **one** criterion under CRITICAL
  (a) Total population $N_e < 500$ (corresponding to actual $N < 2,500$).
  (b) Population fragmented:
     (i) ≤5 subpopulations with $N_e >$ 100 ($N > 500$) with immigration rates <1 per generation, or
     (ii) ≤2 subpopulations with $N_e > 250$ ($N > 1,250$) with immigration rates <1 per generation.
  (c) Census data of >5% annual decline in numbers over past 5 years, or >10% decline per generation over past 2 generations, or equivalent projected declines based on demographic data after

allowing for known cycles.
- (d) Population subject to catastrophic crashes: an average of >20% reduction per 5 to 10 years or 2 to 4 generations, or >50% reduction per 10 to 20 years or 5 to 10 generations, with subpopulations strongly correlated in their fluctuations.
- or (2) Observed, inferred, or projected habitat alteration (i.e., degradation, loss, or fragmentation) resulting in characteristics of (1).
- or (3) Observed, inferred, or projected commercial exploitation or ecological interactions with introduced species (predators, competitors, pathogens, or parasites) resulting in characteristics of (1).

VULNERABLE: 10% probability of extinction within 100 years, or
- (1) Any **two** of the following criteria or any **one** criterion under ENDANGERED.
  - (a) Total population N_e < 2,000 (corresponding to actual N < 10,000).
  - (b) Population fragmented:
    - (i) ≤5 subpopulations with N_e > 500 (N > 2,500) with immigration rates <1 per generation, or
    - (ii) ≤2 subpopulations with N_e > 1,000 (N > 5,000) with immigration rates <1 per generation.
  - (c) Census data of >1% annual decline in numbers over past 10 years, or equivalent projected declines based on demographic data after allowing for known cycles.
  - (d) Population subject to catastrophic crashes: an average of >10% reduction per 5 to 10 years, >20% reduction per 10 to 20 years, or >50% reduction per 50 years, with subpopulations strongly correlated in their fluctuations.
- or (2) Observed, inferred, or projected habitat alteration (i.e., degradation, loss, or fragmentation) resulting in characteristics of (1).
- or (3) Observed, inferred, or projected commercial exploitation or ecological interactions with introduced species (predators, competitors, pathogens, or parasites) resulting in characteristics of (1).

Prior to any general acceptance, we recommend that these criteria be assessed by comparison of the categorizations they lead to in particular cases with the results of formal viability analyses, and categorizations based on existing methods. This process should help to resolve uncertainties about both the practice of, and results from, our proposals. We expect a system such as this to be relatively robust and of widespread applicability, at the very least for most higher vertebrates. For some invertebrate and plant taxa, different kinds of criteria will need to be developed within the framework of the definitions above. For example, many of these species have very high rates of population growth, short generation times, marked or episodic fluctuations in population size, and high habitat specificity. Under these circumstances, it will be more important to incorporate metapopulation characteristics such as subpopulation persistence times, colonization rates, and the distribution and persistence of suitable habitats into the analysis, which are less significant for most large vertebrate populations (Murphy et al. 1990; Menges 1990).

Change of Status

The status of a population or species with respect to risk of extinction should be up-listed (from unlisted to VULNERABLE, from VULNERABLE to ENDANGERED, or from ENDANGERED to CRITICAL) as soon as current information suggests that the criteria are met. The status of a population or species with respect to risk of extinction should be down-listed (from CRITICAL to ENDANGERED, from ENDANGERED to VULNERABLE, or from VULNERABLE to unlisted) only when the criteria of the lower risk category have been satisfied for a time period equal to that spent in the original category, or if it is shown that past data were inaccurate.

For example, if an isolated population is discovered consisting of 500 individuals and no other information is available on its demography, ecology, or the history of the population or its habitat, this population would initially be classified as ENDANGERED. If management efforts, natural events, or both caused the population to increase so that 10 years later it satisfied the criteria of the VULNERABLE category, the population would not be removed from the ENDANGERED category for a further period of 10 years. This time lag in down-listing prevents frequent up-listing and down-listing of a population or species.

Uncertain or Conflicting Results

Because of uncertainties in parameter estimates, especially those dealing with genetics and environmental variability and catastrophes, substantial differences may arise in the results from analyses of equal validity performed by different parties. In such cases, we recommend that the criteria for categorizing a species or population should revert to the more qualitative ones outlined above.

Reporting Categories of Threat

To objectively compare categorizations made by different investigators and at different times, we recommend that any published categorization also cite the method used, the source of the data, a date when the data were accurate, and the name of the investigator who made the categorization. If the method was by a formal viability model, then the name and version of the model used should also be included.

Conclusion

Any system of categorizing degrees of threat of extinction inevitably contains arbitrary elements. No single system can adequately cover every possibility for all species. The system we describe here has the advantage of being based on general principles from population biology and can be used to categorize species for which either very little or a great deal of information is available. Although this system may be improved in the future, we feel that its use will help to promote a more uniform recognition of species and populations at risk of premature extinction, and should thereby aid in setting priorities for conservation efforts.

Summary

1. Threatened species categories should highlight species vulnerable to extinction and focus appropriate reaction. They should therefore aim to provide objective, scientifically based assessments of extinction risks.
2. The audience for Red Data Books is diverse. Positive steps to raise public awareness and implement national and international legislation benefit from simple but soundly based categorization systems. More precise information is needed for planning by conservation bodies.
3. An ideal system needs to be simple but flexible in terms of data required. The category definitions should be based on a probabilistic assessment of extinction risk over a specified time interval, including an estimate of error.
4. Definitions of categories are appropriately based on extinction probabilities such as those arising from population viability analysis methods.
5. We recommend three categories, CRITICAL, ENDANGERED, and VULNERABLE, with decreasing probabilities of extinction risk over increasing time periods.
6. For most cases, we recommend development of more qualitative criteria for allocation to categories based on basic principles of population biology. We present some criteria that we believe to be appropriate for many taxa, but are appropriate at least for higher vertebrates.

Acknowledgments

We would like to acknowledge the support and encouragement of Simon Stuart, Steven Edwards, and Ulysses Seal in the preparation of this paper. We are also very grateful to the many members of the SSC network for the time they put into commenting upon earlier drafts of this paper, and only regret that they are too numerous to mention individually.

Literature Cited

Anderson, R. M., and R. M. May. 1979. Population biology of infectious diseases. Part I. Nature **280**:361–367.

Bean, M. J. 1987. Legal experience and implications. Pages 39–43 in R. Fitter and M. Fitter, editors. The road to extinction. IUCN, Gland, Switzerland.

Belovsky, G. E. 1987. Extinction models and mammalian persistence. Pages 35–57 in M. E. Soulé, editor. Viable populations for conservation. Cambridge University Press, Cambridge, England.

Caughley, G. 1988. A projection of ivory production and its implications for the conservation of African elephants. CSIRO consultancy report to CITES. CSIRO Division of Wildlife and Ecology.

CBSG. 1989. Florida panther: population viability analysis. IUCN/SSC/CBSG: Apple Valley, Minneapolis, Minnesota.

Cumming, D. H. M., R. F. du Toit, and S. N. Stuart. 1989. African elephants and rhinos: status, survey and conservation action plan. IUCN, Gland, Switzerland.

Dobson, A. P., and R. M. May. 1986. Disease and conservation. Pages 345–365 in M. Soulé, editor. Conservation biology—the science of scarcity and diversity. Sinauer Associates, Sunderland, Massachusetts.

Dobson, A. P., and D. Miller. 1989. Infectious diseases and endangered species management. Endangered Species Update **6**(9):1–5.

East, R. 1988. Antelopes: global survey and regional action plans. Part 1. east and north east Africa. IUCN, Gland, Switzerland.

East, R. 1989. Antelopes: global survey and regional action plans. Part 2. southern and south central Africa. IUCN, Gland, Switzerland.

Eudey, A. 1988. Action plan for Asian primate conservation. IUCN/SSC, Gland, Switzerland.

Fitter, R. F. 1974. 25 years on: a look at endangered species. Oryx **12**:341–346.

Fitter, R., and M. Fitter, editors. 1987. The road to extinction. IUCN, Gland, Switzerland.

Fuller, W. A. 1987. Synthesis and recommendations. Pages 47–55 in R. Fitter and M. Fitter, editors. The road to extinction. IUCN, Gland, Switzerland.

Gilpin, M.E. 1987. Spatial structure and population vulnerability. Pages 125–139 in M. E. Soulé, editor. Viable populations for conservation. Cambridge University Press, Cambridge, England.

Gilpin, M. E., and M. E. Soulé. 1986. Minimum viable populations: processes of species extinctions. Pages 19–34 in M. E. Soulé, editor. Conservation biology—the science of scarcity and diversity. Sinauer Associates, Sunderland, Massachusetts.

Ginzburg, L. R., L. B. Slobodkin, K. Johnson, and A. G. Bindman. 1982. Quasiextinction probabilities as a measure of impact on population growth. Risk Analysis **2**:171–181.

Goodman, D. 1987a. The demography of chance extinction. Pages 11–34 in M. E. Soulé, editor. Viable populations for conservation. Cambridge University Press, Cambridge, England.

Goodman, D. 1987b. How do any species persist? Lessons for conservation biology. Conservation Biology **1**:59–62.

Gutiérrez, R. J., and A. B. Carey, editors. 1985. Ecology and management of the Spotted Owl in the Pacific Northwest. General Technical Report PNW-185, USDA Forest Service, Pacific Northwest Station, Portland, Oregon.

Harris, R. B., and F. W. Allendorf. 1989. Genetically effective population size of large mammals: an assessment of estimators. Conservation Biology **3**:181–191.

Heywood, V. H. 1988. Rarity: a privilege and a threat. Pages 277–290 in W. Greuter and B. Zimmer, editors. Proceedings of the XIV International Botanical Congress Koeltz, Konigstein/Taunus.

Holt, S. J. 1987. Categorization of threats to and status of wild populations. Pages 19–30 in R. Fitter and M. Fitter, editors. The road to extinction. IUCN, Gland, Switzerland.

IUCN. 1988. 1988 IUCN red list of threatened animals IUCN, Gland, Switzerland.

King, F. W. 1987. Thirteen milestones on the road to extinction. Pages 7–18 in R. Fitter and M. Fitter, editors. The road to extinction. IUCN, Gland, Switzerland.

Lacy, R. C. 1987. Loss of genetic diversity from managed populations: interacting effects of drift, mutation, immigration, selection and population subdivision. Conservation Biology **1**:143–157.

Lande, R. 1988. Genetics and demography in biological conservation. Science **241**:1455–1460.

Lande, R., and G. F. Barrowclough. 1987. Effective population size, genetic variation and their use in population management. Pages 87–123 in M. E. Soulé, editor. Viable populations for conservation. Cambridge University Press, Cambridge, England.

Lande, R., and S. H. Orzack. 1988. Extinction dynamics of age-structured populations in a fluctuating environment. PNAS **85**:7418–7421.

Leslie, P. H. 1966. Journal of Animal Ecology **25**:291.

Maguire, L. A., U. S. Seal, and P. F. Brussard. 1987. Managing critically endangered species: the Sumatran rhino as an example. Pages 141–158 in M. E. Soulé, editor. Viable populations for conservation. Cambridge University Press, Cambridge, England.

May, R. M., and R. M. Anderson. 1979. Population biology of infectious diseases. Part II. Nature **280**:455–461.

May, R. M., and S. K. Robinson. 1985. Population dynamics of avian brood parasitism. American Naturalist **126**:475–494.

Menges, E. S. 1990. Population viability analysis for an endangered plant. Conservation Biology **4**:52–62.

Munton, P. 1987. Concepts of threat to the survival of species used in Red Data books and similar compilations. Pages 72–95 in R. Fitter and M. Fitter, editors. The road to extinction. IUCN, Gland, Switzerland.

Murphy, D. D., K. E. Freas, and S. B. Weiss. 1990. An environment-metapopulation approach to population viability analysis for a threatened invertebrate. Conservation Biology **4**:41–51.

Nobile, A. G., L. M. Ricciardi, and L. Sacerdote. 1985. Exponential trends of first passage-time densities for a class of diffusion processes with steady-state distribution. J. Appl. Probab. **22**:611–618.

Oates, J. F. 1986. Action plan for African primate conservation: 1986–1990. IUCN/SSC, Gland, Switzerland.

Quinn, J. F., and A. Hastings. 1987. Extinction in subdivided habitats. Conservation Biology **1**:198–208.

Rabinowitz, D., S. Cairns, and T. Dillon. 1986. Seven forms of rarity and their frequency in the flora of the British Isles. Pages 182–204 in M. E. Soulé, editor. Conservation biology—the science of scarcity and diversity. Sinauer Associates, Sunderland, Massachusetts.

Schreiber, A., R. Wirth, M. Riffel, and H. von Rompaey. 1989. Weasels, civets, mongooses and their relations: an action plan for the conservation of mustelids and viverrids. IUCN, Gland, Switzerland.

Scott, P., J. A. Burton, and R. Fitter. 1987. Red Data Books: the historical background. Pages 1–5 in R. Fitter and M. Fitter, editors. The road to extinction. IUCN, Gland, Switzerland.

Seal, U. S., E. T. Thorne, M. A. Bogan, and S. H. Anderson. 1989. Conservation biology and the black-footed ferret. Yale University Press, New Haven, Connecticut.

Shaffer, M. L. 1981. Minimum population sizes for species conservation. Bioscience 31:131–134.

Shaffer, M. L. 1983. Determining minimum viable population sizes for the grizzly bear. Int. Conf. Bear Res. Manag. 5:133–139.

Shaffer, M. L. 1987. Minimum viable populations; coping with uncertainty. Pages 69–86 in M. E. Soulé, editor. Viable populations for Conservation. Cambridge University Press, Cambridge, England.

Shaffer, M. L. 1990. Population viability analysis. Conservation Biology 4:39–40.

Soulé, M. E. 1983. What do we really know about extinction? Pages 111–124 in C. Schonewald-Cox, S. Chambers, B. MacBryde, and L. Thomas. Genetics and conservation. Benjamin/Cummings, Menlo Park, California.

Soulé, M. E., editor. 1987. Viable populations for conservation. Cambridge University Press, Cambridge, England.

Soulé, M. E., and D. Simberloff. 1986. What do ecology and genetics tell us about the design of nature reserves? Biological Conservation 35:19–40.

Comment

Extant Unless Proven Extinct?
Or, Extinct Unless Proven Extant?

JARED M. DIAMOND
Department of Physiology
University of California Medical School
Los Angeles, CA 90024

Whitten, Bishop, Nash, and Clayton (1987; see pp. 42–48) report that recent ornithological surveys of Sangihe Island in Indonesia failed to encounter a previously described species of small bird endemic to that island, the Sangihe flycatcher. Recent surveys of the larger Indonesian island of Sulawesi (Celebes) similarly failed to encounter three endemic species of small fish. The authors conclude that these species may be extinct.

At first, these findings may appear particularistic and unlikely to command headlines in Indonesia, let alone elsewhere. In fact, they are of potentially broad significance, for they illustrate a serious recurrent problem in assessments of the present extinction crisis.

Lists of taxa judged extinct or endangered, as exemplified by the IUCN Red Data books, are assembled as follows. Authorities take a species list for a local fauna or flora (e.g., the birds of North America), initially identify particular taxa which they suspect might be extinct or endangered, and seek more information about those taxa. It will thus be learned that some of those putatively extinct or endangered taxa are really extant or secure, thereby shortening the preliminary "red list." The remaining taxa are assembled into a book of statements of the form that species X is presumed extinct because the last sighting was in 1973 despite particular searches for it, while species Y is presumed endangered because the sole known population numbered 124 individuals as of 1986. An example of such reasoning is the Red Data book *Endangered Birds of the World* (King 1981), which concludes that, out of the ca. 8000 bird species extant in 1600, 87 are now extinct and 283 are endangered.

The implicit assumption behind this approach is that taxa are "extant unless proven extinct." Biologists concerned about an extinction crisis bear the burden of proof. The 8000 − 87 − 283 = 7630 bird species omitted in the Red Data book are presumed extant and secure and are not up for further discussion. Thus, the proportion of the world's bird species for which extinct or endangered status is even suggested is trivial (1% and 4%, respectively). Because there is always some difficulty in establishing a population's true status, advocates of unrestrained economic growth proceed to dispute some of those 370 claims and suggest that the true proportion is even smaller. Biologists are left to defend the weak proposition that they think the proportion will be much higher in the future, despite the low proportion now.

The assumption "extant unless proven extinct" seems appropriate for well-studied taxa in regions with many observers—e.g., birds in North America or Europe. The field marks, geographic ranges, and preferred habitats of all North American and European bird species are well described. Every North American and European bird species is specifically searched for every year, and almost all are observed. No species could escape observation for several consecutive years without becoming a subject of debate. Bird watchers make special efforts to find rare or supposedly extinct species. Hence there is no chance that a species could become extinct without first having been considered endangered for many years. Species not specifically mentioned as being possibly extinct or endangered are known to be extant in good numbers.

The situation is quite different in regions with few observers—e.g., the tropics, where most of the world's species live. The situation is also different for taxa with less popular appeal than birds. Many tropical species described in the nineteenth century have never been observed in recent years, either because a) no observer has recently visited their habitat, b) it is uncertain where to look for them, c) it is unknown how to identify them in the field, or d) there are so few observers and so many species in the area that failure to observe some species is hardly significant and does not attract attention. For example, as one reads through the species accounts in *Walker's Mammals of the World* (Nowak and Paradiso 1983), one repeatedly reads descriptions such as "Known only from the type series of four specimens taken in 1904," "The only observations in life are by Allenby-Smith (1886), who . . . ," "Habits have not been described." Such species could easily have been extinct or endangered for many decades without anyone knowing about it. The assumption "extant unless proven extinct" becomes entirely groundless.

As an example of the differing conclusions that these two assumptions yield for a tropical biota, consider the resident land and freshwater avifauna of the Solomon Islands in the Southwest Pacific. Ornithological exploration of the Solomons began in the nineteenth century and reached a peak between 1927 and 1930, when the Whitney expedition visited every ornithologically significant island. The sole published synthesis of Solomon Island birds is a chapter in a book by Mayr (1945), who suggested that one species (the Solomon Island crowned pigeon) is probably extinct. On this basis the Red Data book (King 1981) lists one extinct and no endangered species for the Solomon Islands.

Since the Whitney expedition there has been much further ornithological activity in the Solomons reported in individual publications but not yet synthesized in published form for the whole archipelago. The Templeton-Crocker expedition, Poncelet, Wolff and Bradley, Cain and Galbraith, Temple and Shanahan, and Schodde each collected on one or a few islands. Baker, Bayliss-Smith, Beecher, Donagghо, Filewood, Hadden, Lang, Parker, Sibley, Tedder, and Virtue each observed on one or a few islands. Between 1969 and 1976 I visited most ornithologically significant Solomon islands. In the course of this field work most of the 154 species recorded previously for the Solomons have been found repeatedly, and 10 species not previously recorded have been obtained.

During my own field work I was puzzled from time to time when I failed to encounter a species previously recorded for a particular island. On re-examining the literature plus unpublished records by modern observers, I discovered that there are 12 species for which no definite records in the Solomons exist since 1953 (Table 1). The significance of these nonrecords varies. At one extreme, the crowned pigeon was discovered in 1904 and has since then been the object of three intensive but unsuccessful searches (by the Whitney expedition, Parker, and me) at the site where it was discovered. Local villagers with detailed knowledge of birds say that their last observations of it were in the 1940s and that it was exterminated by introduced cats (Parker 1972). Similarly, an endemic race of the grey teal was discovered in the lake on Rennell Island in 1928 but has not been found by ornithologists visiting the lake from 1951 onward, possibly as an indirect result of introductions of the fish *Tilapia mossambica* (Diamond 1984a). It is highly probable that these two species are extinct. At the opposite extreme, the Malaita fantail has not been recorded since its discovery in the mountains of Malaita in 1930, but no ornithologist has revisited those mountains, and there is no reason to doubt the fantail's continued existence. Meyer's goshawk and the mustached kingfisher are local (each confined to two islands), rare, and hard to observe, and both probably still exist; there was an uncertain observation of the goshawk in 1967.

The remaining seven species of Table 1 are ground birds, as was the crowned pigeon. Only for Woodford's

Table 1. Resident bird species of the Solomon Islands not definitely recorded since 1953.

Name		Level of Endemism	Last Specimen Collected	Last Reliable Report
Scientific	Vernacular			
Anas gibberifrons	Grey teal	subspecies	1928	1928
Accipiter meyerianus	Meyer's goshawk	—	1927	1967(?)
Nesoclopeus woodfordi	Woodford's rail	species	1936	1968(?)
Columba pallidiceps	Yellow-legged pigeon	species*	1928	1928
Gallicolumba beccarii	Grey-breasted ground dove	subspecies	1953	1978(?)
Gallicolumba jobiensis	White-breasted ground dove	subspecies	1927	1927
Gallicolumba salamonis	Thick-billed ground dove	species	1927	1927
Microgoura meeki	Solomon Island crowned pigeon	genus	1904	1940(?)
Halcyon bougainvillei	Moustached kingfisher	species	1953	1953
Pitta anerythra	Solomon Island pitta	species	1936	1936
Zoothera dauma	Scaly thrush	subspecies	1904	1904
Rhipidura malaitae	Malaita fantail	species	1930	1930

The column "Level of Endemism" denotes whether a taxon is endemic to the Solomon Islands at the subspecies, species, or genus level or else is not endemic (i.e., shared even at the subspecies level with other archipelagoes:—).
*Also in Bismarcks

rail and the grey-breasted ground dove have there been recent claims of observations, in each case uncertain ones. The Solomon Island Pitta was formerly described as common and had a distinctive call; the collector Albert Meek obtained 18 specimens in the first decade of this century. Numerous specimens from early in this century also exist for Woodford's rail, the yellow-legged pigeon, and grey-breasted ground dove, while the white-breasted ground dove, thick-billed ground dove, and scaly thrush are known in the Solomons only from four, two, and one early specimens, respectively. If these seven ground-dwelling species still exist at all, they must be in low numbers and endangered, and some of them may be extinct. They too, like the Solomon Island crowned pigeon and like many other ground bird species on other oceanic islands, may have fallen victims to introduced cats, having evolved in the absence of mammalian predators and thus never having developed behavioral defenses.

In short, according to the assumption "extant unless proven extinct," one bird species is extinct in the Solomons. According to the assumption "extinct unless proven extant," up to 12 species may be extinct or endangered, and 11 of these are not even mentioned in the Red Data book. The true number of extinct or endangered species is probably nine, much closer to the calculation by the latter than the former assumption.

Is the number of extinct or endangered species "hidden" from the Red Data book likely to be unusually high for Solomon Island birds? Quite the reverse: the Solomon figure is surely low compared to avifaunas of other oceanic islands. On remote islands for which we have adequate information, more than half of the bird species have become extinct since the arrival of humans (Olson 1987). The main causes of these extinctions have been habitat destruction, overhunting, and effects of introduced mammals on islands lacking native flightless mammals (Diamond 1984b). The Solomons are less at risk from all three of these factors than are most other oceanic islands. As for habitat destruction, the Solomons are still largely forested, human population density is modest, and the climate is too wet for the fires that denuded Madagascar, Fiji, and lowland Hawaii. No Solomon bird has fallen victim to overhunting because there are no very large species and only two flightless species. While introduced mammals may have caused eight of the nine population crashes in the Solomons, this number of victims is still modest, because the Solomons do have native rats that "immunized" native birds against the extinction waves accompanying arrival of rats on formerly ratless islands such as Lord Howe, Midway, Hawaii, and Big South Cape. Hence the assumption "extinct unless proven extant" applies with even more force to most other oceanic islands.

It will be interesting, for other such tropical biotas, to assemble recent observations and thus to determine how many species are demonstrably "extant and secure." Nearly 3000 bird species are known from South America, many of them described in the last century from mountains of Colombia that are now deforested. I have no idea how many of these 3000 species have been actually recorded in the past decade. The Red Data book lists only two extinct South American bird species; is the true total closer to several dozen, with hundreds more endangered? What about all those mammal species whose status is summarized by a statement such as "known only from the type specimen collected by Swynnerton-Jones in 1882"? For much of the world, as Whitten et al. (1987) have indicated for Sulawesi, species must be presumed extinct or endangered unless shown to be extant and secure.

References and Notes

Diamond, J.M. The avifaunas of Rennell and Bellona Islands. *The Natural History of Rennell Island, British Solomon Islands* 8:127–168;1984a.

Diamond, J.M. Historic extinctions: a Rosetta Stone for understanding prehistoric extinctions. In: P. Martin, R. Klein, eds. *Quaternary Extinctions.* Tucson, AZ: University of Arizona Press; 1984b:824–862.

King, W.B. *Endangered Birds of the World.* The ICBP Bird Red Data Book. Washington, DC: Smithsonian Institution Press; 1981.

Mayr, E. *Birds of the Southwest Pacific.* New York: Macmillan; 1945.

Nowak, R.M.; Paradiso, J.L. *Walker's Mammals of the World,* 4th ed. Baltimore: Johns Hopkins University Press; 1983.

Olson, S.L. Extinction on islands: man as a catastrophe. *Conservation 2100* (in press).

Parker, S.A. An unsuccessful search for the Solomon Islands Crowned Pigeon. *Emu* 72:24–26;1972.

Whitten, A.J.; Bishop, K.D.; Nash, S.V.; Clayton, L. One or more extinctions from Sulawesi, Indonesia? *Conservation Biology* 1:42–48;1987.

Comment

Extant Unless Proven Extinct: The International Legal Precedent

F. WAYNE KING
Department of Natural Sciences
Florida State Museum
Gainesville, FL 32611, U.S.A.

Jared Diamond (1987) correctly described the dilemma of assessing the status of taxa sought but not found during faunal surveys, and of compiling lists of extinct species based on those surveys. Is their extinction implied if they are not found, or must they be presumed still extant until proven extinct? Inherent in any such discussion is the assumption that a survey, in the correct geographic locality and habitat, has been made recently by one or more competent field biologists, or at least by experienced amateurs, who could identify the taxa in question. I agree completely with Diamond's presentation and wonder how many readers realize this is more than an academic problem.

The "extant unless proven extinct" dilemma has had a formal place in international conservation law since 20 March 1979, when it was the subject of debate at the Second Meeting of the Conference of the Parties to the Convention on International Trade in Endangered Species of Wild Fauna and Flora (CITES), convened in San José, Costa Rica.

Because Article II of CITES requires that Appendix I contain all species threatened with extinction and affected by trade, and that Appendix II contain species that might become threatened, it is clear that CITES deals with living species and excludes extinct taxa. However, when CITES was drafted in 1973, species were listed in the appendices on the basis of diverse data, some exhaustive and some frivolous. As a consequence, some taxa were inappropriately listed. In 1976, at the First Meeting of the Conference of the Parties to CITES in Berne, Switzerland, an attempt was made to improve the listing process. Criteria were adopted that set a biological standard for adding and deleting taxa from the appendices. In addition to data on the actual or potential negative impact from trade, the criteria set forth in Resolution of the Conference 1.1 (Berne 1976) for addition of taxa to Appendix I require biological data on its status from any of a variety of sources. In descending order of preference, these are (1) hard scientific data collected over a number of years that demonstrate trends in population size and geographic range; or lacking that, (2) data from a single survey of population size and range; or if that is not available, (3) reports from reliable, nonscientific observers; or in the absence of any of the above types of data, (4) reports on habitat destruction, excessive trade, or other potential causes of extinction.

The companion Resolution Conf. 1.2 for deletion of taxa from Appendix I requires more rigorous data than are used for addition of species, data that "must transcend informal or lay evidence of changing biological status and any evidence of commercial trade which may have been sufficient to require the animal or plant to be placed on an appendix initially; and should include at least a well-documented population survey, indication of the population trend sufficient to justify deletion, and an analysis of the potential for commercial trade in the species or population."

After these criteria were developed in Berne, the Parties decided to go back and review the appropriateness of the taxa that had been listed in the Appendices prior to their adoption. That review was initiated at the Special Working Session of the Parties to CITES, held in Geneva, in 1977, and has continued at every subsequent meeting.

The continuing review process led Australia to worry about several possibly extinct species that earlier had been placed on Appendix I at the insistence of that nation. At the Costa Rica meeting in 1979, the Australian delegation questioned how appendix species that were presumed to be extinct should be treated. They pointed out that the documented data on species believed to be extinct vary widely between taxa. The extinction of some taxa will be presumed though not yet proven and the available data will not satisfy the Berne criteria for either addition or deletion of taxa. The extinction of others will be well documented and easily would satisfy the criteria for deletion. As a consequence, the Australians proposed the adoption of a standard approach to the question of extinction of listed taxa.

The Australians summarized these initial discussions in Document 2.14 to give all the delegations a chance to consider the various sides of the issue before the discussion was renewed in Plenary Session a day later. That document asserts that when a population of a wild species consists of

> a small number of individuals with a restricted distribution it is possible to monitor population fluctuations. The decline and extinction of such species can be readily documented and recognized. Another category of taxa threatened with extinction includes species distributed sparsely over an extensive geographic range or consisting of discrete, discontinuous populations in remote inhospitable areas. Documentation of the dynamics of such taxa is difficult and there will always be doubt about the status of such species. The acquisition of adequate records may be hampered by such factors as the trappability and cryptic habits of the species.

Doc. 2.14 further proposed that Parties seeking the deletion of presumed extinct taxa should demonstrate that an effort has been made to confirm the existence of such taxa (e.g., one or more recent surveys should have been conducted), and that it might be wise to wait until after a set period of time had passed before a species is considered extinct. The potential for trade in species previously believed extinct but that have been rediscovered was a sufficient worry that the Australians proposed immediate reinstatement of the taxa if such trade developed, or if that could not be accomplished, retention on Appendix I of those taxa that have great trade value but are presumed extinct.

The lively debate that followed in the Plenary Session in Costa Rica revealed that most delegations were sympathetic to removing extinct taxa from the appendices. However, like the Australians, many delegates worried about removing taxa that were presumed, but not proved, to be extinct, since removal of a species from the CITES Appendix also terminates the protection from excessive commercial trade provided by that listing. The Parties to CITES only meet in conference every second year to debate and vote on additions and deletions in the Appendices. Between conferences, changes in the Appendices are difficult to achieve because the two-thirds majority of the Parties needed for approval of any change must be accomplished by postal vote. Clearly if a presumed-to-be-extinct species were rediscovered after being removed from the Appendix, it could be exploited by unscrupulous traders for one or two years before protection could be reinstated.

Following two days of discussion in Plenary, the Parties adopted Resolution of the Conference 2.21 (San José 1979), which reads:

> NOTING that the definitions of the appendices exclude extinct species;
>
> NOTING also that criteria adopted for developing amendment proposals to the appendices ensure that extinct species are not added to the appendices;
>
> RECOGNIZING the difficulty of determining without doubt that some species presently on the appendices are extinct;
>
> NOTING finally that such species if rediscovered could become items of great albeit short-lived trade interest;
>
> THE CONFERENCE OF THE PARTIES TO THE CONVENTION
>
> RECOMMENDS that no action be taken to remove such species from the appendices and that species not observed for at least 50 years despite repeated surveys be annotated in the appendices as p.e. (possibly extinct).

In the nine years since the resolution was adopted, a total of eight species (six of which are Australian) have been annotated 'p.e.':

Thylacinus cynocephalus, the thylacine of Tasmania, has not definitely been seen in the wild since 1930, and the last captive individual died in 1934, though rumored sightings continue to surface;

Caloprymnus campestris, the desert rat kangaroo of northeastern South Australia and southwestern Queensland, was last recorded in 1935 and not found by surveys in 1961 and subsequent years;

Chaeropus ecaudatus, the pig-footed bandicoot of western New South Wales, Victoria, South Australia, and southern Northern Territory, and a small area of southern Western Australia, was last seen in 1926;

Rhodonessa caryophyllacea, the pink-headed duck of eastern India, was last collected in the wild in the 1920s, and the last captive specimens died in 1942;

Anodorhynchus glaucus, the glaucous macaw of southeastern Brazil, Paraguay, Uruguay, and northern Argentina, was believed extinct in the 1969 and has not been seen since (interestingly, although *A. glaucus* is annotated p.e., the listing disappeared when the entire genus *Anodorhynchus* was added to Appendix I at the 5th Meeting of the Conference of the Parties at Buenos Aires in 1985);

Geopsittacus occidentalis, the night parrot of South Australia and Western Australia, has not been definitely seen in the last 65 years although unconfirmed sightings were reported in 1967;

Psephotus pulcherrimus, the paradise parrot of central Queensland and northern New South Wales, has not been sighted since 1922; and

Dasyornis broadbenti littoralis, the rufous bristlebird

of extreme southwestern Western Australia, was last seen in the wild in 1940.

Despite attempts to find them, six of the eight species have not been seen in over 50 years as required by Resolution Conf. 2.21.

In the absence of hard data proving their extinction, the Parties to CITES have chosen to provide maximum protection from international trade to species annotated 'p.e.' by leaving them on Appendix I. The result is that functionally, within the context of CITES, species are considered extant until proven extinct. However, as suggested by Diamond (1987), for the majority of species that have not been seen in many years and for which recent surveys have not been conducted, there is little choice but to treat them as extinct unless proven extant. This is especially true when such a treatment would provide needed protection for species that are not listed on CITES.

Literature Cited

Diamond, J. 1987. Extant unless proven extinct? Or, extinct unless proven extant? Conservation Biology **1(1)**:77–79.

What Exactly Is an Endangered Species? An Analysis of the U.S. Endangered Species List: 1985–1991

DAVID S. WILCOVE
MARGARET MCMILLAN
KEITH C. WINSTON

Environmental Defense Fund
1875 Connecticut Avenue, N.W.
Washington, D.C. 20009, U.S.A.

Abstract: *Critics of the Endangered Species Act have asserted that it protects an inordinate number of subspecies and populations, in addition to full species, and that the scientific rationale for listing decisions is absent or weak. We reviewed all U.S. plants and animals proposed for listing or added to the endangered species list from 1985 through 1991 to determine the relative proportion of species, subspecies, and populations, and their rarity at time of listing. Approximately 80% of the taxa added to the list were full species, 18% were subspecies, and 2% were distinct population segments of more widespread vertebrate species. The proportion of subspecies and populations was considerably higher among birds and mammals than among other groups. The median population size at time of listing for vertebrate animals was 1075 individuals; for invertebrate animals it was 999. The median population size of a plant at time of listing was less than 120 individuals. Earlier listing of declining species could significantly improve the likelihood of successful recovery, and it would provide land managers and private citizens with more options for protecting vanishing plants and animals at less social or economic cost.*

Que es exactamente una especie en peligro de extinción? Un análisis de la lista de especies en peligro de extinción de U.S.A.

Resumen: *Los críticos del Acta de Especies en Peligro de Extinción han afirmado que, además de proteger especies, esta proteje un número excesivo de subespecies y poblaciones, y que la justificación científica en el proceso de elección de las especies a listar está ausente o es débil. Hemos revisado todas las plantas y animales de USA que fueron propuestos para listar o adicionados a la lista de especies en peligro de extinción entre 1985 y 1991 a los efectos de determinar la proporción relativa de especies, subespecies y poblaciones, y su rareza en el momento del listado. Aproximadamente un 80 porciento de los taxones adicionados a la lista fueron especies; 18 porciento fueron subespecies y 2 porciento fueron segmentos poblacionales diferentes de las especies de vertebrados más comunes. La proporción de subespecies y poblaciones fue considerablemente más alta entre pájaros y mamiferos que entre otros grupos. La mediana del tamaño poblacional en el momento del listado fue de 1.075 individuos; para animales invertebrados fue de 999. La mediana del tamaño poblacional de las plantas en el momento del listado fue menor que 120 individuos. Listados de especies en declinación anteriores podrían mejorar significativamente la posibilidad de una recuperación exitosa, y proveería a los administradores de tierras y ciudadanos privados con más opciones para proteger plantas y animales en proceso de desaparición a un costo social o económico menor.*

Address correspondence to D. S. Wilcove.
Paper submitted August 10, 1992; revised manuscript accepted November 9, 1992.

Introduction

Celebrated by some people and reviled by others, the Endangered Species Act of 1973 is one of the strongest and most controversial environmental laws in the United States. Its stated purpose is to prevent the extinction of plants and animals by conserving the ecosystems upon which they depend. The act authorizes the executive branch to compile a list of threatened and endangered taxa, and it directs all federal departments and agencies to "utilize their authorities in furtherance of the purposes of this Act." The conservation measures it requires include "all methods and procedures which are necessary to bring any endangered species or threatened species to the point at which the measures provided pursuant to this Act are no longer necessary" (statutory language from U.S. Fish and Wildlife Service 1988).

Most of the controversy surrounding the act stems from a clash of values, a belief that the economic and social costs associated with protecting imperiled species exceed whatever benefits are gained by such measures. Some concerns, however, are directed towards the science underlying the law. For example, the act permits the listing of species and subspecies of plants and animals as well as "distinct population segments" of vertebrates. Critics contend that the growing roster of endangered taxa reflects the addition of numerous subspecies and populations, rather than full species, to the list. This perception has led to repeated calls to revise the law so that only full species are eligible for protection. Former Interior Secretary Manuel Lujan has been the most prominent supporter of this idea, as demonstrated by his well-publicized comments on the endangered Mount Graham red squirrel, *Tamiasciurus hudsonicus grahamensis* (Sward 1990): "Nobody's told me the difference between a red squirrel, a black one or a brown one. Do we have to save every subspecies?"

A second controversy centers around the standards by which taxa are listed (Dennis et al. 1991). Some have argued that the act protects taxa that are not truly endangered. In a 1992 Senate hearing, one witness spoke of the need to "set some higher standards for those things we list" (Gordon 1992), while in a cover story in *The Atlantic*, Mann and Plummer (1992) described the science underlying the Endangered Species Act as "so incomplete that it verges on the fraudulent." Alternatively, others have argued that by the time many species are officially listed, their numbers are so low that prospects for recovery are poor (Endangered Species Coalition 1992). Rohlf (1991) has commented that the distinctions between threatened and endangered taxa appear to have no uniform biological meaning.

Despite these concerns, few studies have examined the endangered species list in a systematic manner. We reviewed all U.S. taxa proposed for listing or added to the list between 1985 and 1991. Our objectives were to determine the relative proportion of species, subspecies, and populations, the breakdown between plants and animals, and the rarity of these organisms at time of listing.

The Listing Process

Prior to discussing our methodology, a brief review of the listing process is useful (see Bean 1983 for more details). The decision to list a species, subspecies, or population is made by the Secretary of the Interior (via the U.S. Fish and Wildlife Service [FWS]) or, in the case of certain marine species, by the Secretary of Commerce (via the National Marine Fisheries Service [NMFS]). The law permits them to list a plant or animal for any one of five reasons: (1) present or threatened destruction of habitat; (2) overutilization for commercial, recreational, scientific, or educational purposes; (3) losses due to disease or predation; (4) the inadequacy of existing laws and regulations to protect the organism in question; and (5) "other natural or manmade factors affecting its continued existence" (U.S. Fish and Wildlife Service 1988). For the purposes of the act, all listed taxa, including subspecies and populations, are considered "species," and they may be listed as either endangered or threatened. An endangered species is "any species which is in danger of extinction throughout all or a significant portion of its range"; a threatened species is "any species which is likely to become an endangered species within the foreseeable future throughout all or a significant portion of its range" (U.S. Fish and Wildlife Service 1988).

If FWS or NMFS determines that listing a particular plant or animal is warranted, it publishes a notice of its intent to do so in the *Federal Register* and in local newspapers. The proposal notice states the reasons for listing and solicits public comments. A final listing determination must be made within a year of publishing the proposal, although under certain circumstances this deadline may be extended by six months. According to the law, a final listing decision must be made "solely on the basis of the best scientific and commercial data" available to the secretary. Notice of a final decision is published in the *Federal Register* and elsewhere, and once again the rationale for the decision is given.

Methods

Data Collection and Analysis

We reviewed all proposed and final listings from 1985 through 1991 to determine: (1) the breakdown of listed organisms in the following categories: mammals, birds, reptiles, amphibians, fishes, mollusks, arthropods, and plants (plants were subdivided into annuals, biennials, and perennials); (2) the proportion of species, subspecies, and populations within each category; (3) the total number of individuals (equal to total population size) of each listed taxon at time of listing; and (4) the total number of known populations of each listed plant or animal at time of listing. This information was obtained from either the *Federal Register* or the *Endangered Species Technical Bulletin,* a publication of FWS. Wherever possible, we used data from the final listing notice. Data were taken from the listing proposal when adequate information was not available in the final notice or when the final notice had not been published. In a small number of cases, we contacted FWS or outside experts to clarify confusing statements in the *Federal Register* or the *Endangered Species Technical Bulletin.* For the desert tortoise (*Gopherus agassizii*), population data were taken from an unpublished report by the Bureau of Land Management. We used this report because the data were unavailable in published sources. We wanted to include the tortoise in our study because it has the largest extant population among listed vertebrates.

We restricted our analysis to wild populations of species known to occur in the United States (including Puerto Rico, the U.S. Virgin Islands, and Pacific Trust territories). If the range of a listed species, subspecies, or population extended beyond the borders of the U.S., we included population data from its entire range. Captive populations were not counted in determining the total number of individuals or the total number of known populations.

Where a range of values was given for either the total number of individuals or the total number of populations, we always chose the highest number for analysis. When the population was presented as an upper limit (such as "<200"), we chose the next highest whole number as the actual population size ("199"). Both rules are conservative in that they yield the largest possible population tallies for listed species. Seedlings and saplings of listed plants were not included in population tallies. Nor did we attempt to distinguish between clonal and nonclonal individuals in any of the analyses.

The Piping Plover (*Charadrius melodus*) and Roseate Tern (*Sterna dougallii*) were listed under a split designation, in which some populations were classified as endangered and others as threatened. For overall comparisons of numbers of individuals and populations in different taxonomic groups, each of these birds was treated as a single entity, with threatened and endangered populations combined. For the purposes of comparing threatened versus endangered taxa, however, the populations were treated separately.

Finally, we excluded obviously outdated population data. For example, when the Golden-cheeked Warbler (*Dendroica chrysoparia*) was listed in 1991, the only available population estimates dated back to 1976. In the intervening years, massive losses of warbler habitat had occurred, triggering the listing decision. Therefore, the 1976 population estimates were not used in this study.

Data were analyzed using SYSTAT (Wilkinson 1988). Statistical comparisons were made using the nonparametric Mann-Whitney U Test.

Data Caveats

As species become rarer and more localized, they become easier to census. Thus, population data may be more readily available for the rarest species. Data estimates also become more precise and accurate as the species becomes rarer. On the other hand, below a certain level, a plant or animal may become too scarce to census.

The listing of a species may also spur people to search for new populations, thereby boosting the population tally. When the running buffalo clover (*Trifolium stoloniferum*) was listed in 1987, for example, FWS knew of only a single population consisting of four plants. Within two years after listing, biologists had discovered seven additional populations (Lowe et al. 1990). Alternatively, we have noted instances in which populations of a plant or animal were destroyed between the time it was proposed for listing and the finalization of that decision. Half of the remaining population of the prickly-ash (*Zantholyxum thomasianum*) on St. Thomas Island was destroyed by a developer a few days before the species was officially listed as endangered. We emphasize, therefore, that data presented in this study reflect FWS's assessment of the status of the taxon, *based on the best available information at time of listing.*

Finally, it can be difficult to determine from the literature how many populations of a particular plant or animal exist. Phrases such as "known from only three sites" or "present in two localities" are frequently used to describe the spatial distribution of a species, and one must exercise judgment in determining whether these terms refer to multiple populations. We took a conservative approach and omitted species about which we were uncertain. Nonetheless, the data on number of populations should be viewed cautiously.

Results

In total, 492 plants and animals were listed or proposed for listing from 1985 through 1991 (Table 1). Plants comprised by far the largest portion of this total (68%), followed by vertebrate animals (19%) and invertebrate animals (13%). Overall, 20% of the taxa proposed or listed during this period were subspecies or populations rather than full species (18% subspecies, 2% populations; see Table 1). The proportion of listings involving subspecies or populations differed markedly among

Table 1. Breakdown of U.S. plants and animals listed or proposed for listing under the Endangered Species Act, 1985–1991.

Taxonomic group	n	Species	Subspecies	Populations[1]	% Subspecies	% Populations
Mammals	23	7	16	0	70	0
Birds	15	3	8	4	53	27
Reptiles	10	6	2	2	20	20
Amphibians	3	3	0	0	0	0
Fishes	43	30	11	2	26	5
Arthropods	23	18	5	N.A.	22	N.A.
Mollusks	43	41	2	N.A.	5	N.A.
Plants	332	286	46	N.A.	14	N.A.
Total	492	394	90	8	18	2

[1] Populations of invertebrate organisms and plants cannot be listed under the Endangered Species Act; only species and subspecies of plants and invertebrate animals are eligible for listing.

taxa. In general, a higher proportion of the vertebrates represented subspecies or populations than was true for other taxa. For example, 80% of the birds and 70% of the mammals proposed for listing or actually listed represented subspecies or populations. The percentage of subspecies dropped to 5% for mollusks and 14% for plants.

Data on total number of individuals and total number of populations at time of listing were available for 235 and 373 taxa respectively (Figs. 1 and 2). Vertebrate and invertebrate animals did not differ significantly with respect to either total number of individuals (median values of 1075 and 999 respectively; Mann-Whitney U Test, $p = 0.385$) or number of populations (median values of 2 and 3 respectively; $p = 0.521$). Listed plants had significantly fewer individuals than listed animals (median values of 119.5 and 999, respectively; $p < 0.001$). These fewer individuals were spread among more populations, however (median values of 4 and 2.5 for total number of populations for plants versus animals; $p = 0.006$).

Among listed plants, perennials had significantly fewer individuals than did annuals (median = 99 versus 3254.5; $p < 0.001$); perennials also had significantly fewer populations (median = 4 versus 9; $p = 0.004$).

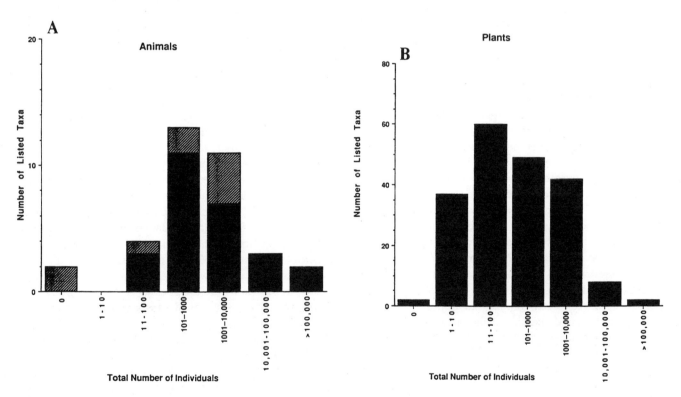

Figure 1. (A) Distribution of total population sizes (total number of individuals) at time of listing for animals. Solid bars denote vertebrates; hatched bars denote invertebrates. Median population size for all animals = 999. See text for details. (B) Distribution of total population sizes (total number of individuals) for plants. Median population size = 119.5.

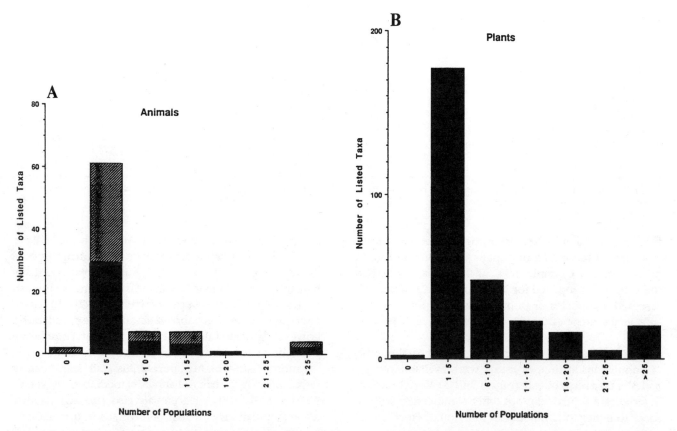

Figure 2. (A) Distribution of total number of populations at time of listing for animals. Solid bars denote vertebrates; hatched bars denote invertebrates. Median number of populations for all animals = 2.5. (B) Distribution of total number of populations at time of listing for plants. Median number of populations = 4.

Biennial plants, totalling only three species, were excluded from this analysis.

Significant differences emerged between species classified as threatened versus endangered for both animals and plants. Vertebrate animals listed as endangered had significantly fewer individuals (median = 407.5) than those listed as threatened (median = 4161; $p = 0.005$); there were no significant differences in numbers of populations. Our sample size was inadequate to compare threatened and endangered invertebrate animals. When vertebrate and invertebrate taxa were combined, animals listed as endangered had significantly fewer individuals (median = 515) than those listed as threatened (median = 4161; $p = 0.006$) but did not differ significantly with respect to number of populations. Compared to threatened plants, endangered plants had fewer individuals (median = 99 versus 2499.5; $p < 0.001$) and fewer populations (median = 3 versus 9; $p < 0.001$).

Discussion

The contention that the overall growth of the endangered and threatened species list has stemmed from the addition of numerous subspecies and populations is unfounded. Populations constitute less than 2% of recent additions to the list, while subspecies comprise 18%. However, the overwhelming majority of mammals and birds listed in recent years have been subspecies and populations rather than full species, which may contribute to the misperception. It should be noted that the ability to list individual populations has been cited by environmental groups as an example of the Endangered Species Act's flexibility (Endangered Species Coalition 1992). For example, populations of the Bald Eagle (*Haliaeetus leucocephalus*) can be protected in the coterminous 48 states without imposing the act's restrictions in Alaska, where the species is much more numerous. Moreover, removal of subspecies and populations from the endangered species list would eliminate special protection for some of the most charismatic endangered animals in the United States, including grizzly bear (*Ursus arctos horribilis*), Peregrine Falcon (*Falco peregrinus*), Florida panther (*Felis concolor coryi*), and gray wolf (*Canis lupus*).

Early intervention is critical to the success of endangered species recovery efforts. Yet our analysis indicates that most species, subspecies, and populations protected under the Endangered Species Act are not receiving that protection until their total population size and number of populations are critically low. Obvious examples would include the 39 plants that were listed when 10 or fewer individuals were known to exist, or animals such as the freshwater mussel *Quadrula fragosa*, which was listed when only a single nonreproducing population remained.

Recent studies and opinions in the field of population viability analysis support our contention that protection under the Endangered Species Act is coming much too late for most species. The International Union for Conservation of Nature and Natural Resources (IUCN) Captive Breeding Specialist Group (IUCN 1987), for example, recommends that captive populations be established for vertebrate species with wild populations below 1000 individuals, a reflection of their poor prospects for long-term survival in the wild. The median total population size of a vertebrate at time of listing—1075 individuals—is perilously close to this number. Median values for both total population size and number of populations of vertebrates are also below the levels that would qualify a species as "endangered" under the IUCN classification scheme proposed by Mace and Lande (1991)—the second most serious of their three categories of endangerment.

Soulé (1987) suggests that a "bottom line" value for a viable population of a vertebrate species would be "in the low thousands," with viability defined as a 95% probability of persistence, without loss of fitness, for several centuries. He cautions, however, that each species should be evaluated individually because " '[f]ew thousand' is not a rule-of-thumb. Rather, it is a possible, order of magnitude lower boundary." Extinction models developed by Belovsky (1987) indicate that for small to mid-sized mammals, populations of several thousand to several tens of thousands of individuals are necessary to achieve a 95% chance of persistence for 100 years. Thomas (1990), reviewing data on species survival rates on islands, arrived at similar values. He recommends minimum conservation targets of 1000–10,000 individuals for birds and mammals, depending upon the magnitude of the species' natural population fluctuations (more variable populations require higher conservation goals). Virtually all of these values greatly exceed the median population size of listed vertebrates.

It should also be noted that effective population sizes (N_e) of vertebrates are typically less than one half and sometimes as little as one quarter of total population size (see Barrowclough & Coats 1985; Harris & Allendorf 1989). Census populations of several hundred to a few thousand individuals therefore represent much smaller effective populations, which may be vulnerable to the loss of genetic diversity associated with small effective population sizes. Moreover, the values for total population size used in this study usually represent the sum of two or more separate populations; in many cases, N_e values for the separate populations would range from a few dozen to a few hundred.

Total population sizes of invertebrate species did not differ significantly from those of vertebrate species. With a median total population size of 999 individuals, listed invertebrate species may be more vulnerable to extinction than listed vertebrates if, as predicted, their smaller body sizes and shorter individual lifetimes make them especially vulnerable to environmental fluctuations (see Diamond 1984; Murphy et al. 1990). Thomas (1990) suggests that to ensure comparable viability, population targets for rare insects should be at least one order of magnitude greater than population targets for vertebrates. No such margin of safety is apparent in the limited population data available for listed invertebrates.

Population viability models for plants are almost nonexistent (Menges 1990). Nonetheless, the strikingly small population sizes of most listed plants (median = 119.5 individuals) suggest a high risk of extinction and a low probability of recovery. Moreover, depending upon the species, some or all of the individuals may be clones, thereby limiting the potential for genetic diversity. The presence of seed banks may increase the survival prospects for many of these plants, but we have no data on seed banks. Perennials had significantly smaller total population sizes and fewer populations than annuals. We hypothesize that perennials may be less vulnerable to environmental fluctuations and therefore able to persist for longer periods of time at very low population levels. This hypothesis obviously requires further testing and is not meant to suggest that protection of perennials should be delayed until population sizes are extremely small.

Critics of the Endangered Species Act have questioned whether the distinction between threatened and endangered taxa has any biological basis. Our study shows that, in general, endangered species are rarer than threatened ones. Among animals, endangered taxa have significantly fewer individuals than threatened taxa; among plants, endangered species and subspecies have significantly fewer individuals and fewer populations. However, the low population levels of some threatened taxa (such as the Inyo California Towhee [*Pipilo crissalis eremophilus*], listed as threatened with fewer than 200 individuals) strongly suggest they should have been classified as endangered.

Finally, our results may help to explain why to date only five species have recovered to the point where listing was no longer necessary (U.S. General Accounting Office 1988). The Fish and Wildlife Service and the National Marine Fisheries Service are simply not protecting imperiled taxa soon enough. If rare and declining species, subspecies, and populations were listed before they reached such low numbers, prospects for successful recovery would improve. Earlier listing might also provide these agencies with more options for protecting vanishing plants and animals at less social or economic cost.

Acknowledgments

We thank Michael Bean, Deborah Crouse, Rod Fujita, Bruce Manheim, Dennis Murphy, Reed Noss, Katherine Ralls, J. Michael Scott, Mark Shaffer, Tim Tear, and an anonymous reviewer for their helpful comments on earlier drafts of this paper. John Fay reviewed the manuscript and graciously helped us to classify plants as annuals, perennials, and biennials. Our work would not have been possible without the generous support of the Pew Scholars Program in Conservation and the Environment.

Literature Cited

Barrowclough, G. F., and S. L. Coats. 1985. The demography and population genetics of owls, with special reference to the conservation of the Spotted Owl (*Strix occidentalis*). Pages 74–83 in R. J. Gutierrez and A. B. Carey, editors. Ecology and management of the Spotted Owl in the Pacific Northwest. General Technical Report PNW-185. USDA Forest Service, Pacific Northwest Forest and Range Experiment Station. Portland, Oregon.

Bean, M. J. 1983. The evolution of national wildlife law. Second edition. Praeger Publishers, New York, New York.

Belovsky, G. E. 1987. Extinction models and mammalian persistence. Pages 35–57 in M. E. Soulé, editor. Viable populations for conservation. Cambridge University Press, Cambridge, England.

Dennis, B., P. L. Munholland, and J. M. Scott. 1991. Estimation of growth and extinction parameters for endangered species. Ecological Monographs 61:115–143.

Diamond, J. M. 1984. "Normal" extinctions of isolated populations. Pages 191–246 in M. H. Nitecki, editor. Extinctions. University of Chicago Press, Chicago, Illinois.

Endangered Species Coalition. 1992. The Endangered Species Act: a commitment worth keeping. The Wilderness Society, Washington, DC.

Gordon, R. E. 1992. Testimony presented before the Senate Environment and Public Works Committee, April 10.

Harris, R. B., and F. W. Allendorf. 1989. Genetically effective population size of large mammals: an assessment of estimators. Conservation Biology 3:181–191.

IUCN Captive Breeding Specialist Group. 1987. The IUCN policy statement on captive breeding, as approved by the 22nd meeting of the IUCN Council, Gland, Switzerland.

Lowe, D. W., J. R. Matthews, and C. J. Moseley. 1990. The official World Wildlife Fund guide to endangered species of North America. Beacham Publishing, Washington, DC.

Mace, G. M., and R. Lande. 1991. Assessing extinction threats: toward a reevaluation of IUCN threatened species categories. Conservation Biology 5:148–157.

Mann, C. C., and M. L. Plummer. 1992. The butterfly problem. The Atlantic 269(1):47–70.

Menges, E. S. 1990. Population viability analysis for an endangered plant. Conservation Biology 4:52–62.

Murphy, D. D., K. E. Freas, and S. B. Weiss. 1990. An environment-metapopulation approach to population viability analysis for a threatened invertebrate. Conservation Biology 4:41–51.

Rohlf, D. J. 1991. Six biological reasons why the Endangered Species Act doesn't work—and what to do about it. Conservation Biology 5:273–282.

Soulé, M. E. 1987. Where do we go from here? Pages 175–183 in M. E. Soulé, editor. Viable populations for conservation. Cambridge University Press, Cambridge, England.

Sward, S. 1990. Secretary Lujan and the squirrels: Interior chief calls Endangered Species Act "too tough." San Francisco Chronicle, May 12.

Thomas, C. D. 1990. What do real population dynamics tell us about minimum viable population sizes? Conservation Biology 4:324–327.

U.S. Fish and Wildlife Service. 1988. Endangered Species Act of 1973 as amended through the 100th Congress. U.S. Fish and Wildlife Service, Washington, DC.

U.S. General Accounting Office. 1988. Endangered species: management improvements could enhance recovery program. GAO/RCED-89-5. U.S. General Accounting Office, Washington, DC.

Wilkinson, L. 1988. SYSTAT: the system for statistics. SYSTAT, Inc., Evanston, Illinois.

Note

Habitat Protection Under the Endangered Species Act

JOHN G. SIDLE

DAVID B. BOWMAN

Endangered Species Division
U.S. Fish and Wildlife Service
2604 Saint Patrick Avenue
Grand Island, Nebraska U.S.A. 68803

Although some species have been hunted into extinction, the major cause of extinctions has been the destruction of natural habitat. One of the purposes of the United States Endangered Species Act of 1973, as amended, is "to provide a means whereby the ecosystems upon which endangered species and threatened species depend may be conserved."

However, protection of habitat of species listed as endangered or threatened pursuant to the Act is incomplete. Although not often done, the federal government can acquire habitat and create refuges for species under Section 5 of the Act. Also infrequent is the designation or labeling of public or private land (not acquisition) as critical habitat under Section 4 for subsequent protection under Section 7 (Sidle 1987). Under Section 7, federal agencies must ensure that any action authorized, funded, or carried out by such agencies will not jeopardize the continued existence of endangered or threatened species, or result in the destruction or adverse modification of designated critical habitat. However, Section 7 does not apply to state agencies, private citizens, and organizations unless federal permits or funds are involved. Sadly, in spite of the Act, the habitats of endangered and threatened species can slowly or quickly disappear.

Section 9 details prohibitions regarding the taking, possession, and importation of listed species. These activities generally involve direct human contact with listed species. For some listed species, the Act may appear to convey an inadequate conservation standard: for example, shooting an endangered bird is a crime, but bulldozing its habitat may not be. However, the Act defines "take" very broadly to include "harass, harm, pursue, hunt, shoot, wound, kill, trap, capture or collect, or to attempt to engage in any such conduct" (16 *United States Code* 1532 [19]). Section 9 does not specifically mention the destruction of habitat. Prior to November 1981, regulations promulgated by the U.S. Department of the Interior defined harm in the above definition of take as such (50 *Code of Federal Regulations* § 17.3):

> "Harm" in the definition of "take" in the Act means an act or omission which actually injures or kills wildlife, including acts which annoy it to such an extent as to significantly disrupt essential behavioral patterns, which include, but are not limited to, breeding, feeding or sheltering; significant environmental modification or degradation which has such effects is included within the meaning of "harm."

Environmental modification and destruction often occur over a long period, and species slide into oblivion without a corpus delecti. An example is the endangered palila (*Loxioides bailleui*), a honeycreeper that numbered about 3444 ± 956 in 1987 (Sparling 1987) and is endemic to the remaining mamane (*Sophora chrysophylla*) and naio (*Myoporum sandwicense*) forests on the slopes of Mauna Kea on the island of Hawaii between the elevations of 2000 and 2900 meters—about 10 percent of the bird's historical range (U.S. Fish and Wildlife Service 1986).

Mamane trees provide the palila with food, shelter, and nest sites. Naio trees also provide nest sites and food. Feral goats and sheep, which were introduced in the late eighteenth century, browse mamane seedlings and shoots, curtail forest regeneration, and have caused the decline of the palila's habitat (U.S. Fish and Wildlife Service 1986). For many years hunting interests prevailed against proposals to eradicate sheep and goats managed by the state of Hawaii for sport hunting purposes.

Although critical habitat has been designated for the palila (50 *Code of Federal Regulations* § 17.95), most of it lies within the state of Hawaii's Mauna Kea Game Management Area. Because there is no federal activity involved, Section 7 cannot be invoked to address the destructive and adverse browsing in the palila's habitat. However, in 1979, U.S. District Court for Hawaii ruled that the Hawaii Department of Land and Natural Resources was violating the Act because its management of feral sheep and goats on Mauna Kea constituted an unlawful taking of palila (*Palila v. Hawaii Department of Land and Natural Resources,* 471 F. Supp. 985 [1979]) (Palila I). Feral sheep and goats were modifying habitat and disrupting the palila's feeding behavior. Palila I was affirmed by the Ninth Circuit Court of Appeals (*Palila v. Hawaii Department of Land and Natural Resources,* 639 F.2d 495 [1981]).

The ecological effect of palila habitat destruction is unequivocal. Even small numbers of feral sheep and goats can denude a large area and eliminate forest regeneration. Unless stopped, feral goats and sheep would eventually eliminate palila habitat and palila numbers would decline accordingly. By allowing goats and sheep to browse, the state of Hawaii was effectively taking (harming) palila.

In 1981, in response to Palila I, the Department of the Interior proposed to remove any mention of environmental modification or disruption from the definition of harm (46 *Federal Register* 29490) but was persuaded to promulgate the following redefinition of harm (46 *Federal Register* 54748):

> "Harm" in the definition of "take" in the Act means an act which actually kills or injures wildlife. Such act may include significant habitat modification or degradation where it actually kills or injures wildlife by significantly impairing essential behavioral patterns, including breeding, feeding, or sheltering.

This redefinition of harm was similar to the previous definition, and the Department of the Interior explained that harm was not limited to:

> direct physical injury to an individual member of the wildlife species.... The purpose of the redefinition was to preclude claims of a Section 9 taking for habitat modification alone without any attendant death or injury of the protected wildlife. Death or injury, however, may be caused by impairment of essential behavioral patterns which can have significant and permanent effects on a listed species.

Harm clearly means detrimental actions to species' habitat as well as the standard meaning of directly killing individuals. Bean (1983) states that Palila I clarified the dilemma of habitat destruction when a species is not currently present. Vital nesting habitat of a migratory bird, for example, cannot be harmed regardless of whether the birds are present or wintering elsewhere.

In November 1986 the U.S. District Court for Hawaii issued another ruling in the same case—this time involving another introduced ungulate browser, the European mouflon (*Ovis musimon*) (*Palila v. Hawaii Department of Land and Natural Resources and Sportsmen of Hawaii,* 649 F. Supp. 1070 [1986]) (Palila II). Expert witnesses at trial, including those of the defendants, agreed that the mouflon sheep are currently degrading the mamane forest, that this degradation is irreversible because it is suppressing the forest's regeneration, that palila depend on mamane for their existence, and that continued degradation could drive the palila into extinction. Palila II refused to accept the defendants' claim that any effect the mouflon has on mamane and indirectly on palila is only a potential injury and does not fall within the Department of the Interior's revised definition of harm. The defendants also maintained that an actual decrease of palila numbers must be shown. However, Palila II concluded that the Department of the Interior's redefinition of harm was not substantially different from its previous definition (Bean 1987) and stated:

> A finding of "harm" does not require death to individual members of the species; nor does it require a finding that habitat degradation is presently driving the species further toward extinction. Habitat destruction that prevents the recovery of the species by affecting essential behavioral patterns causes actual injury to the species and effects a taking under Section 9 of the Act.

Palila II also stated that a "showing of harm does not require a decline in population numbers." The palila is already endangered, exists at a critical population level, and needs to recover. The U.S. Fish and Wildlife Service (1986) has determined that the palila will be recovered and removed from the endangered species list when the population has achieved a density of 25 birds per square kilometer (5000 birds) throughout the 200 sq km of critical habitat continuously for five years. Regarding harm and recovery, Palila II stated:

> The key to the Secretary's [of the Interior] definition is harm to the species as a whole through habitat destruction or modification. If the habitat modification prevents the population from recovering, then this causes injury to the species and should be actionable under Section 9.

Like the browsing feral goats and sheep in Palila I, the grazing mouflon sheep in Palila II were decreasing food and nesting sites and thus suppressing the palila population and preventing it from expanding to recovery.

Palila II also had something to say about the enforcement of Section 9. Oftentimes compromise or accommodation is sought in endangered species confrontations. Defendants in Palila II sought a multiple-use approach to accommodate their sport hunting and endangered species protection responsibilities. However,

Palila II stated that the Act "does not allow a 'balancing' approach for multiple use considerations." All the mouflon sheep must go. Besides, there are other areas in Hawaii where they can be hunted. Section 9 does not allow other management strategies or policies for placative purposes that would play "Russian roulette with a critically endangered species."

Support for the palila decisions reaches back to the famous snail darter (*Percina tanasi*) case (*Tennessee Valley Authority v. Hill,* 437 U.S. 153 [1978]). Aside from Section 7 violations, the U.S. Supreme Court also indicated that environmental effects (loss of spawning habitat, siltation, and low oxygen content of the reservoir) of a completed Tellico Dam might amount to a taking of snail darters under Section 9. Snail darters floating belly up were not required.

The 1982 amendments to the Act, however, allowed for incidental taking in Section 7 consultations. Section 7(b)(4) permits taking of some individuals of an endangered species if the taking is incidental to the federal action under consultation, and if such incidental taking is not to the degree that the continued existence of the species is likely to be jeopardized (50 *Code of Federal Regulations* 402.14[i]). For example, the U.S. Fish and Wildlife Service (1987) has determined that construction and operation of the proposed Two Forks Dam on the South Platte River in Colorado will result in a nonjeopardizing incidental take of the endangered interior population of the least tern (*Sterna antillarum*) and threatened piping plover (*Charadrius melodus*) downstream on the Platte River, Nebraska. Because the project sponsor has agreed to implement reasonable and prudent measures to minimize the incidental take of least terns and piping plovers, the project may proceed without violating Section 9.

Incidental taking aside, the palila decisions enunciated a broad taking concept that potentially subjects many private landowners to risk of prosecution or civil suit against land development activities for taking through habitat destruction. Habitat status is usually the determining factor in a species' road to recovery. Stronger habitat protection is the key to the effectiveness of most endangered species conservation (Vaughn 1987). According to Palila I and II, the Department of the Interior's definition of "harm" gives great leverage to the enforcement of habitat protection under Section 9 of the Act. Such enforcement may require considerable political will (Bean 1983), and some believe that federal agencies are not inclined to invoke Section 9 to curtail either drastic or subtle long-term habitat modifications (Vaughn 1987).

Literature Cited

Bean, M. J. 1983. The evolution of national wildlife law. Praeger, New York, New York, USA.

Bean, M. J. 1987. The federal endangered species program. Pages 147–160 *in* R. L. DiSilvestro, editor. Audubon wildlife report 1987, National Audubon Society, New York, New York, USA.

Sidle, J. G. 1987. Critical habitat designation: is it prudent? Environmental Management **11**:429–437.

Sparling, O. W. 1987. The endangered palila of Hawaii. Endangered Species Technical Bulletin **12**(9):6.

U.S. Fish and Wildlife Service. 1986. Revised palila recovery plan.

U.S. Fish and Wildlife Service. 1987. Biological opinion on the Platte River off-site effects of the Denver Water Department's Two Forks Dam project. Letter from U.S. Fish and Wildlife Service, Denver, Colorado to U.S. Army Corps of Engineers, Omaha, Nebraska dated October 14, 1987.

Vaughn, D. C. 1987. The whooping crane, the Platte River, and endangered species legislation. Nebraska Law Review **66**:175–211.

Essay

Six Biological Reasons Why the Endangered Species Act Doesn't Work—And What to Do About It

DANIEL J. ROHLF

Natural Resources Law Institute
Lewis and Clark Law School
10015 S.W. Terwilliger Boulevard
Portland, OR 97219, U.S.A.

Abstract: *Law plays an important role in shaping land management decisions. The success of efforts to conserve biodiversity thus depends to a large degree on how well scientific knowledge is translated into public policy. Unfortunately, the Endangered Species Act, the United States's strongest legal tool for conserving biodiversity, contains serious biological flaws. The statute itself, as well as agency regulations and policies that implement the law, include provisions that fail to account accurately for important biological concepts such as ecosystem conservation, patch dynamics, and the probabilistic nature of stochastic threats to a species' persistence. Moreover, the procedures of federal agencies charged with implementing the Endangered Species Act in some cases make it difficult for interested outside reviewers to evaluate the agencies' scientific findings and methodology. However, the Endangered Species Act also gives interested individuals and groups several opportunities to provide input into the process of managing threatened and endangered species. Conservation biologists should practice focused advocacy by taking advantage of such opportunities to steer law in a more biologically sound direction.*

Resumen: *Las leyes juegan un papel importante al darle forma a las decisiones sobre el manejo de la tierra. El éxito de los esfuerzos para conservar la biodiversidad depende en buena parte de que tan bién sea traducido el conocimiento científico dentro de las reglamentaciones públicas. Desafortunadamente, el Acta sobre las Especies en Peligro, la herramienta más fuerte dentro de los Estados Unidos para conservar la biodiversidad, contiene serias faltas biológicas. El estatuto en sí mismo, así como las regulaciones de las agencias y los reglamentos que implementan la ley, incluyen aspectos que fallan para tomar en cuenta certeramente importantes conceptos biológicos como la conservación de los ecosistemas, la dinámica de ecosistemas fragmentados y la naturaleza probabilística de amenazas al azar en contra de la existencia de una especie. Por otra parte, los procedimientos de las agencias federales a cargo de implementar el Acta de las Especies en Peligro, en algunos casos, hacen difícil la intervención de revisores externos interesados en evaluar los hallazgos y metodolgías científicas de la agencias. Aún así, el Acta de las Especies en Peligro da, a los individuos y grupos interesados, varias oportunidades para ofrecer impulsos durante el proceso del manejo de las especies amenazadas y en peligro. Los biólogos de la conservación deben de enfocar su apoyò al tomar ventaja de estas oportunidades para orientar a la ley en una dirección que contenga bases biológicas sólidas.*

Introduction

Conservation biology is a unique discipline in that its practitioners not only study biodiversity, they also work to slow and ultimately prevent its erosion. As such, conservation biologists must wear many hats. Thomas and Salwasser (1989) noted that to more effectively influence land and wildlife management, conservation biologists must also serve as teachers, biopoliticians, and even "gladiators" in the arena of land-use planning. This article adds another calling to that list—lawyers. Law plays a major role in shaping natural resource management, particularly (but by no means exclusively) on public land. Even the best biological advice has little practical effect if it does not fit within the legal con-

Paper submitted July 6, 1990; revised manuscript accepted October 10, 1990.

straints imposed upon resource managers. Therefore, in addition to their other duties, conservation biologists must work to ensure that lawmakers and administrators accurately translate scientifically based conservation recommendations into public policy.

The Endangered Species Act remains the United States's strongest and most comprehensive species conservation strategy, as well as a model for other nations' protection efforts (Rohlf 1989). However, the Act has had very limited success in achieving its stated goal of halting and reversing the trend toward species extinctions. In the seventeen years since passage of the Act in 1973, only a handful of protected species have recovered to the point where they no longer face extinction. Meanwhile, Wilson (1988) estimates that human actions extirpate 17,500 species each year. Although its extremely limited influence over resource management outside the United States hampers the Act's ability to conserve global biodiversity, other factors contribute to the statute's domestic ineffectiveness. Politics and economic debates have interfered with efforts to conserve some species, as the recent bitter controversies surrounding shrimpers' use of turtle excluder devices and efforts to curtail logging in Spotted Owl habitat demonstrate. More fundamentally, however, some provisions of the statute itself, as well as key interpretations of the Act by administrators charged with its implementation, conflict with sound biological principles. Biological flaws in the law itself significantly contribute to its ineffectiveness in conserving biodiversity.

This article examines the interplay between the law and science of conserving imperiled species. After briefly describing how the Endangered Species Act works, I analyze provisions and interpretations of the Act that are inconsistent with basic tenets of conservation biology. I conclude with comments on how conservation biologists can and should influence policymakers in an effort to make law consistent with science.

How the Endangered Species Act Works

Congress enacted the Endangered Species Act to "provide a means whereby the ecosystems upon which endangered species and threatened species depend may be conserved, [and] to provide a program for the conservation of such ... species. ..."(16 U.S.C. Section 1531 [b]). To qualify for Endangered Species Act protections, a species must appear on the official list of endangered species, defined by law as those likely to become extinct within all or a significant portion of their range, or of threatened species, those likely to become endangered in the foreseeable future. The Secretaries of Interior and Commerce (acting through the U.S. Fish and Wildlife Service and National Marine Fisheries Service, respectively) have authority to add to and delete from these lists based on whether a species faces extinction due to any variety of natural or human-caused factors (see 16 U.S.C. Section 1533 [a] [1]). The Secretaries must also draw up "recovery plans" for each listed species, which set forth conservation goals and specify actions necessary to achieve them. The current lists include over 600 species that occur within the United States and over 500 species that occur elsewhere in the world.

Section 7 contains some of the Act's principal substantive protections for listed species. It directs federal land managers and other federal agencies to insure that their activities do not jeopardize the continued existence of listed species or adversely modify habitat critical to those species. To provide federal agencies with expert biological advice to help them comply with this mandate, the statute requires agencies considering a specific action to consult with the Fish and Wildlife Service or the National Marine Fisheries Service before committing to a course of conduct. After this consultation, the service issues a written "biological opinion," which details the proposed activity's probable influence on protected species, suggests project alternatives or modifications that would avoid or lessen adverse effects, and sets forth the biological information upon which the opinion is based. Although the action agency makes the final decision on whether to proceed, agencies seldom go forward with a project if the biological opinion reports a likelihood of jeopardy to the species or adverse modifications of its critical habitat. Congress amended section 7 in 1978 after the famous snail darter case to allow a committee composed of Cabinet-level officials to grant exemptions to the absolute protections of this section. Rarely convened, the so-called "God Committee" has granted only one exemption in its history.

Unlike section 7, whose provisions apply only to federal agencies, section 9 applies to private individuals, corporations, and state and local governments as well. It prohibits anyone from "taking" a species listed as endangered. "Taking" includes not only direct infliction of physical harm on a member of an endangered species, but also alteration of an endangered species' habitat that in turn kills or injures members of the species. In addition, section 9 prohibits any sort of domestic or international commerce in endangered species and products made from those species, a particularly important provision for conservation of listed species that occur exclusively outside the United States. The Secretaries of Interior and Commerce have issued regulations that apply the above protections to the vast majority of threatened species as well.

Regulations and official interpretations of the Act play an important role in shaping the law. Regulations, issued by Fish and Wildlife Service and National Marine Fisheries Service, define terms in the Act that the statute does not explicitly define. They also interpret the Act's directives and set forth specific procedures for implementing its provisions. Although a federal court has authority to set aside regulations it finds in conflict with the statute itself, this seldom happens. In practice, therefore, regulations greatly influence the scope and implementation of the Act's protections. Additionally, Fish and Wildlife Service and National Marine Fisheries Service receive legal advice to help them interpret the Act from the Solicitor's Office and Office of General Counsel within the Departments of Interior and Commerce, respectively. Regulations, solicitor's opinions, and internal Fish and Wildlife Service and National Marine Fisheries Service policies shape and define legal protections for endangered and threatened species.

Six Biological Reasons Why the Endangered Species Act Is Not an Effective Tool for Conserving Biodiversity

1. The Act Primarily Protects High-Profile Individual Species Rather Than Overall Biodiversity.

Legal experts as well as biologists have criticized the Act's single-species approach to biodiversity conservation. Smith (1984) notes that this approach has been a traditional element of conservation regulations due to the historical fact that overhunting and other forms of direct exploitation depleted or extirpated many species; he goes on to point out that species currently face greater threats due to habitat reduction, making ecosystem conservation preferable to single-species protection. Scott et al. (1987) argue that "Emergency Room Conservation" expends inordinate effort and resources on a few species that, by the time they are finally listed as endangered or threatened, may be too far gone to save.

Despite its focus on single species, the Endangered Species Act could play a significant role in protecting biodiversity on a broader scale. The Act could serve as an extremely useful tool for preserving keystone species, thus indirectly benefiting the many other life forms in some way dependent upon those species. Additionally, the law could systematically extend protections to indicator species whose relative abundance provides a measuring stick for overall health of entire ecosystems. Such strategies could substantially mitigate many of the shortcomings inherent in the Act's single-species approach.

Unfortunately, despite congressional statements favoring ecosystem approaches, policy-makers have tended to emphasize Endangered Species Act protection for high-profile single species rather than incorporate umbrella protections for biodiversity into the statute. In response to a 1979 amendment to the statute requiring Fish and Wildlife Service and National Marine Fisheries Service to develop priority systems to guide their listing decisions, Fish and Wildlife Service adopted a scheme that favored what it termed "higher" life forms such as mammals and birds. Congress specifically disapproved of this priority system in 1982 when it again amended the Act. Moreover, in committee reports accompanying the 1982 amendments, lawmakers made clear that even though the Act focuses on individual species, its purposes are "far broader." Amplifying the Act's reference to ecosystem conservation, a conference committee report noted that protected species "must be viewed in terms of their relationship to the ecosystem of which they form a constituent element" (House of Representatives 1982), while a Senate report argued that it is biologically prudent to place emphasis on listing species that "form the basis of ecosystems and food chains" (U.S. Senate 1982). These pronouncements notwithstanding, Fish and Wildlife expressly refused to incorporate species' importance within ecosystems as a listing priority criterion when it rewrote its listing priority guidelines in 1983. The current guidelines instead concentrate primarily on the magnitude and immediacy of threats facing species, thus officially adopting an "emergency room" approach to biodiversity conservation. The National Marine Fisheries Service adopted a similar scheme in its listing priority system.

Species recovery priorities also reflect little consideration for maximizing overall biodiversity. Both Services base their guidelines for allocating resources for recovery efforts primarily on degree of threat as well as a subjective determination of "recovery potential." These agencies also have historically focused their recovery expenditures on a handful of high-profile species. Between 1982 and 1986, Fish and Wildlife Service spent almost half of the funds available to it for developing and implementing recovery plans on twelve species—only six of which the agency considered highly threatened. The General Accounting Office (1988) found that this imbalance resulted from the government's attempt to maintain a positive public perception of its recovery efforts by placing special emphasis on species with high public appeal. In an attempt to force more even resource allocations between listed species, Congress in 1988 amended the Endangered Species Act to require that recovery plan development and implementation proceed "without regard to taxonomic classification" (16 U.S.C. Section 1533 [f] [1] [A]). Although this amendment may encourage the Services to consider species other than "charismatic megafauna," it did not change the emphasis on degree of threat as a primary recovery priority criterion.

2 The Act Lacks Clearly Defined Thresholds to Delineate Endangered, Threatened, and Recovered Species.

Shaffer (1987) listed agreement on an appropriate level of security as an important element in systematic biodiversity conservation. The parameters one chooses to define a "secure" population have tremendous influence on decisions concerning management of that population and its habitat. Greater security for a given species or population generally requires larger and more numerous habitat areas and in some cases more extensive monitoring and management.

Although the Act implicitly sets a standard for separating relatively secure species and populations from those facing extinction, the statutes does not clearly define or even specifically describe its security standard. The threshold separating species listed as threatened and endangered from those considered "recovered" or not eligible for protection in effect constitutes a standard of security because it extends special protections to species on one side of this line. The Act and its regulations, however, simply describe endangered species as those in danger of extinction throughout all or a significant portion of their range; threatened species include those likely to become endangered. The law makes no reference to quantitative or even qualitative parameters of what constitutes a "danger" of extinction.

Ironically, advances in conservation biology have demonstrated that defining a "secure" population involves making policy choices as well as determining required habitat and population sizes. To describe a viable population, Shaffer (1981, 1987) noted that one must first establish a time frame of reference and desired degree of certainty of continued existence. As an example, Shaffer arbitrarily defined a viable population as the smallest isolated population with a 99% chance of remaining extant for 1000 years in the face of stochastic threats to its existence. However, he acknowledged that establishing these explicit criteria is a policy decision rather than a biological question; society in essence must choose the amount of "insurance" it wishes to purchase against the risk of extinction.

Although Congress's pronouncements about the importance of protecting imperiled species suggest that lawmakers made a policy decision to afford species a high degree of security, the Endangered Species Act's lack of explicit biological criteria leaves species security determinations to the Services charged with implementing the statute. However, the Services have failed to set specific time and certitude standards for systematically differentiating "secure" species from those facing extinction and thus eligible for protection under the Act. Instead, the Services have attempted to draw this distinction on a case-by-case basis in reference to qualitative factors such as a species' historic abundance and threats to its existence. As a result, the terms "endangered" and "threatened" have no uniform biological meaning. Although scientific uncertainty always limits the precision of objective standards, we use them in other forms of environmental regulation. The Environmental Protection Agency, for example, uses an objective estimate of the number of cancers caused to determine whether to allow use of specific pesticides. This allows decision-makers, as well as the public, to perceive a clear policy choice: should we define "safety" as one additional cancer death per ten thousand people or one additional death per one million? Use of an objective standard of species security would permit a similarly clear choice.

Without explicit criteria to define a secure population, however, the degree of security afforded to species by the Act varies according to discretionary ad hoc determinations by the Services. This creates a risk that the security afforded to particular species in listing decisions or section 7 consultations may not correspond to the high value Congress placed on biodiversity protection. Absence of objective standards makes it very difficult to challenge the Services' decisions. Courts, which lack biological expertise, give these agencies a great deal of deference; without specific criteria by which to measure agency determinations, courts are unlikely to question the services' listing decisions and biological opinions.

Moreover, making species security decisions on a case-by-case basis without reference to objective standards necessarily injects political and economic considerations into the process of making what by law are supposed to be biological decisions. This occurs because without previously set, objective security parameters, the task of defining species security on a case-by-case basis involves making policy as much as making a scientific determination. In making a listing decision, for example, the service must first define the point at which a particular species is "in danger" of extinction—which Shaffer notes is essentially a policy choice—and then determine biologically whether the species' population has reached that point. In other words, determining which biological questions to ask involves making a policy choice. In the controversial Spotted Owl case, for instance, the GAO (1989) concluded that Fish and Wildlife had refused to list owls in part due to political pressure from Interior Department officials' concern over effects on the Northwest's timber industry. This lobbying probably played a significant role in how Fish and Wildlife chose to define "danger of extinction" in that particular case, thus influencing the agency's supposedly biologically based decision not to afford Spotted Owls protection under the Endangered Species Act.

Finally, the agencies' current ad hoc approach to determining species security under the Act virtually insures that security standards for different species will not be uniform. Such disparities raise problems similar to those inherent in the single-species approach to biodiversity conservation discussed above. For example, a Service may choose to give a highly visible or popular species a relatively high degree of security even though it does not play a particularly important ecological role. Differential standards of species security thus undercut efforts to systematically protect biodiversity.

3. The ESA Does Not Adequately Protect Metapopulations.

Metapopulation dynamics play an important role in the persistence of many species. The existence of several or many populations is critical for species that inhabit patches in a shifting mosaic of habitats (Picket & Thompson 1978). Multiple populations also serve as a source of colonists and thus as a hedge against environmental stochasticity. Additionally, even minimal interbreeding between relatively isolated populations can be a key factor in maintaining such populations' overall genetic fitness (Gilpin 1987). Finally, metapopulation and patch dynamics are likely to become increasingly important as habitat areas become fragmented. In cases of extreme habitat fragmentation, artificial breeding exchanges between populations that have become completely reproductively isolated could be a vital future management tool.

The Endangered Species Act permits protection of populations as well as entire species by authorizing the Secretaries to list distinct population segments of vertebrate fish and wildlife as threatened or endangered. For example, grizzly bears in the lower 48 states are listed as threatened, whereas the relatively healthy Alaska grizzly population receives no Endangered Species Act protections. Thus, the Act prohibits actions that jeopardize the continued existence of grizzlies in the contiguous states even if such actions would not jeopardize grizzlies as a species.

When distinct population segments of a given species are not separately listed, however, they generally do not enjoy this type of protection. Under Fish and Wildlife Service policy, an action that jeopardizes or even wipes out a population of a listed species is not considered a section 7 violation of the Act unless the action jeopardizes the entire species. For example, assume that an imperiled species has five distinct population segments. If Fish and Wildlife Service separately listed each of the five populations as threatened or endangered, a federal action that threatened to destroy only one of the populations would violate section 7, since as a result of its separate listing the population would legally be considered a separate species. Separate listings for populations are exceptional, however. On the other hand, if the agency had declared the entire species—including all five populations—threatened or endangered in one listing decision, an action resulting in destruction of one population would probably not be precluded under section 7 since four others would still remain. In the latter case, each population would not legally constitute a separate species and thus would not be eligible for separate protection.

This legal shell game with the definition of species adversely affects conservation efforts by ignoring metapopulation dynamics. If an entire biotic species has declined to the point where it is considered threatened or endangered, each remaining population could play a critical role in improving the species' chances for long-term persistence. However, barring limited exceptions and relatively rare instances of separate listings for each population, the Act does not protect distinct populations of listed species.

4. Many Biological Determinations Under the Act Are Not Adequately Documented, Preventing Meaningful Scrutiny and Participation from the Public and Scientific Community.

The Endangered Species Act directs federal agencies to use the "best scientific and commercial data available" to fulfill their responsibilities under the Act. Other provisions of the statute are designed to reinforce this requirement. Section 7 requires agencies to prepare written biological assessments that evaluate how proposed actions are likely to affect listed species. Additionally, the Services must include in their biological opinions summaries of the information upon which the opinions' conclusions are based.

Since biological assessments and biological opinions are public documents, they permit public as well as independent scientific scrutiny of federal decisions that affect threatened and endangered species. Such scrutiny plays an important role in assuring that agencies' decisions are biologically sound. Congress encouraged interested parties outside the government to take an active role in enforcing the Act's provisions by granting anyone the right to challenge in court agency actions alleged to violate the statute. Taking advantage of this right, plaintiff groups have overturned several federal agency decisions made without sufficient biological data.

However, current administrative interpretations of the Act have reduced the documentation required to accompany federal decisions that affect listed species, correspondingly limiting the public's and scientific community's opportunity to independently evaluate these decisions. In 1986, the Services adopted regulations substantially limiting application of the Act's consultation procedures. Prior to 1986, federal agencies were required to comply with section 7's "formal" consultation procedures—which result in preparation of a biological opinion—whenever they determined that a proposed activity could affect a listed species. Under current practice, however, agencies need not go through this process if they decide that a proposal will not "adversely affect" protected species. Predictably, the number of consultations resulting in biological opinions immediately dropped dramatically. In 1979, 1980, and 1981, the Services together conducted an average of about 3500 consultations and issued around 650 written biological opinions each year (U.S. House of Representatives 1982). However, while the number of consultations in 1986 alone soared to almost 11,000, the number of biological opinions issued dropped to 421.

Such a trend is troubling. Although the Services must concur with other federal agencies' "no adverse affect" determinations in writing, such concurrence statements—unlike biological opinions—need not discuss or even mention the information upon which the Services based their findings. This forecloses any outside scrutiny of the methods used to determine that a proposal will not adversely affect listed species, or of the information upon which such a determination is based. Further, Fish and Wildlife Service also greatly reduced the number of activities for which agencies must prepare biological assessments.

Public disclosure of biological conclusions and the information upon which those conclusions are based promotes decision-making based on sound science. Though the Act requires agencies to use the best scientific information available when making decisions—and encourages outside parties to enforce this provision—reduction of Act documentation requirements threatens to erode scientifically credible decision-making by driving science behind closed doors.

5. The Act Does Not Protect Habitat Reserves Sufficiently to Sustain "Recovered" Populations.

The Act ultimately strives to bring populations of listed species to the point where they are no longer endangered or threatened with extinction. To do so, the law recognizes that adequate habitat must exist to sustain so-called recovered population levels. Section 7 of the Act prohibits federal agencies from destroying or adversely modifying habitat that either service formally declares to be "critical" to protected species. Critical habitat, as defined by the Act, means specific geographical areas that contain those physical or biological features necessary for recovery of listed species. In light of this definition, section 7's prohibition against destroying critical habitat provides listed species with an important legal protection above and beyond the Act's prohibition against jeopardizing the continued existence of listed species, that is, legal protection for species' chances of recovery to healthier population levels.

For a few years after the Act was enacted, the services actively implemented the statute's protection of critical habitat, occasionally even declaring large areas critical habitat for listed species. In 1975, for example, Fish and Wildlife Service declared over 100,000 acres in Mississippi as critical habitat for sandhill cranes, a designation that later played a key role in forcing the U.S. Department of Transportation to reroute a highway planned to be built through the area. By the late 1970s, however, fierce opposition to critical habitat designations set off legislative as well as administrative actions to weaken Endangered Species Act habitat protections.

Congress dealt with the critical habitat issue when it amended the Act in 1978. Though lawmakers required the Services to define—the extent "prudent" and determinable—critical habitat for a species at the same time it was added to the protected list, Congress also allowed these agencies to exclude areas from critical habitat status on economic or other grounds. This provision marked a significant departure from the Act's emphasis on biologically based decision-making.

The agencies charged with implementing the Act essentially responded to the controversy over critical habitat designation by avoiding the issue altogether. In 1979, Fish and Wildlife Service withdrew several proposed critical habitat designations, included a 10-million-acre critical habitat designation for grizzly bears. Moreover, by broadly interpreting Congress's exception that critical habitat need not be established if not "prudent" or determinable, the Services avoided designating critical habitat concurrent with listings. In 1986, for example, Fish and Wildlife Service listed 45 species as threatened or endangered, but made concurrent critical habitat designations for only four species.

Also in 1986, the Services made critical habitat designations essentially moot by reading out of the law section 7's protections for habitat sufficient to support recovered populations of listed species. The agencies interpreted section 7 to prohibit only those actions that diminish the value of critical habitat for *both* the survival and recovery of listed species. In other words, if a federal action hurts a species' chances for recovery but does not imperil its bare survival, the action does not violate section 7. Although this view of the law is questionable in light of the Act's language and intent, it has never been successfully challenged.

Webster (1987) notes that a similar standard apparently applies to habitat of listed species in private ownership. Although the Act's section 7 standards govern only federal or federally controlled activities, section 9 also applies to private as well as public land. It prohibits "taking" of listed species, and has been broadly construed to ban habitat alterations on private land that kill or injure protected species. However, in 1982 Congress created a process that allows private parties to apply for a permit to "incidentally" take threatened and endangered species in the course of otherwise lawful activities. To obtain such a permit, a party must submit a "habitat conservation plan"; Fish and Wildlife Service or National Marine Fisheries Service may grant the permit if it finds that the plan will not appreciably reduce the survival and recovery of listed species. As in their interpretation of the phrase "survival and recovery" in the context of section 7's critical habitat provision, the services apparently feel free to issue incidental "taking" permits unless a proposal threatens to appreciably diminish a species' survival—a deleterious impact on recovery alone is not sufficient grounds for permit denial. For example, Fish and Wildlife Service approved a habitat conservation plan for the threatened Coachella Valley fringe-toed lizard that calls for development of 75% of the lizards' remaining habitat, with the remainder to be placed in reserves. Elimination of three-quarters of the species' habitat clearly will adversely affect its recovery chances. However, the permit was approved because it did not threaten both the lizard's survival and recovery.

6. Bodies Charged with Implementing and Enforcing the Act Tend to Discount Uncertain or Nonimmediate Factors in Their Decision-Making Processes.

Uncertainty plays a critical role in scientific study of the extinction process. Stochastic factors substantially influence population persistence; estimates of population persistence must therefore be expressed in terms of probabilities. Additionally, scientists do not completely understand many biological and ecological processes. Researchers often express this scientific uncertainty as a factor of error, which they report along with their conclusions.

Those charged with making decisions under the Act also must often deal with uncertainty. Rather than treating uncertainty in a probabilistic manner, however, decisions involving conservation of threatened and endangered species often ignore or discount uncertain threats to these species or use the existence of uncertainty to justify inaction.

Environmental stochasticity—chance events such as forest fires, drought, floods, and similar habitat disruptions—is an important factor influencing population persistence. Although scientists typically study stochastic natural events that affect habitat, environmental stochasticity has a human-related component as well. For example, a given population existing in a riparian habitat faces the threat of chance environmental events such as flooding. In addition, the population may be affected by future human-caused environmental changes such as construction of a dam or water pollution. However, entities that implement and enforce the Act's protections tend to overestimate species' chances of survival by discounting or ignoring natural as well as human-related stochastic threats to species' environments; as a result, listed species often receive less protection than is necessary to ensure their continued existence.

For example, Endangered Species Act regulations refer to future activities to be conducted by private entities or state and local governments in the same area as a proposed federal action as "cumulative effects." However, when either service prepares a biological opinion analyzing a federal action's effect on protected species, it considers only those cumulative effects that are "reasonably certain" to occur. Under this standard, the service accounts only for planned activities that have cleared all legal and financial hurdles and thus presently give every indication of taking place. Rather than considering all risks based on their probability of occurrence, this type of analysis ignores all human-related stochastic threats to a species save those that are virtually sure to occur. In effect, therefore, this procedure discounts the role environmental stochasticity plays in species persistence. Consequently, the services overestimate species' chances of survival and thus underprotect listed species.

These agencies have demonstrated a similar tendency when considering nonimmediate threats to listed species. For example, when Fish and Wildlife Service removed brown pelicans from protected status under the Act, it dismissed threats to the birds' habitat posed by manganese mining, not on the grounds that such operations posed no threat to the species, but because it determined that mineral development was unlikely to take place in the near future (USFWS 1985).

Courts too sometimes discount nonimmediate threats to listed species. In a case challenging offshore oil and gas leasing, for example, plaintiffs argued that leasing violated the taking prohibition in section 9 of the Act, which forbids actions that "harm" endangered species. The court, despite an admission by the government that future activities stemming from the leases could harm protected whales, refused to interfere with leasing because the harm was not "sufficiently imminent or certain" (*North Slope Borough* v. *Andrus* [1979]).

Such refusals to consider future risks to listed species' survival are particularly dangerous. Most human planning horizons span at most a few decades, a very short period from the standpoint of biological evolution. Thus, risks that seem far in the future from a human perspective can loom as significant threats to species' persistence. Conservation biologists take such risks into account when calculating a species' time to extinction by considering the biological consequences of an event, discounted by that event's likelihood of occurring. Ignoring uncertain future threats causes agencies and courts to overestimate species' chances of long-term survival and thus to underprotect those species in present-day decision-making.

A Note on Biopolitics

Reviewers who provided valuable comments on an earlier draft of this essay wondered whether reforms to address the Act's shortcomings are politically feasible. Their point in essence is that our efforts to protect biodiversity come up short due more to a lack of political will than to the Act's technical deficiencies. At first glance, this view seems incorrect. Congress apparently has already made a clear policy choice in favor of protecting species, even in the face of serious economic consequences. Lawmakers stressed that the Act gives listed species "the benefit of the doubt" and, despite persistent efforts by groups opposed to the Act's constraints, have consistently refused to significantly weaken its provisions. Moreover, the U.S. Supreme Court construed the ESA as protecting listed species "at any cost" (*TVA* v. *Hill* [1978]).

Despite these pronouncements, efforts to strengthen protections for threatened and endangered species will indeed face serious political hurdles. At least initially, the greatest lies in separating policy from science. As is common in other contexts, Congress has said one thing about species protection yet actually done another. The Act makes general commitments to preserve biodiversity but transfers important policy decisions to those not directly accountable to the electorate. This permits politicians to point to their solid environmental voting record while at the same time pressuring administrative agencies responsible for implementing the Act not to make decisions that significantly curtail economic activities, particularly in their districts. Reduced protections for biodiversity are then passed off as "science" rather than conscious, politically driven policy choices.

Conservation biologists and others need to redouble their efforts to impress upon elected officials and the public the worth of saving imperiled species. However, until policy decisions—the degree of security to give listed species, for example—are taken away from administrative agencies and given to politically accountable decision-makers, such efforts will have limited influence.

A Call for Scientific Involvement in the Legal Process

This section encourages conservation biologists and other scientists to take on a role as focused advocates working to make the Act more effective at preventing human-caused species extinctions, and provides specific suggestions on how to do so. Salzman (1989) describes focused advocacy as a person (or group) reporting data concerning an area in which he or she has expertise as well as deeply held convictions, and pressing to ensure that the information is interpreted correctly and acted upon. It is entirely appropriate—and crucial—for scientists to become focused advocates for strengthening the Endangered Species Act. Congress has already made a policy decision to protect species facing extinction, even if such protection carries substantial economic or other costs. Conservation biologists have a strong incentive to support and further the Act's policy of preventing human-caused species extinctions. They also possess the biological knowledge and data crucial to strengthen current efforts to reach that goal. What are urgently needed are increased efforts to ensure that such knowledge and information are translated into the law and regulations that govern resource management.

The following are specific suggestions to facilitate such efforts.

1. Scientists should develop a degree of legal sophistication.

Law exerts a strong influence over wildlife and public land management, and to a generally lesser degree over management of private land, in two ways. First, it sets substantive standards for management activities and establishes incentives to comply with, or penalties for violating, those standards. Additionally, laws and regulations specify the procedures public agencies must follow in making decisions and biological determinations, often including procedures that give the public opportunities to participate.

It is important for scientists to understand the legal meaning of substantive terms, as well as how that meaning affects management. The legal meaning of a technical term can substantially differ from the term's scientific definition. For example, biologists generally broadly define "cumulative effects" acting on a species to include synergisms, "nibbling" or incremental effects, indirect effects, effects that overlap in time or space, and even delayed or remote effects (National Research Council 1986). However, as defined by regulations implementing the Act, cumulative effects include only those state and private activities that are reasonably certain to occur in the vicinity of a federal project (50 C.F.R. Section 402.02). Knowledge of such distinctions is critical to understand, evaluate, and influence efforts to protect threatened and endangered species. It is also vital for scientists to understand how legal terms actually affect management activities. As outlined above, for example, regulatory revisions in 1986 changed the standard for triggering section 7 consultation from "may affect" to "may adversely affect." This seemingly insignificant change drastically reduced the public's access to information for outside evaluation of federal agencies' biological determinations.

Attorneys have done a good job of convincing people that they play an indispensable role in making and deciphering the law. Unfortunately, this has tended to discourage scientists from participating in the legal process of formulating and implementing guidelines for managing threatened and endangered species. Left to their own devices, lawyers and policy-makers have committed critical biological errors that limit the effectiveness of such guidelines. To recognize and successfully work to correct these biological errors, conservation biologists need to increase their knowledge of the law. One need not attend three years of law school to develop a good working knowledge of laws that deal with biodiversity conservation. Simon (1988), for example, contains an excellent introduction to many federal land management statutes, as well as an excellent basic legal primer for nonlawyers.

2. Perform directed research.

The Endangered Species Act is "information forcing" in much the same way that antipollution laws are "technology forcing." The Clean Air Act, for example, requires some pollution sources to employ the "best available control technology." Part of the thrust of this requirement is to encourage technological advances in pollution control by guaranteeing private developers of superior technology a market for their product. Similarly, the Endangered Species Act requires federal agencies to use the "best scientific and commercial data available" in fulfilling their responsibilities under the Act. This requirement simply refers to the best data available, not just to data produced by government scientists.

This legal requirement provides conservation biologists with broad opportunities to perform "directed research" to benefit species conservation. Salzman (1989) described how the Point Reyes Bird Observatory intentionally directed research efforts toward specific biologically sensitive areas and species. This research encouraged biopolitical decisions to protect the Point Reyes biota by providing decision-makers with data on bird and seal populations as well as biological information on threats to those populations. Similar research directed at specific endangered and threatened species and the threats they face could have significant influence over federal decisions that affect listed species and their habitat.

Federal agencies most commonly use outside biological studies in the context of section 7 consultation between Fish and Wildlife Service or National Marine Fisheries Service and other federal agencies proposing particular actions. Accordingly, the influence of a particular study varies in proportion with its specificity in relation to the project under consideration. For example, a study examining the effect of open-road density on grizzly bears' use of specific habitat types will have a greater impact on national forest management decisions than general studies of human-bear interactions. Researchers should thus practice focused advocacy by directing their research toward determining the likely impacts of ongoing or future federal actions on listed species of concern. By law, federal agencies must consider the resulting information.

3. Take full advantage of opportunities for participation.

The Endangered Species Act encourages public participation in the federal government's efforts to conserve endangered and threatened species by providing several opportunities for interested parties to give their input on questions of both biology and policy. Traditionally, environmental organizations, local, state, and federal agencies, and representatives of affected industries have comprised the overwhelming majority of outside participants in Endangered Species Act proceedings and policy-making. However, rectifying current biological deficiencies in legal guidelines for species protection will require widespread and effective participation from the scientific community.

The Act permits any interested party to petition either service to add or delete species or populations from the threatened and endangered lists. Whenever a service concludes that a petitioned action may be warranted, or is considering listing a species or population on its own initiative, it conducts a status review. Notices of ongoing status reviews appear in the *Federal Register*. Immediately after such a notice is published, the listing agency solicits written comments from the public concerning the proposed action, and, if requested, a public hearing. Through use of the petition process and participation in status reviews, conservation biologists can exert powerful and desperately needed influence over the process of determining which elements of biodiversity need the Act's protections.

The agencies responsible for implementing the Act practice so-called "notice and comment" rule-making when formulating regulations and policies relating to conservation of listed species. Under these procedures, the agencies publish a notice of proposed regulations or policies in the *Federal Register* and invite public comments. Scientists, other than those affiliated with interest groups, seldom take advantage of these opportunities to provide input. As detailed in this article, however, biologically unsound regulations and policies detrimentally affect species conservation efforts. The participation of the scientific community in the Services' regulatory and policy-making procedures is thus particularly crucial.

Finally, the Act requires that a recovery plan for threatened and endangered species be prepared. Recovery plans set forth and prioritize actions deemed necessary to increase the numbers and security of listed species to the point where they no longer require protection under the Act. Since the law specifies that the Services base recovery plans solely on biological considerations, and that plans contain objective, measurable criteria, the public participation procedures for recovery planning also give scientists an important avenue for focused advocacy.

Conclusion

Because law wields considerable influence over the actions of resource managers, biological deficiencies in legal guidelines can adversely affect efforts to conserve species facing extinction. Both the Act itself and the regulations and policies of agencies charged with implementing the statute contain significant biological deficiencies. Fortunately, however, the Act's extensive public involvement provisions give conservation biologists many opportunities to practice focused advocacy for biodiversity conservation. Given the current precarious state of the earth's biological resources, conservation biologists can no longer afford to leave law solely to lawyers.

Acknowledgment

The author wishes to thank the Natural Resources Law Institute at Northwestern School of Law, Lewis and Clark College for a Fellowship which made work on this article possible.

Literature Cited

General Accounting Office. 1988. Management improvements could enhance recovery program. Report No. RCED-89-5.

General Accounting Office. 1989. Spotted owl petition evaluation beset by problems. Report No. RCED-89-79.

Gilpin, M. E. 1987. Spacial structure and population vulnerability. Pages 125–140 in M. E. Soulé, editor. Viable populations for conservation. Cambridge University Press, Cambridge, England.

National Research Council. 1986. Ecological knowledge and environmental problem-solving. National Academy Press, Washington, D.C.

North Slope Borough v. *Andrus,* 486 F. Supp. 326 (D.D.C. 1979), *aff'd & rev'd,* 642 F. 2d 610 (1980).

Pickett, S. T. A., and J. N. Thompson. 1978. Patch dynamics and the design of nature reserves. Biological Conservation 13:27–37.

Rohlf, D. R. 1989. The Endangered Species Act: a guide to its protections and implementation. Stanford Environmental Law Society, Stanford, California.

Salzman, J. E. 1989. Scientists as advocates: the Point Reyes Bird Observatory and gill netting in central California. Conservation Biology 3(2):170–180.

Scott, J. M., B. Csuti, J. D. Jacobi, and J. Estes. 1987. Species richness: a geographical approach to protecting future biodiversity. Bioscience 37(11):782–788.

Shaffer, M. 1981. Minimum population sizes for species conservation. Bioscience 31(2):131–134.

Shaffer, M. 1987. Minimum viable populations: coping with uncertainty. Pages 69–86 in M. E. Soulé, editor. Viable populations for conservation. Cambridge University Press, Cambridge, England.

Simon, D. J., editor. 1988. Our common lands: defending the national parks. Island Press, Washington, D.C.

Smith, E. M. 1984. The Endangered Species Act and biological conservation. Southern California Law Review 57(1):361–413.

Thomas, J. W., and H. Salwasser. 1989. Bringing conservation biology into a position of influence in natural resource management. Conservation Biology 3(2):123–127.

TVA v. *Hill,* 437 U.S. 153 (1977).

U.S. Fish and Wildlife Service. 1985. Endangered and threatened wildlife and plants; removal of the Brown Pelican in the southeastern United States from the list of endangered and threatened wildlife. Federal Register 50:4938–4956.

U.S. House of Representatives. 1982. House Conference Report No. 97-835.

U.S. Senate. 1982. Senate Report No. 418.

Webster, R. E. 1987. Habitat conservation plans under the Endangered Species Act. San Diego Law Review 24(1):243–271.

Wilson, E. O. 1988. The current state of biological diversity. Pages 3–18 in E. O. Wilson, editor. Biodiversity. National Academy Press, Washington, D.C.

Comments

Response to: "Six Biological Reasons Why the Endangered Species Act Doesn't Work and What to Do About It"

MICHAEL O'CONNELL

World Wildlife Fund
1250 24th Street, NW
Washington, D.C. 20037, U.S.A.

Introduction

In his essay in the September 1991 issue of *Conservation Biology,* Daniel J. Rohlf proposes six biological reasons why the Endangered Species Act supposedly doesn't work and suggests remedies. Although on the surface the assertions in the essay appear correct, they are a reflection less of problems with the legislation than of failure to carry it out. The Endangered Species Act is in fact a remarkably prescient statute that has been plagued since its adoption by ineffective implementation. Further, the essay overlooks the overriding problem of the endangered species program: a disappointing lack of commitment by Congress to fund the provisions of the Act. Even more important, Rohlf's indictment comes at a particularly inopportune time, as pressure mounts in many areas to dismantle the Act. Propitiously, however, the active role the essay recommends for conservation biologists in the debate is not only correct, it is crucial.

Although not commonly acknowledged, the Endangered Species Act of 1973 was not a singular event. Rather, it was the product of an evolution of sorts, a succession of environmental statutes that grew more comprehensive as scientific knowledge about biological diversity grew. In 1966, the passage of the Endangered Species Preservation Act (P. L. 89–669, Sections 1–3) initiated federal efforts to protect species from extinction. Three years later the Endangered Species Conservation Act of 1969 (P. L. 91–135, 83 Stat. 275) attempted to remedy deficiencies recognized in the earlier legislation, such as lack of habitat protection and applicability only to native wildlife. A number of shortcomings in the 1969 Act were recognized, which again Congress attempted to rectify. In 1973, the political climate was right for Congress to enact a much more comprehensive effort to protect imperiled species, and the Endangered Species Act (ESA) was passed.

The drafters of the Act in 1973 were operating with the best scientific knowledge available at the time. Given the legislation's provisions for protecting imperiled species and "the ecosystems upon which [they] depend" (ESA Section 2[b]) and for designating critical habitat, it went fairly far toward protecting biodiversity. However, since 1973 the scientific process has far outpaced the political one. While we now know that a species-based approach to biodiversity conservation falls short, an ecosystem basis has not been incorporated into ESA very rapidly. Attempts to enact additional legislation tailored to recent revelations in conservation biology and biodiversity have failed, not because of poor science but because competing interests, such as economic concerns, are powerful and well organized. A bill currently before Congress to establish a biodiversity conservation program suffers from the same problems.

Problems Lie with Implementation

The criticisms in Rohlf's essay center on the Act's supposed failure to conserve biodiversity. An individual examination of those criticisms shows that in fact if implemented as written, the ESA goes much further toward conserving biodiversity than generally recognized.

As the essay asserts, regulations and official interpre-

Paper accepted November 6, 1991.

tations play a vital role in shaping the implementation of the ESA. These implementing provisions are often the result of political pressures that undermine the clear mandate of the ESA and the original intention of Congress. In the past, policies such as the Fish and Wildlife Service's (and society's) favoring of so-called "higher" life forms over the often equally critical ecosystem links such as invertebrates and plants have flown in the face of the Act's clear mandate to consider a far broader array of organisms. In addition, Congress has been sluggish in appropriating funds for general recovery programs; nearly one-half of recovery funds are earmarked specifically for 12 species that are either of low conservation priority or have high public appeal (GAO 1988). Thus, the emphasis on protecting high-profile species, as evidenced by the millions of dollars poured into efforts such as California condor recovery, are not the fault of the Act itself, as implied in the essay; rather, they result from the political climate under which the Act must operate.

The essay cites the lack of a clearly defined threshold for recovery as a primary problem with the Act. This, too, is the residue of politics rather than a weakness in the legislation. It is true that lack of explicit biological criteria make determinations of recovery success difficult. However, their absence from the Act is perhaps more a measure of flexibility than oversight. On several occasions—most notably in 1982—Congress has both implied in legislation and made clear in report language that it has in mind strong thresholds to determine both jeopardy and recovery. For example, in the conference committee report for authorization of Section 10(a), Congress directed the Fish and Wildlife Service to consider the "extent to which [a habitat conservation plan] is likely to *enhance* the habitat of listed species or *increase* the long term survivability of the species and its ecosystem" (House of Representatives 1982; emphasis added). Fish and Wildlife Service regulations themselves state that incidental takings will be permitted only if they "will not appreciably reduce the likelihood of survival and recovery of the species" (FWS 1982).

An attempt to define thresholds explicitly biologically, while clearly optimal from a purely scientific standpoint, would be in reality foolhardy. Shaffer's (1987) selection of 99% probability of persistence for 1000 years, though attractive, is fundamentally an arbitrary goal. Others can just as easily cite 50% probability for 100 years as acceptable. Neither has a sound scientific basis, because the appropriate threshold, as Shaffer was astute enough to recognize, is a purely political question. Given the reality of today's political climate, it is fairly certain where the line would be drawn. It seems appropriate at this point to pursue a general objective such as "significant improvement in survival probability," determining specific objectives on a case-by-case basis.

A further criticism in the essay is the supposed inadequacy of metapopulation protection under the Act. Research into metapopulation dynamics (Gilpin 1987, for example) continually reveals more and more about the crucial role of these mechanisms in maintaining species. Contrary to Rohlf's oversimplification, nothing in the ESA compels the Fish and Wildlife Service to tolerate continued loss of populations until only one remains. Not only can loss of individual populations or genetically distinct units constitute jeopardy to a listed species, but individual populations themselves may be separately listed and protected.

Another failure cited in the essay lies with documentation of biological determinations. The Act states that determinations of listing and other decisions shall be made by the Secretary "solely on the basis of the best scientific or commercial data available" and encourages public participation in the process through a number of mechanisms (ESA Section 4(b)1A). The 1986 administrative reductions in the amount of documentation necessary to substantiate Section 7 findings and subsequent failure on the part of the Fish and Wildlife Service to adequately document its determinations, fell during a decade filled with attempts to weaken most environmental regulations. While it is indisputable that public disclosure of information is critical to sound biological decisions, faulting the Endangered Species Act for such politically influenced regulations is mistaken.

Perhaps the most unfortunate charge in Rohlf's essay, given the current political hysteria surrounding the Act's supposed "inflexibility" with regard to species protection, is his condemnation of the Section 10(a) habitat conservation planning process. These plans attempt to reconcile economic activity with species conservation and are an increasingly important valve to release the pressure of mounting dissatisfaction with the economic constraints imposed by the ESA. Bean et al. (1991) examined this process in detail and concluded that while some fine tuning was necessary, the HCP process had significant potential to improve the situation for a number of listed species.

The essay's comments regarding the Coachella Valley HCP in California in particular reflected a lack of familiarity with the history of that plan and resulted in a misleading interpretation. While the HCP did in fact leave open 75% of the remaining potential habitat of the Coachella Valley fringe-toed lizard to development, most of that area was questionable as a reserve. The lizard depends on shifting fine-grained sand dunes for survival, which in turn rely on an uninterrupted sand source. The sand source for most of the remaining habitat was already sufficiently altered by flood control facilities and highways that the only remaining defensible reserve system comprised but 25% of the remaining lizard habitat. Further, the essay discounts the positive actions the HCP accomplished on behalf of the lizard, primarily involving limiting factors other than habitat loss that were contributing to the species' decline. For example, elimination of off-road vehicle use in lizard habitat and control of invasive dune-stabilizing plants have had important positive effects on lizard abundance.

It is also important to note that the highly fragmented ownership pattern in the Coachella Valley meant that increasing the percentage of habitat protected could have easily pushed the cost of the HCP from the actual $25 million to over $100 million. Development interests made it clear that they would have rather spent the money challenging the constitutionality of ESA in court than taking action to conserve the lizard if the HCP's cost approached $100 million. This, in combination with the fact that it was virtually impossible for the Fish and Wildlife Service to enforce the ESA's taking prohibitions in the valley, indicate that the HCP was arguably a much better outcome for the fringe-toed lizard than at first glance.

The essay's final criticism of the ESA lies with the Fish and Wildlife Service's poor record on considering cumulative effects and future planned impacts on species and habitat. This reasonable accusation plainly acknowledges, however, that fault lies in implementation of the statute, not in the Act itself. It is ironic that while the Fish and Wildlife Service has promulgated regulations requiring consideration of cumulative effects, stochastic events and future human-related threats, these are rarely fulfilled in practice. In particular, cumulative effects considerations have been constrained by a dubious departmental legal opinion dating from the early 1980s. The problem goes even deeper. Federal court decisions have varied widely with regard to what constitutes fulfillment of such obligations. Given this scenario, it is not surprising that consideration of cumulative effects has been problematic in implementing the Act.

In perhaps its ultimate oversight, the essay neglects the fact that inadequate funding is largely responsible for the six deficiencies it cites. Despite a clear need, Congress has continually failed to commit adequate funding to the provisions of the Act. Current funding of the endangered species program stands at about $38 million per year, far from sufficient to evaluate which of the more than 3800 candidate species deserve listing, much less to provide for their recovery. A recent audit by the Office of the Inspector General of the Department of the Interior concluded that existing levels of funding are pathetically inadequate to carry out the Act's requirements with regard to listing and recovery (Dept. of Interior 1990). This does not include consultation and administration provisions, which are underfunded as well. Given this obstacle, creative solutions are necessary for the Act to continue to be effective. Alternatives such as listings on an ecosystem basis and multispecies recovery planning are two of many that have been proposed.

Six Positive Aspects of ESA

As I have attempted to demonstrate, the Act's shortcomings in biodiversity conservation lie with the political process of funding and implementation. Although it may not go far enough to remain in stride with what we now know about biodiversity, the ESA is still one of the most important and farsighted environmental regulations in existence, and it is under rapidly growing negative pressure. In that spirit, I offer six brief positive aspects of the ESA.

- The Act is the only U.S. legislation protecting imperiled fauna and flora and their habitat, as well as genetic diversity.
- The ESA is sufficiently clear, uncomplicated, and concise that its goals can be achieved, given adequate funding and political will.
- The Act currently is flexible enough to resolve most conflicts. The Section 7 consultation process and HCPs (Section 10a), the private land counterpart, have proven effective in reconciling economic activity with endangered species conservation.
- The ESA is a forceful example for the rest of the world.
- The Act provides protection for an important and widely popular public good.
- Despite adoption of the ESA in 1973, more than a dozen species have disappeared from the wild. Without a strong Act, many more may join them.

Biopolitical Reality

Despite demonstrating impressive knowledge of conservation biology and pointing out several areas where ESA implementation could be improved, the recent essay reflects a dangerous misinterpretation of the current political climate surrounding the Endangered Species Act. This indictment of the Act comes just at the moment its supporters are struggling, in the face of rapidly organizing opposition, to retain the integrity of even the existing version. Meetings are being held across the country, often financed by industry, to discuss anti-ESA strategy. An almost constant flow of weakening amendments floods the halls of Congress. They appear as add-ons to highway bills, as language in agency appropriations amendments, and even buried in unemployment legislation. Even the most effective action that could be taken to shore up the Act—increasing its funding level—will be an uphill battle.

In the face of such political challenges it is important not to create the erroneous impression that the Act itself is the culprit, an impression that carries the real danger of becoming even more fuel for the fire of those who, claiming the Act "doesn't work," would like to scale back its provisions. Given mounting political pressure and uncertain support for the ESA from the current administration, efforts to protect this important legislation must be redoubled. Fortunately, that is where Mr. Rohlf's call for scientific involvement in the policy process hits the nail right on the head. Few scientists can profess the understanding to plunge headlong into the political process. Yet their increased involvement in the process of legislating sound science, and attempts to develop a small degree of political sophistication, are critical. Conservation biologists also can and should play a pivotal role in helping the Fish and Wildlife Service implement the ESA and provide new direction for biodiversity conservation efforts. Many members of the Society for Conservation Biology echoed these sentiments at the most recent annual meeting in Madison, Wisconsin. The need is clear. The opportunity to contribute to a healthy and strengthened Endangered Species Act should not be missed.

Literature Cited

Bean, M. J., S. G. Fitzgerald, and M. A. O'Connell. 1991. Reconciling conflicts under the Endangered Species Act: the habitat conservation planning experience. World Wildlife Fund, Washington, D.C. 110 pp.

Gilpin, M. E. 1987. Spacial structure and population vulnerability. Pages 125–140 in M. E. Soulé, editor. Viable populations for conservation. Cambridge University Press, Cambridge, England.

Rohlf, D. J. 1991. Six biological reasons why the Endangered Species Act doesn't work—and what to do about it. Conservation Biology 5:273–282.

Shaffer, M. 1987. Minimum viable populations: coping with uncertainty. Pages 69–86 in M. E. Soulé, editor. Viable populations for conservation. Cambridge University Press, Cambridge, England.

U.S. Department of Interior. 1990. Audit report: endangered species program, U.S. Fish and Wildlife Service. Report no. 90–98. Office of the Inspector General, Washington, D.C. 40 pp.

U.S. Fish and Wildlife Service. 1982. Regulations implementing section 10(a) of the Endangered Species Act: 16 U.S.C. 1539(a)(2)(B)(iv).

U.S. General Accounting Office. 1988. Endangered species: management improvements could enhance recovery program. Resources, Community and Economic Development Division, Washington, D.C. 100 pp.

U.S. House of Representatives. 1982. Endangered Species Act amendments of 1982, Conference committee report no. 97–835. 97th Congress, Second Session at 31.

Response to O'Connell

DANIEL J. ROHLF

Natural Resources Law Institute
Lewis and Clark Law School
10015 S.W. Terwilliger Boulevard
Portland, OR 97219, U.S.A.

American experiences, in particular over the past decade, unquestionably validate Michael O'Connell's central point: there is an important relationship between politics and federal efforts to conserve biodiversity. Curiously, however, while accurately pointing out that political high jinks have severely limited the Endangered Species Act's effectiveness, O'Connell himself characterizes the ESA's salvation largely in political terms. This view is misguided for two reasons. First, policy makers and proceedings directly accessible to the public—not executive branch officials adhering to prevailing political views—should establish federal biodiversity conservation strategy. Second, there is no shortage of groups and individuals attempting, through political channels, to strengthen public policy on biodiversity. Conservation biologists, therefore, should not place high priority on heeding O'Connell's call to develop "political sophistication." Instead, scientists can most effectively sway federal policy by using their technical expertise to inform both the public and decision makers of the causes and consequences of earth's losses of biodiversity.

O'Connell creates confusion by failing to clearly distinguish between federal policy making and politics. The former consists principally of legislative decisions made by Congress. While the process of arriving at such decisions is by definition "political," it is open to all who care to participate as voters, activists, or lobbyists. On the other hand, the executive branch of the federal government in theory simply carries out the wishes of Congress, but as a practical matter the views of the incumbent administration often influence interpretation and implementation of federal legislation. In contrast to the legislative process, this sort of "playing politics" with federal policy occurs with little or no public participation or even public knowledge. O'Connell correctly notes that agency regulations interpreting the ESA, as well as agency actions implementing the statute, account for many of the reasons why the ESA does not work. Yet he misses the mark by failing to recognize that it is the ESA's generality that permits politically driven officials to interpret and implement the Act almost as they wish.

O'Connell cites approvingly the ESA's "flexibility." Yet, whether this flexibility advances or hinders conservation of biodiversity depends of course on who is doing the flexing. For example, the ESA's vaguely defined listing/recovery threshold essentially constitutes the sort of general standard O'Connell finds attractive. Rather than insulating from political influence the question of when to add species to the ESA's protective lists, this lack of specificity has allowed administrative agencies under Reagan and Bush to define restrictively the circumstances under which species become eligible for listing. The National Marine Fisheries Service (Thompson 1991) finally quantified these circumstances when it recently suggested a listing/recovery threshold of a 95% chance of persistence for only 100 years. Aside from an act of Congress, advocates of greater protection of biodiversity have little recourse to change such an interpretation. Courts must defer to *reasonable* administrative interpretations of the law; the vague legislative history O'Connell cites in arguing that lawmakers had in mind "strong thresholds" would almost certainly prove insufficient to convince a court to overturn the Service's interpretation. Of course, one could simply hope that the next administration pushes agencies to adopt a more protective interpretation. However, such flip flops in federal policy hardly further a long-term biodiversity conservation strategy. General standards within the ESA itself thus open the door to the sort of political meddling O'Connell criticizes.

Rather than allow prevailing political whims to dictate biodiversity conservation policy, publicly accountable officials should openly and explicitly make these decisions. If in the ESA itself Congress established a biologically explicit listing/recovery threshold, for example, this standard would remain constant regardless of the political philosophy of the White House occupant. Moreover, a statutory expression of precise congressional intent would provide courts a measuring stick, enabling them to strike down agency interpretations inconsistent with this intent.

Although he faults politics for the ESA's shortcomings, O'Connell applauds what he sees as the Act's ability to broker what can only be termed political deals, or in his words, "valve[s] to release the pressure of mounting dissatisfaction with the economic constraints imposed by the ESA." As an example, O'Connell praises the Habitat Conservation Plan which permits development of three-quarters of threatened Coachella Valley fringe toed lizards' remaining habitat. Yet O'Connell makes no attempt to evaluate the HCP's biological impacts upon the lizard population and its chances for recovery. Instead, he favors the Plan because it protects a "defensible reserve system," and because it holds economic costs to a level that avoided a legal challenge from developers. However, in approving the Coachella Valley HCP, the U.S. Fish and Wildlife Service (1986) did not maintain that the Plan protects all remaining high quality lizard habitat; it merely noted that the Plan's "three reserves total 15.2% of the historic habitat of [fringe-toed lizards] and 26.8% of the remaining habitat not yet subject to sand stabilization." Indeed, O'Connell openly acknowledges that property values and legal threats played a role in determining the size of "defensible" reserve areas. Even though the ESA's incidental taking provision provided a mechanism to avoid confrontation with development interests, this political harmony becomes irrelevant if the protected area ultimately proves insufficient to sustain the lizard population. The Coachella Valley situation thus typifies many of the ESA's problems. When economics and expediency replace biological criteria as indicia of a conservation strategy's success, federal biodiversity conservation policy has clearly gone awry.

O'Connell seems reluctant to ask Congress to improve the ESA, and federal biodiversity conservation strategy generally, because he fears lawmakers will not reach "correct" decisions. He implies that the influence of powerful commercial interests has torpedoed biodiversity legislation, criticizes congressional funding for ESA implementation, and even labels as "dangerous" commentary that discusses the ESA's flaws. In voicing such concerns, however, O'Connell indicts the United States' political system as well as politics. Would he propose as solutions a ban on lobbying by groups deemed antithetical to biodiversity conservation, or censorship of opinions critical of the ESA? Further, by characterizing inadequate funding levels as a "failure" on the part of Congress, O'Connell does not allow for the possibility

that lawmakers have not allocated more funds to ESA implementation because proponents of such action have failed to make their case forcefully, both to Congress and to the public. American political theory holds that the best policy decisions result when publicly accountable representatives choose from a "marketplace of ideas" open to all. O'Connell, on the other hand, seems dangerously close to advocating a sort of biological Big Brother which simply dictates the "best" way to protect biodiversity.

Currently, the drive to strengthen the ESA and other federal laws on biodiversity most lacks activities of an empirical rather than a political nature. At a meeting I attended in April 1991, numerous well-organized and experienced environmental groups agreed to launch a major political campaign to strengthen or at least maintain the ESA. Rather than merely adding their voices to this chorus, conservation biologists should use their expertise to supply a critical tool for such efforts—information. Atmospheric scientists have played a key role in instigating restrictions on activities that destroy stratospheric ozone, not through these scientists' political activities, but by providing the public and decision makers with empirical data detailing the causes and dire consequences of ozone depletion. Economic impacts of banning CFCs have taken a back seat to action because everyone realizes the very real danger of failing to address the problem. Until the public and Congress have a similar appreciation of the causes and serious consequences of eroding biodiversity, federal policies such as the ESA will be seen by many as luxuries our economy cannot afford, and thus continue to suffer political difficulties.

In sum, conservation biologists should not form a PAC and wade into the political fray. Instead, armed with an appreciation of the biological shortcomings of federal laws dealing with biodiversity conservation, they should set off for the field and laboratory, making certain to communicate the results of their "directed research" (Salzman 1989) to policymakers and the public. Ironically, this is the most politically sophisticated strategy of all.

Literature Cited

Thompson, G. G. 1991. Determining minimum viable populations under the Endangered Species Act. NOAA Technical Memorandum NMFS F/NWC-198.

Salzman, J. E. 1989. Scientists as advocates: the Point Reyes Bird Observatory and gill netting in central California. Conservation Biology **3**:170–180.

U.S. Fish and Wildlife Service. 1986. Record of decision for issuance of endangered species permit to allow incidental take of the Coachella Valley fringe-toed lizard. Federal Register **51**:15, 702–715, 704.

Comments

Rejoinder to Rohlf and O'Connell: Biodiversity as a Regulatory Criterion

JOSEPH P. DUDLEY

Environmental Planning Division (Natural Resources)
Headquarters United States Air Force
Washington, D.C. 20332, U.S.A.
and
University of Tennessee—Knoxville
Waste Management Research and Education Institute
Knoxville, TN 37996, U.S.A.

The points made by both Rohlf (1991, 1992) and O'Connell (1992) regarding the reauthorization of the U.S. Endangered Species Act (ESA: 16 U.S.C. 1531–1543) are valid and well-taken in context. Nonetheless, the authors fail to examine a key issue underlying their point of contention: the usefulness of biodiversity per se as a criterion for government regulatory action. While conservation of biological diversity should undoubtedly take precedence over single-species issues as a strategic goal, I question its ultimate usefulness as a tactical device in enforcing governmental conservation action.

This is particularly true for the developing countries, where biodiversity issues are often considered manifestations of continued northern colonialist attitudes and double standards pertaining to economic development opportunities for Third World nations. Scientific and philosophical concerns for global biodiversity hold little meaning for government officials concerned with developing national industrial capacity or for rural communities facing possible famine as the result of crop destruction by elephants, locusts, and other problem wildlife. Where governments and local communities are concerned, health issues such as malaria and schistosomiasis provide far more convincing arguments for controlling hydropower development projects in tropical rainforest regions than do scientific and philosophical fears for conservation of global biodiversity (Robert Goodland, personal communication).

Biological diversity has been slowly coming of age as a legislative issue in the United States Congress over the past decade (Thomas & Salwasser 1989). The first explicit U.S. federal legislative mandate to fund "biological diversity" programs was passed in the month following the "National Forum on Biodiversity" sponsored by the American Academy of Sciences and the Smithsonian Institution (September 1986; see Wilson 1988), and it targeted the Agency for International Development (Special Foreign Assistance Act 1986). This program was reauthorized under the Foreign Operations Appropriations Act of 1990. The Congressional biodiversity mandate has expanded in recent years to include the U.S. Department of Defense ("Legacy Resource Management Program": Department of Defense Appropriations Act 1991). The executive branch has been much more reluctant to adopt a firm position on biodiversity issues at either the national or the global level.

Rohlf (1991, 1992) has recognized the limitations of legislation for coping with biological diversity issues, and this topic merits closer inspection. In order to be amenable to effective governmental regulation, objective criteria for evaluation of compliance with biodiversity mandates must be established. We must therefore address the following issues: (1) How shall biodiversity be defined for the purposes of government regulation? (2) How shall different aspects of biodiversity be classified in terms of regulatory oversight and actions? (3) What metrics and standards shall be used in establishing baselines and determining noncompliance with laws and legal statutes?

I predicted in a manuscript (Dudley, unpublished manuscript) submitted to this journal two years ago that summary measures of biodiversity might in future become the primary input provided by conservation biologists within the decision frameworks of proposed major federal actions (*sensu* Section 1508.18: National Environmental Policy Act Regulations 1978). Those academicians who consider this issue passé must recognize that in natural resources management there can be a lag-time of decades to incorporate changes in major conservation paradigms into governmental policies and regulations. I must therefore emphasize my concern that the usefulness and limitations of biodiversity indices need careful and thorough reevaluation in light of this pending development.

Once having defined parameters for quantifying biodiversity within regulatory contexts, how are we to rank different indices and/or values? How shall evolutionary diversity (i.e., phyletic diversity) be addressed, and where will the trade-offs fall? For example, in managing for maximum biodiversity within a forest habitat, which has priority: three obligate old-growth bird species with evolutionary lineages distinct through the family/ordinal level, or six edge-adapted bird species of a single genus? Unfortunately, this simplistic example reflects the type of management decisions that government officials and land managers must ultimately resolve by administrative fiat.

The limitations of indices of species richness and diversity (such as Simpson's Index) in this context are readily appreciated (Noss & Harris 1986). Indices are mathematical constructs that describe numbers of species on site and/or in what proportions they occur, but they provide no qualitative data regarding how many of what species are actually present (Noss 1983). Landscape fragmentation and invasion of species-poor and endemic biotas (such as desert oasis and oceanic island communities) by weedy exotics may cause increases in local species diversity while an actual erosion of biological diversity at the regional/global scale is occurring (Noss 1983; Noss & Harris 1986). In these instances, increases in species diversity indices are more useful for assessing net ecological degradation than for establishing the vitality and integrity of regional biotas (see Noss 1983).

It is not enough to use our "technical expertise to inform both the public and decision makers of the causes and consequences of earth's losses of biodiversity" (Rohlf 1992). Nor is the participation in the process of developing legislation and "political sophistication" advocated by O'Connell (1992) and Rohlf (1992) adequate to the task before us. Neither of these processes gets us to the true cutting edge: the regulatory definitions and administrative directives under which actual implementation will occur (Rohlf 1991). Laws exist only within the parameters under which they are actually interpreted and implemented. As the proverb states: "Do not fear the law—fear the judge."

Rohlf (1992) has stressed that scientists must appreciate how poorly the intricacies of biological systems fit into government legal and regulatory contexts. Unfortunately, one cannot adequately encompass in regulatory language comprehensible to nonbiologists the full range of factors underlying the concept of biodiversity within a single brief paragraph. (Any and all efforts to refute this statement are greatly encouraged.) Yet ultimately this is what must happen if biodiversity is to be effectively established as a criterion for compliance with laws and governmental regulations. I submit that an accurate regulatory definition of biological diversity may be beyond our grasp.

I agree in principle with O'Connell's (1992) assessment of the benefits of a strong united front supporting reauthorization of the ESA, despite its serious biological flaws. I would expand Rohlf's (1992) target audience for biodiversity awareness actions to include policy implementers along with policy makers and the public. In the long run, perhaps far more important than communicating research results (data, conclusions) is our ability to demonstrate their applicability to the actual management decisions faced by members of government administrative and regulatory guilds (see Thomas & Salwasser 1989).

Conservation biologists should recognize participation in the development and implementation of biodiversity-related governmental regulations and policies as an inherent responsibility (Thomas & Salwasser 1989; Rohlf 1992), even a duty (Noss 1989), of their avowed profession. We must make certain that government regulatory measures implementing biodiversity legislation reflect the true spirit, if not the state-of-the-art science, of conservation biology precepts. Legal compliance with the current "no net loss of wetlands" policy of the United States government is a case in point: under the "no net loss" criterion, prairie potholes and cypress domes can be replaced by cattle tanks and artificial water impoundments of equal area. Is "no net loss of species" in this sense an adequate measure of merit or an improvement over ESA?

Therefore, with all its faults and inadequacies, ESA (or some permutation thereof) must probably remain as one of our most powerful tools in the struggle to preserve biodiversity. Given the difficulties inherent to translating biodiversity into a regulatory context and the fact that for the moment ESA is all we have, I must follow O'Connell (1992) in preferring an ESA devil we know to a nebulous biodiversity angel that we know not.

Acknowledgments

Helpful critiques of an earlier draft were provided by David Ehrenfeld, Reed Noss, Michael O'Connell, and Robin O'Malley. Assistance in tracking biodiversity legislation references was provided by Carl Gallegos, Frank Harris, Joanne Roskoski, and Sy Sohmers. The views expressed herein are those of the author and do not necessarily represent official Department of Air Force or Department of Defense policy.

Literature Cited

Department of Defense Appropriations Act. 1991. Making appropriations for the Department of Defense for the fiscal year ending September 30, 1992, and for other purposes. Congressional Record-House November 18, 1991:H10417, paragraph (*a*).

Dudley, J. P. Unpublished Manuscript. Megadiversity at macroevolutionary scales: a paleoecological perspective on neotropical biodiversity.

Foreign Operations Appropriations Act. 1990. Committee Report for Public Law 101–167. House Committee Report 101–344, Item 67, Amendment No. 161.

O'Connell, M. 1992. Response to: "Six biological reasons why the Endangered Species Act doesn't work—and what to do about it." Conservation Biology **6**:140–143.

National Environmental Policy Act Regulations. 1978. Council on Environmental Quality: regulations for implementing the procedural provisions of the National Environmental Policy Act [42 U.S.C. 4321 et seq.] 40 CFR Parts 1500–1508.

Noss, R. F. 1983. A regional landscape approach to maintain diversity. BioScience **33(11)**:700–706.

Noss, R. F. 1989. Who will speak for biodiversity? Conservation Biology **3**:202–203.

Noss, R. F., and L. D. Harris. 1986. Nodes, networks, and MUMs: preserving diversity at all scales. Environmental Management **10(3)**:299–309.

Rohlf, D. J. 1991. Six biological reasons why the Endangered Species Act doesn't work—and what to do about it. Conservation Biology **5**:273–282.

Rohlf, D. J. 1992. Response to O'Connell. Conservation Biology **6**:144–145.

Special Foreign Assistance Act. 1986. Title III—Protecting tropical forests and biological diversity in developing countries, Section 302. Protecting biological diversity. Public Law 99–529. 100 STAT 3014 (also 22 U.S.C. 2151q).

Thomas, J. W., and H. Salwasser. 1989. Bringing conservation biology into a position of influence in natural resource management. Conservation Biology **3**:123–127.

Wilson, E. O., editor. 1988. Biodiversity. National Academy Press, Washington, D.C.

Editorial

On Reauthorization of the Endangered Species Act

We have reached a pivotal moment in history, a time when public understanding and appreciation of biotic diversity is at an all-time high. Yet, paradoxically, the forces that seek to weaken environmental laws are stronger than ever before. Many special interests have united in an attempt to weaken the Endangered Species Act in the United States, and they are spreading disinformation far and wide. Long-time observers of the political processes related to the protection of endangered species in the U.S. believe that the Act, now being considered for reauthorization, has never been in greater danger.

The United States assumed world leadership in recognizing the value of conserving vanishing plants and animals when Congress passed the Endangered Species Act in 1973. But much has changed since then. The last 20 years have witnessed unprecedented degradation of our environment. Human population growth and resource consumption have led to more pollution, atmospheric changes, the destruction of wildlands, and an accelerated loss of species, subspecies, and populations. Concurrently, the past two decades have witnessed tremendous advances in the sciences of ecology, genetics, and conservation biology. Our knowledge of the relationship between human civilization and nature has grown apace. Now we know how important living diversity is to us. From this knowledge has emerged a powerful and broad consensus among scientists that we must redouble our commitment to saving vanishing species. More than ever, the United States needs a strong and effective Endangered Species Act.

The vast majority of biological scientists agrees fundamentally about the importance of conserving the diversity of life on Earth. Although legitimate disagreements exist about some scientific details of endangered species protection, these differences should not obscure the widespread consensus of most scholars on the following major points:

1. The rate of extinction of species has increased dramatically over the past four centuries and continues to escalate. The overwhelming majority of extinctions in recent centuries has been caused by human activities. This is true for the world as a whole and for the United States in particular.
2. Losses of species are irreversible. Even under the most optimistic scenarios, the technological obstacles that stand in the way of recreating species from their DNA will not be overcome in the foreseeable future.
3. We cannot reliably predict how many or which species can be removed from ecosystems without significantly altering those ecosystems or perhaps causing their collapse. In some cases, species have been lost or added to ecosystems without causing obvious, major disruptions. In other cases, such additions or losses have resulted in disastrous consequences for biodiversity and ecosystem integrity.
4. Mycorrhizal fungi and grizzly bears may be equally important to the healthy functioning of an ecosystem. Overwhelming evidence shows that the role of a species within an ecosystem is not necessarily proportional to its size, abundance, position in the food web, or charisma.
5. The long-term survival of a species depends on its ability to adapt to changing environmental conditions. The ability to adapt depends in large measure on the genetic diversity within a species. Different subspecies and populations of a given species are often genetically distinct and represent genetic blueprints for survival in the face of environmental challenges.
6. We cannot predict which populations or subspecies will prove to be critical to the long-term survival of a species. Such information is unknowable except in retrospect for very specific cases. Moreover, we can-

not anticipate which genes within populations or subspecies will ultimately prove useful for agriculture, medicine, or industry.
7. Delay in action to protect imperiled species greatly increases the duration, expense, and chance of failure of recovery efforts. In general, endangered species recovery efforts in the United States have begun much later than would have been desirable to offer reasonable likelihoods of success and to minimize financial expense.
8. Recovery efforts for endangered species have not only been too late in many cases, they have also been too little. Official population goals for species recovery are sometimes lower than population sizes at the time species were listed, and often fall far short of levels that viability analyses indicate are necessary for long-term population persistence.
9. Captive propagation or breeding in zoos, aquariums and game parks has been instrumental in saving some species. It will continue to be a useful conservation tool. Captive breeding, however, is not an effective general strategy to save rare species. Because it is intensive, it also is expensive. Zoos, aquariums, and game parks do not have the resources or expertise to handle more than a tiny fraction of the species in need. Moreover, species cannot remain in captivity forever without losing crucial genetic and behavioral adaptations to the wild. Captive breeding also does not protect the habitats in which these species live, and continued habitat loss likely means that even more species eventually must enter captivity.
10. Habitat protection is the key to species conservation. Species have been shaped by, and themselves shape, the habitats in which they live. Conservation strategies that try to restore and maintain natural habitats offer greater promise than strategies that attempt to conserve species apart from their habitats. Habitat-based strategies also increase the chances that other species occupying the same areas will not become endangered.
11. Imperiled species tend to be concentrated in relatively few ecosystems. We expect this pattern to hold even as our knowledge of the flora and fauna of the United States increases. This has two very important and reassuring implications regarding the Endangered Species Act. First, the amount of land needing special management to ensure recovery of endangered species will not increase nearly as fast as the list of endangered species. Second, protection of endangered ecosystems and "hot spots" of biodiversity where many rare and common species are concentrated will eliminate the need to list many species separately.
12. Conservation of habitat for endangered species will, in the long term, bring economic benefits by preserving natural goods and services. But even in the short term it need not cause economic harm; conservation is often compatible with, and can enhance, the variety and value of economic uses of land and water.

All of these points demonstrate the need for a strong and effective Endangered Species Act, to affirm our nation's commitment to ensure the survival of all aspects of our biological endowment. But to do the job effectively, the Endangered Species Act needs modification and improvement. During the debate over its reauthorization, the following recommendations should be kept elevated and in focus.

First, listing of species as threatened or endangered under the Endangered Species Act should be expedited to protect vanishing plants and animals while chances of recovery are still high. Congress should not burden the listing process with additional procedural requirements. Because adequate quality control mechanisms are already built into the listing process, additional peer review of listing decisions would only impede implementation of the Act. Congress must provide greater funding to the Departments of Interior and Commerce to enable these agencies to respond promptly to species in need. In addition, those agencies should be encouraged to list multiple species from imperiled habitats where such listing will better serve to achieve overall conservation goals.

Second, conservation strategies should emphasize the protection and restoration of natural habitats. Strategies based on intensive manipulation of individual plants and animals should only be relied upon in cases where no other options are available. Protection of ecosystems upon which endangered and threatened species depend is optimized by giving priority to the listing of species that would extend protection to other species in the same habitats. These "umbrella species" tend to have large area requirements or low population densities. The listing of multiple species from single habitat types will help assure that adequate amounts of habitat are conserved to meet their many needs.

The identification of critical habitat should be completed as close as possible to the time of listing to assure that consultations and permitting processes are based on the best available information, and to avoid increasing the costs of eventual conservation efforts. To achieve that objective, Congress would be well-advised to amend the Act to unburden the critical habitat designation from the requirement of economic impact assessment, an exercise that rarely assists efficient conservation planning; more often it serves to delay an otherwise straightforward science-based determination.

Third, more effective and timely recovery planning should be encouraged. Recovery planning should emphasize strategies that maintain and restore species and ecological processes in natural habitats; recovery plans should rarely rely on heroic manipulations and artificial ex situ techniques. The federal government should seek the best available expertise both inside and outside of government to assist in the development of those plans. Mandated deadlines for recovery plans are necessary to ensure timely completion and implementation, and recovery plans must mandate specific actions. Again, recovery plans that target multiple species, where such species coexist, can greatly enhance the efficiency of conservation planning. In addition, jeopardy opinions should be made in cases where actions might threaten recovery *or* survival of a listed species, as opposed to the current policy of only issuing such opinions when proposed actions threaten both the survival *and* the recovery of a listed species.

Fourth, vertebrates, invertebrates, and plants warrant equal treatment with respect to listing and protection. The authority of the Secretaries of Interior and Commerce to list and conserve vertebrate populations should be continued and expanded to include invertebrates and plants. Similarly, penalties for the taking of listed plants on public or private lands should be the same as for the taking of listed animals.

While incorporating these recommendations into a reauthorized Endangered Species Act will strengthen it greatly, the Act cannot bear full responsibility for the conservation of species and the ecosystems upon which they depend. Existing environmental laws, such as the National Forest Management Act and Clean Water Act, also must be strictly enforced. A number of highly contentious endangered species controversies would never have arisen had these other laws been faithfully obeyed.

We urge our colleagues to speak out in defense of the Endangered Species Act and the scientific principles upon which it rests, and against the deliberate, highly organized disinformation campaign being waged against it. This task is urgent. In recent trade agreement activities and proposals for health care reform, the United States has sent a clear message to the rest of the world that economic development and the well-being of people are important. If at the same time the Endangered Species Act is weakened, we send an adjunct message that the loss of biodiversity is nothing to be concerned about. But if the Act is strengthened, we signal hope for an historic turning point from which the process of biotic impoverishment can at last be reversed.

Dennis Murphy
Stanford University

David Wilcove
Environmental Defense Fund

Reed Noss
Oregon State University and University of Idaho

John Harte
University of California, Berkeley

Carl Safina
National Audubon Society

Jane Lubchenco
Oregon State University

Terry Root
University of Michigan

Victor Sher
Sierra Club Legal Defense Fund

Les Kaufman
New England Aquarium

Michael Bean
Environmental Defense Fund

Stuart Pimm
University of Tennessee

Correspondence should be addressed to Dennis D. Murphy, Department of Biological Sciences, Stanford University, Stanford, CA 94305-5020. This editorial is an abstracted version of an open letter to Congress drafted and cosigned by these and additional Pew Scholars in Conservation and the Environment. Support to the Pew Scholars Committee on Reauthorization of the Endangered Species Act has been generously provided by the Pew Scholars Program in Conservation and the Environment.

Contributed Papers

Biological Integrity and the Goal of Environmental Legislation: Lessons for Conservation Biology

JAMES R. KARR

Department of Biology
Virginia Polytechnic Institute and State University
Blacksburg, VA 24061-0406, U.S.A.

Abstract: *Passage of environmental legislation often creates a state of euphoria among supporters such that implementation programs are not rigorously evaluated. Endangered species and water resource legislation are two examples of sound environmental legislation with major weaknesses in their implementation. Regulations to implement the U.S. Clean Water Act, for example, emphasize water quality (physical and chemical properties of water) by calling for uniform standards for contaminants rather than the broader goal of improving the quality of water resources. As a result, improvements in the quality of water resource quality have been limited despite massive expenditures. Natural resource agencies' narrow emphasis on harvested and threatened and endangered species has similar consequences. Unless a comprehensive and rigorous definition of the goals of biodiversity legislation is developed and adhered to by regulatory agencies, efforts to implement biodiversity legislation could lead to similar problems. Protection of biodiversity should be considered a subset of the need to protect the biological integrity of natural resource systems and the ecological health of the biosphere. Programs to protect biodiversity should reflect that more holistic goal and include provisions for evaluating success at attaining stated goals and for making midcourse adjustments in programs when resources are not being adequately protected.*

Resumen: *La aprobación de legislación ambiental con frecuencia crea tal estado de euforia entre los seguidores que los programas de implementación se siguen con ménos rigidez. La legislación sobre las especies en peligro y los recursos de agua son dos ejemplos de legislación con bases ambientales sólidas y con grandes debilidades en su implementación. Las regulaciones para implementar el Acta de Agua Potable de los Estados Unidos, por ejemplo, enfatiza la calidad del agua (propiedades físicas y químicas) haciendo un llamado para uniformizar los estándards de los contaminantes más que la meta más amplia de mejorar la calidad del recurso agua. Como resultado, los mejoramientos en la calidad del agua han sido limitados a pesar de los gastos masivos que se han hecho. El estrecho énfasis de las agencias de los recursos naturales acerca de las especies cosechadas, amenazadas y en peligro tiene consecuencias similares. A menos que se desarrolle una definición comprensiva y riguroza de las metas de la legislación sobre la biodiversidad y se le adhieran las agencias regulatorias, los esfuerzos para implementar la legislación de la biodiversidad pueden dar lugar a problemas similares. La protección de la biodiversidad debe de ser considerada como parte de la necesidad de proteger la integridad biologica del sistema de recursos naturales y la salud ecológica de la biosfera. Los programas para proteger la biodiversidad deben reflejar esta meta más comprensiva e incluir previsiones para evaluar el éxito al ir alcanzando las metas enunciadas y crear mecanismos para hacer ajustes sobre la marcha en los programas cuando los recursos no estén siendo protegidos adecuadamente.*

Introduction

Just as biological systems are being fragmented by diverse human activities throughout the world, the environmental legislation of the past two decades is a fragmented approach to a broad range of problems. Most environmental legislation and regulations in support of legislation are reactive (damage control) rather than proactive; problems are treated only after degradation is obvious. In addition, duplicate or competitive programs result from the fragmentation of responsibilities among agencies. A recent addition to the legal landscape is a biodiversity bill, an effort to establish a comprehensive United States policy on biological diversity. This legislation comes on the heels of a major review of U.S. programs in biological conservation by the Office of Technology Assessment (OTA 1987). Major goals of the legislation include development of a national policy on biodiversity conservation and a federal strategy to maintain biodiversity, creation of a National Center for Biological Diversity and Conservation Research, and evaluation of project impacts on biological diversity when environmental impact statements are prepared.

Passage of biodiversity legislation would be a major environmental landmark. But as is so often the case, the program to implement that legislation may be more important in determining the success of the law than its explicit provisions will be. Regulations to implement a new law may be inadequate because of ambiguities in the law or in judicial interpretations, or because agency efforts to draft regulations are inappropriate. Effective implementation depends upon clear and concise definition of goals and proper choice of indexes to measure environmental health. The latter must be based on rigorous application of ecological principles.

Biological diversity can be defined as *"the variety and variability among living organisms and the ecological complexes in which they occur"* (OTA 1987). This definition and recent usage of the term "ecological integrity" (Scheuer 1989), recognizes the importance of biological integrity and ecological health as explicit goals of biodiversity legislation. These terms can be defined as follows:

Biological integrity — "The capability of supporting and maintaining a balanced, integrated, adaptive community of organisms having a species composition and functional organization comparable to that of natural habitat of the region" (Karr & Dudley 1981).

Ecological health — "A biological system — whether it is a human system or a stream ecosystem — can be considered healthy when its inherent potential is realized, its condition is [relatively] stable, its capacity for self-repair when perturbed is preserved, and minimal external support for management is needed" (Karr et al. 1986).

Unfortunately, these concepts are not central to implementation programs in many environmental situations. Examples of this problem include soil erosion programs initiated in the Dust Bowl of the 1930s, which evolved to focus primarily on enhancing crop production; endangered species legislation that bogged down in a bureaucratic listing process; and water resource legislation that concentrated on control of point sources of pollution through construction of wastewater treatment plants. In this paper, I illustrate potential problems with the implementation of biodiversity legislation using examples from my experience with other environmental legislation — the Endangered Species Act (ESA) and Clean Water Act (CWA).

Endangered Species Act

Endangered species legislation at state and federal levels and through international treaties and conventions has been instrumental in protecting species, in part because it provides easily defined biological goals. By protecting species threatened with extinction, ESA extends legal rights to nonhuman species for their own sake, a clear expansion of the ethical landscape in western culture. Regrettably, that legislation is not sufficient for exactly the same reason — narrowly defined goals centered on endangered species do not protect species from human impacts until their populations are reduced to the point that they are endangered. Further, projects that endanger whole ecosystems are not subject to review, except in special cases, such as when wetlands are affected.

Individuals concerned about endangered species are heartened by recent advances in the implementation of that legislation: (1) expansion of endangered species concerns beyond warm cuddly creatures to include cold-blooded vertebrates, invertebrates, and plants whose populations are in jeopardy, and (2) increased emphasis on "critical habitat" protection. Both strengthen the protection afforded to endangered species.

On the other hand, implementation of endangered species legislation has created a bureaucratic quagmire that diverts attention from larger goals. To some extent, that quagmire derives from the use of esthetic and moral arguments to support the existence of endangered species programs (the warm cuddly issue noted above). While those arguments are not inappropriate, one could argue that they are not sufficient, because many persons do not share concern for fellow travellers on spaceship Earth (Karr 1990b).

Often, protection of endangered species is viewed as an obstruction to progress rather than part of an overall strategy to protect Earth's biological integrity. In projects like the Tellico Dam (snail darter), logging of Northwest old-growth forest (spotted owl), and astronomical development on Mt. Graham in Arizona (Mt. Graham red squirrel), ESA goals confront major projects of public concern. Plans to build over a dozen astronomical facilities on Mt. Graham met resistance on several environmental grounds, including the presence of the red squirrel (*Tamiasciurus hudsonicus grahamensis*), a unique, insular subspecies recognized by the U.S. Fish and Wildlife Service as an endangered species. A number of other species and subspecies may warrant similar recognition (U.S. Forest Service 1988). The University of Arizona, however, used political muscle to solicit the aid of the Arizona congressional delegation to push through an exception for the first phase of astronomical development. The action, passed as an attachment to unrelated legislation, circumvented both the National Environmental Policy Act and the Endangered Species Act. In recent action, a U.S. District Court judge temporarily banned observatory construction pending a review of the project's impact.

That endangered species legislation and the regulations used to implement it have evolved is a positive step. The responsibility of practicing conservation biologists, however, must be to ensure that such evolution proceeds rapidly. Robust biological insight must be incorporated into the planning process at the earliest possible stage.

Clean Water Act

The first major legislation designed to protect water resources was passed nearly 100 years ago in 1899 (Karr, 1991b). Since then, a series of water pollution control acts have expanded the federal government's role in maintaining water quality, with particular emphasis on funds for constructing wastewater treatment facilities, developing technologies to improve those facilities, expanding lists of pollutants to be regulated, and increasing enforcement to control point sources of pollution.

Major advances were made in controlling domestic effluents, but a growing array of toxic materials produced by human society and the pervasive, more-difficult-to-control non–point sources (diffuse runoff from urban, agricultural, and forest lands) remained largely untreated. Atmospheric deposition of toxic materials is yet another problem that has not been adequately addressed that threatens both terrestrial and aquatic environments. The magnitude of the degradation over the past century is illustrated by data from the Illinois and Maumee rivers, where 66 percent and 43 percent, respectively, of resident fish faunas have experienced major population declines or been extirpated (Karr et al. 1985b). As is so often true, no single factor is responsible for this degradation. The primary culprits in the Illinois and Maumee were agriculture, impoundments and levees, navigation, toxics, consumption of water, and introduction of exotics.

This breadth of factors illustrates the folly of trying to restore water resources with single-minded approaches, the dominant theme of agencies dealing with the nation's waters. Environmental agencies often have a primary focus on control of point sources of chemical pollutants through construction of wastewater treatment plants, whereas conservation departments deal primarily with physical habitat degradation in fishable streams. (Paradoxically, wholesale introduction of exotics by those same agencies often tears the fabric of native ecosystems.) In recent years, programs to protect minimum critical flows have gained considerable attention. All treat serious water resource problems but none is sufficient alone.

As a result, the biological integrity of water resources continues to decline in most areas. A few exceptions exist (e.g., Lake Washington near Seattle (NAS 1986), where a single stress, eutrophication, was identified and controlled by a creative interaction of science and political will). More commonly, however, water engineers or planners do not acknowledge the existence, let alone the importance, of the biological communities associated with water resources. Thus, the quality of water resources has not been protected because the primary regulatory approach concentrates on water quality (or water quantity in some western areas) rather than the quality of the water resource. A major lesson emerges from three decades of water resource law and regulations for its implementation — *environmental legislation must be both well conceived and carefully implemented through sound regulations.*

Perhaps the clearest call for attention to Earth's life support system occurs in the Water Quality Act Amendments of 1972 (PL 92-500) and later revisions of that Act with the mandate "to restore and maintain the physical, chemical, and biological integrity of the Nation's waters." The inclusion of biological integrity was thoughtful (Ballentine & Guaraia 1977), but no provisions were made to define that goal precisely and develop integrative ways to measure successful attainment of that goal. A simple surrogate for biological integrity, water chemistry, was used to evaluate the degradation that stimulated the passage of the Clean Water Act (Karr & Dudley 1981; Karr 1987). As a result, expensive treatment plants were built with relatively little or no effort to measure their impact on the quality of the water resource. In one study (Karr et al. 1985a), biotic integrity was degraded by the chlorine added in secondary treatment, and addition of a territory denitrification plant did not improve biotic integrity. Simply put, if water resources are to be protected, a quantitative, ecologically sophisticated method is needed to monitor the biotic integrity of all waters. No nonbiological techniques exist that can serve as a surrogate for the direct measurement of biological conditions in a stream.

Biologists have intuitively known and agreed with this position for over two decades, and a variety of methodologies for achieving it have been proposed (Worf 1980; Taub 1987; Ford 1989; Fausch et al. 1990; Karr 1990a). Laboratory studies of acute toxic effects dominated early work with the goal of establishing criteria for specific pollutants (USEPA 1976), an approach that was challenged by many (Thurston et al. 1979; Levin et al.

1989). Focusing solely on acute toxicity in the laboratory misses chronic effects in the field and the synergistic effects of combinations of chemical pollutants. Field monitoring of selected (indicator) taxa was also tried using fish, benthic invertebrates, and diatoms, but by focusing exclusively on assessing the presence of pollution-tolerant forms, researchers miss the opportunity to evaluate other aspects of biotic integrity such as individual health, sizes of populations of component species, and the trophic structure of the community. Because limits to the biological integrity of a water resource vary in space and time, no single approach can be expected to detect and reverse all degradation. To protect water resources, five sets of variables (Fig. 1) that may be affected by human action must be evaluated, and the factor or combination of factors that is responsible for degradation must be treated (Karr et al. 1986).

The need for a more integrative approach to evaluate this diversity of human impacts stimulated the development of an index (Karr 1981; Karr et al. 1986) to assess biological conditions in a river or stream using fish communities. It is called the Index of Biotic Integrity (IBI). This multiparameter index uses attributes of fish communities to evaluate human effects on a stream and its watershed. Use of IBI in both research and regulatory contexts (Karr et al. 1986; Hughes & Gammon 1987; Karr et al. 1987; Hirsch et al. 1988; Hite 1988; Miller et al. 1988; Steedman 1988; Ohio EPA 1988; Plafkin et al. 1989) has spread throughout North America and even to Europe. IBI has been used to evaluate chemical and habitat degradation affecting both in-stream and watershed-level activities. Further, the conceptual framework of the fish IBI has been adopted by benthic biologists in efforts to develop robust methods to measure degradation by monitoring invertebrate communities (Ohio EPA 1988; Plafkin et al. 1989). Ecologically sophisticated biological monitoring is now widely recognized as an essential part of monitoring programs. No other approach provides direct, integrative information about biological conditions at a sample site. The advantages of rigorous analysis of biotic integrity are widely recognized, as are the disadvantages of lack of biotic assessment (Ohio EPA 1988; Plafkin et al. 1989; USEPA 1987, 1988, 1989).

The IBI was conceived to provide a broadly based and ecologically sound tool to evaluate biological conditions in a stream. A single sample from a stream reach is evaluated using 12 metrics (Table 1) to determine the extent to which the resident community diverges from that expected of an undisturbed site in the same geographic area and of the same stream size. Unlike efforts to define chemical criteria that do not take variation by geographic region into account, this approach explicitly recognizes natural variation in water resource conditions. The twelve metrics are grouped into three classes that evaluate the extent to which the fish community has reduced species richness and/or altered species composition, the trophic composition is altered due to human effects on the energy base and trophic dynamics of the resident biota, and the abundance and condition of fish within the community are affected by humans. The IBI can be used to assess local conditions or to evaluate a basin through a number of samples within a watershed (Steedman 1988). Similarly, simultaneous samples at a series of sites can be used to compare areas or multiple samples from a single site can be used to evaluate trends over time (Karr et al. 1986; Hughes & Larsen 1988). Finally, when degradation is detected, judicious use of the IBI can provide information about the factors responsible for degradation.

Monitoring and assessment efforts using the conceptual approach of IBI are more likely to protect the quality of water resources than are more conventional programs that ignore or limit the use of direct biological assessments because the IBI incorporates all the stressors shown in Figure 1.

Discussion

Environmental legislation inevitably has narrow, specific goals, although an umbrella goal that motivates virtually all environmental legislation is the protection of biological integrity and ecological health. It is essential that framers of legislation and those responsible for implementing legislation keep both these issues in mind. Congressional testimony on the 1972 Water Quality Act convinces even the skeptical reader that some advocates of that bill had similar insights. History, however, shows that the larger umbrella was lost with the development of regulations and water resource programs (Anonymous 1981a, b, 1983). Some published material on biodiversity protection and recent discussion at meetings reinforce my concern about the need to avoid

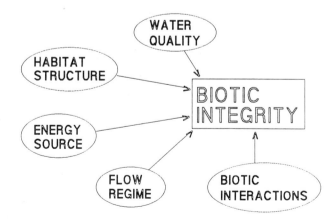

Figure 1. Five major classes of environmental factors that affect the integrity of an aquatic biota.

Table 1. Metrics used to assess fish communities in the midwestern United States (modified from Karr 1981 and Fausch et al. 1984).

Category	Metric[a]	Scoring criteria[b]		
		5	3	1
Species richness and composition	1. Total number of fish species	Expectations for metrics 1–5 vary with stream size and region.		
	2. Number and identity of darter species			
	3. Number and identity of sunfish species			
	4. Number and identity of sucker species			
	5. Number and identity of intolerant species			
	6. Proportion of individuals as green sunfish	<5%	5–20%	>20%
Trophic composition	7. Proportion of individuals as omnivores[c]	<20%	20–45%	>45%
	8. Proportion of individuals as insectivorous cyprinids	>45%	45–20%	<20%
	9. Proportion of individuals as piscivores (top carnivores)	>5%	5–1%	<1%
Fish abundance and condition	10. Number of individuals in sample	Expectations for metric 10 vary with stream size and other factors.		
	11. Proportion of individuals as hybrids	0%	>0–1%	>1%
	12. Proportion of individuals with disease, fin damage, and skeletal anomalies	0–2%	>2–5%	>5%

[a] Tabulated metrics are for the original IBI developed in the Midwest. More general metrics applicable outside the Midwest include the following: Metric 1—Native fish species, 2—Benthic species, 3—Water-column species, 4—Long-lived species, 6—Tolerant species, 8—Proportion insectivores, 11—Proportion exotics (see Miller et al. 1988).
[b] Ratings of 5, 3, and 1 are assigned to each metric according to whether its value approximates, deviates somewhat from, or deviates strongly from the value expected at a comparable site that is relatively undisturbed.
[c] Omnivores are defined here as species with diets composed of ≥25% plant material and ≥25% animal material.

a narrowing of perspective with respect to the biodiversity bill. Biodiversity, for example, has been defined as "the *elements* of the biosphere" in contrast to ecological processes, "the *interactions* among species and between species and their environments" (Reid & Miller 1989). What if the framers of biodiversity regulations err in focusing only on the former?

Neither protection of endangered species nor the maintenance of clean water will assure that biological integrity and ecological health are protected. Similarly, neither cataloging biological diversity through a national biological inventory nor protecting fragments of natural ecosystems will assure that biological integrity and ecological health will be preserved under biodiversity legislation. Conservation biologists must guard against the allocation of resources to aspects of the biodiversity problem that are too limited or to the development of measuring devices that are too narrow to protect the larger fabric of biological integrity and ecological health. The biodiversity bill must be interpreted as a mandate to protect genes within populations, populations within species, species within communities, communities/ecosystems within landscapes, and landscapes within the biosphere.

Four other problems must be dealt with if biodiversity legislation is to attain its potential. (1) Environmental protection should not be held hostage to political maneuvering (as happened in the case of telescopes versus squirrels on Mt. Graham). (2) Jurisdictional disputes often break out between state and federal authorities and among agencies. These may involve defense of territory or efforts to avoid responsibility. For many years, I heard USEPA officials say that protecting biotic integrity was not their mandate and then cite the implementation regulations of the CWA. (3) Many do not recognize the biological foundations of most environmental problems. (4) Finally, a move away from the ethical, theological, political, economic, and management perspectives that place human life and products above the "less useful" nonhuman life would aid attainment of biodiversity goals.

The disciplines of conservation biology and ecology are at a major threshold. Never before have the challenges and the opportunities been greater and the consequences of inaction more severe. Throughout history, the human-environment interaction has generally been significant only at relatively small spatial and temporal scales. Disease, acts of predation, and accidents were the major decimating factors. Because of population growth and the expanding influence of technology, man's interactions with his environment have shifted such that we face a growing array of difficult challenges, including many that we have not evolved to deal with (Orenstein & Ehrlich 1989). The very difficult problems created by global warming are examples of the consequences of our actions.

Mankind's principal interaction with his environment is no longer at the scale of individuals, and mankind's normal solutions, medicine and technology, can no longer be viewed as panaceas to resolve the ills of modern society. Our biological problems are no longer centered on the health of the individual. Rather, because of our actions, the health of the biosphere is threatened, and we need a new generation of "medicine" men. They should be people with ecological background, trained "physicians of the environment." Like medical pathologists, environmental pathologists must have knowledge of ecologically healthy as well as degraded biological communities if their prescriptions are to be taken seriously.

Literature Cited

Anonymous. 1981a. WPCF roundtable discussion — congressional staffers take a retrospective look at PL 92-500. Journal of the Water Pollution Control Federation **53**:1264–1270.

Anonymous. 1981b. Congressional staffers take a retrospective look at PL 92-500. Part 2. Journal of the Water Pollution Control Federation **53**:1370–1377.

Anonymous. 1983. Changing the Clean Water Act: reflections on the long-term. Journal of the Water Pollution Control Federation **55**:123–129.

Ballentine, K., and L. J. Guaraia, editors. 1975. The integrity of water. United States Environmental Protection Agency, Washington, D.C.

Fausch, K. D., J. Lyons, J. R. Karr, and P. L. Angermeier. 1990. Fish communities as indicators of environmental degradation. Bioindicators of stress in fish. Special Symposium Publication, American Fisheries Society. Bethesda, Maryland. In press.

Ford, J. 1989. The effects of chemical stress on aquatic species composition and community structure. Pages 99–144 in S. A. Levin, M. A. Harwell, J. R. Kelly, and K. D. Kimball, editors. Ecotoxicology: problems and approaches. Springer-Verlag, New York.

Hirsch, R. M., W. M. Alley, and W. G. Wilber. 1988. Concepts for a national water-quality assessment program. United States Geological Survey Circular 1021. Washington, D.C.

Hite, R. L. 1988. Overview of stream quality assessments and stream classification in Illinois. Pages 98–119 in T. P. Simon, L. L. Holst, and L. J. Shepard, editors. Proceedings of the first national workshop on biocriteria, 1987. Lincolnwood, Illinois. EPA 905/9-89/003. United States Environmental Protection Agency, Chicago, Illinois.

Hughes, R. M., and J. R. Gammon. 1987. Longitudinal changes in fish assemblages and water quality in the Willamette River, Oregon. Transactions of the American Fisheries Society **116**:196–209.

Hughes, R. M., and D. P. Larsen. 1988. Ecoregions: an approach to surface water protection. Journal of the Water Pollution Control Federation **60**:486–493.

Karr, J. R. 1981. Assessment of biotic integrity using fish communities. Fisheries **6**(6):21–27.

Karr, J. R. 1987. Biological monitoring and environmental assessment: a conceptual framework. Environmental Management **11**:249–256.

Karr, J. R. 1990a. Bioassessment and non–point source pollution: an overview. Pages 4-1–4-18, in Second national symposium on water quality assessment. Office of Water, United States Environmental Protection Agency, Washington, D.C.

Karr, J. R. 1991b. Biological integrity: a long-neglected aspect of water resource management. Ecological Applications. **1**:1:in press.

Karr, J. R. 1990b. Endangered species: an overview of problems and needs. Pages in Virginia's endangered species. McDonald and Woodward, Blacksburg, Virginia.

Karr, J. R., and D. R. Dudley. 1981. Ecological perspective on water quality goals. Environmental Management **5**:55–68.

Karr, J. R., K. D. Fausch, P. L. Angermeier, P. R. Yant, and I. J. Schlosser. 1986. Assessing biological integrity in running waters: a method and its rationale. Special Publication 5, Illinois Natural History Survey, Urbana, Illinois.

Karr, J. R., R. C. Heidinger, and E. H. Helmer. 1985a. Sensitivity of the index of biotic integrity to change in chlorine and ammonia levels from wastewater treatment facilities. Journal of the Water Pollution Control Federation **57**:912–915.

Karr, J. R., L. A. Toth, and D. R. Dudley. 1985b. Fish communities of midwestern rivers: a history of degradation. BioScience **35**:90–95.

Karr, J. R., P. R. Yant, K. D. Fausch, and I. J. Schlosser. 1987. Spatial and temporal variability on the index of biotic integrity in three midwestern streams. Transactions of the American Fisheries Society **116**:1–11.

Levin, S. A., M. A. Harwell, J. R. Kelly, and K. D. Kimball, editors. 1989. Ecotoxicology: problems and approaches. Springer-Verlag, New York.

Miller, D. L., P. M. Leonard, R. M. Hughes, et al. 1988. Regional applications of an index of biotic integrity for use in water resource management. Fisheries **13**(5):12–20.

National Academy of Sciences. 1986. Ecological knowledge and environmental problem-solving: concepts and case studies. National Academy Press, Washington, D.C.

Office of Technology Assessment. 1987. Technologies to maintain biological diversity. OTA-F-330. Office of Technology Assessment, Congress of the United States, Washington, D.C.

Ohio Environmental Protection Agency. 1988. Users manual for biological field assessment of Ohio surface waters. Three volumes. Surface Water Section, Division of Water Quality Monitoring and Assessment, Ohio Environmental Protection Agency, Columbus, Ohio.

Ornstein, R., and P. R. Ehrlich. 1989. New world, new mind. Simon and Schuster, New York.

Plafkin, J. L., M. T. Barbour, K. D. Porter, S. K. Gross, and R. M. Hughes. 1989. Rapid bioassessment protocols for use in streams and rivers: benthic macroinvertebrates and fish. EPA/444/4-89-001. United States Environmental Protection Agency, Washington, D.C.

Reid, W. V., and K. R. Miller. 1989. Keeping options alive: the scientific basis for conserving biodiversity. World Resources Institute, Washington, D.C.

Scheuer, J. H. 1989. The National Biological Diversity Conservation and Environmental Research Act (H.R. 1268): a new approach to save the environment. Bulletin of the Ecological Society of America **70**:194–195.

Steedman, R. J. 1988. Modification and assessment of an index of biotic integrity to quantify stream quality in southern Ontario. Canadian Journal of Fisheries and Aquatic Sciences **45**:492–501.

Taub, F. B. 1987. Indicators of change in natural and human-impacted ecosystems: status. Pages 115–144 in S. Draggan, J. J. Cohrssen, and R. E. Morrison, editors. Preserving ecological systems — the agenda for long-term research and development. Praeger, New York.

Thurston, R. V., R. C. Russo, C. M. Fetterolf, Jr., T. A. Edsall, and Y. M. Barber, Jr., editors. 1979. A review of the EPA Red Book: quality criteria for water. Water Quality Section, American Fisheries Society, Bethesda, Maryland.

U.S. Environmental Protection Agency. 1976. Quality criteria for water. United States Environmental Protection Agency, Washington, D.C.

U.S. Environmental Protection Agency. 1987. Surface water monitoring: a framework for change. Office of Water Policy Planning and Evaluation, Office of Water, United States Environmental Protection Agency, Washington, D.C.

U.S. Environmental Protection Agency. 1988. WQS draft framework for the water quality standards program. Office of Water, United States Environmental Protection Agency, Washington, D.C. Draft, 11-8-88.

U.S. Environmental Protection Agency. 1989. Water quality standards for the 21st century. Office of Water, United States Environmental Protection Agency, Washington, D.C.

U.S. Forest Service. 1988. Final environmental impact statement, proposed Mt. Graham astrophysical area, Pinaleno Mountains, Coronado National Forest. Southwestern Region, United States Department of Agriculture Forest Service, Tucson, Arizona.

Worf, D. L. 1980. Biological monitoring for environmental effects. Lexington Books, D. C. Heath and Company, Lexington, Massachusetts.

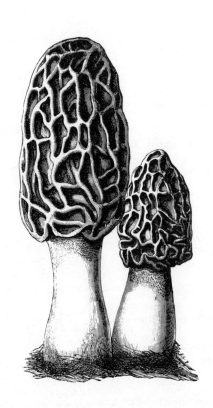

Conservation Education

The Question of Management

There is a growing presumption that humans must now take control of the planet and its life processes. The September 1989 issue of *Scientific American* devoted an entire issue to the subject of "Managing Planet Earth." In it one finds pleas for "adaptive" policies for planetary management that require "improving the flow of information," "technologies for sustainable development," and "mechanisms... to coordinate managerial activities" (Clark 1989). Economists of similar mindset intend to "find the right policy levers" to manage "all assets, natural resources, and human resources" (Repetto 1986). Language like this conjures images of economists and policy experts sitting in a computerized planetary control room, coolly pushing buttons and pulling levers, guiding the planet to something called "sustainable growth."

I do not doubt that something needs to be managed. I would like to raise questions, however, about what and how we manage. For would-be planet managers, it is a matter of no small consequence that God, Gaia, or evolution was doing the job nicely until human population, technology, and economies got out of control. This leads me to think that it is humans that need managing, not the planet. This is more than semantic hairsplitting. "Planetary management" has a nice ring to it. It places the blame on the planet, not on human stupidity, arrogance and ecological malfeasance, which do not have a nice ring. The term avoids the messy subjects of politics, justice, and the discipline of moral choice. Planetary management, moreover, appeals to our desire to be in control of things. It appeals to our fascination with digital readouts, computer printouts, dials, gauges, and high tech of all sorts. Management is mechanical, not organic, and we like mechanical things, which reinforce our belief that we are in control.

Plans to manage the Earth are founded on the belief that ignorance is a solvable problem, that with enough research, satellite data, and computer models, we can take command of spaceship earth. There is a great deal more to learn about the Earth, no doubt. But there are good reasons to believe that its complexity is permanently beyond our comprehension. A square meter of topsoil several inches deep teems with life forms that still have not been studied; we know nothing about their relationship to the planet. The same can be said of most of the "machinery of nature." The salient fact is not our knowledge, but our ignorance. To the complexity of nature, from soil bacteria to planetary biogeochemical cycles, add human impacts with all kinds of synergies, feedback loops, leads, and lags, and the idea of managing the planet takes on a different prospect altogether. It would require a level of knowledge that we are not likely to reach. Even if we could, other limits would appear.

I am referring to the problem of human evil, recalcitrance, and our capacity to rationalize almost anything. There is a limit to how well we can comprehend the good, and there are other limits to our willingness to do it. The world is not simply a set of problems solvable by adjustments in prices or more efficient technologies. There can be no good argument against efficiency in the use of resources or against prices that include all environmental costs. But I think it is a mistake to believe that we face only problems solvable by painless market adjustments and better gadgets, not dilemmas that will require wisdom, goodness, and a rationality of a higher sort. The word "hubris," meaning overweening arrogance, is not heard much any more, but it applies, I think, to the belief that we can sufficiently understand the complexities of the Earth to manage it for good purposes. It is also hubris to believe that the technologies, mindset, and methods of management are always benign. They are not. The attempt to manage anything carries with it political, social, ecological, and moral consequences. We may assume, I think, that these rise with scale.

Another approach to management was expressed by Wendell Berry, who wrote in the September 1989 issue of *Harpers* that "we are not smart enough or conscious enough or alert enough to work responsibly on a gigantic scale" (Berry 1989). His answer has nothing to do with management, unless by that term we include self-management. He says, "we must acquire the character and the skills to live much poorer than we do. We must waste less. We must do more for ourselves." There are points of agreement between Berry and the planet managers — the need to reduce waste, for example. But where Berry talks openly about living more poorly, the planet managers talk of sustainable growth as if life can continue as it has for the past four decades. Berry's emphasis on self-reliance skills and character, which have no market value, strikes an odd note in a time when we think that people act only from economic motives, not from a sense of duty, rightness, or even righteousness. Elsewhere Berry has described the environmental crisis as one of character that cannot be improved by technology or market adjustments, although it can be made worse by them.

The difference between the planet managers and Berry arises from different judgments about the causes of the crisis. The *Scientific American* authors are silent about the origins of our plight, as if the crisis has no historical causes, no antecedents, no roots in science, technology, economics, or politics. I think it no coincidence that the advertisements of IBM, McDonnell Douglas, General Motors, and Boeing are scattered throughout the issue. It is difficult to avoid the conclusion that planet managers have made their peace with the powers that be and have joined the movement to extend human domination of nature to the fullest extent.

Berry's point that living poorly requires both character and skills strikes at the root of the matter. For planet managers the only skills required are scientific expertise, technological ingenuity, and economics. In contrast, the kind of skills Berry has in mind are those of an ecologically competent and active citizenry who know how to do for themselves and have the character traits of frugality, truthfulness, and goodheartedness necessary to be a good neighbor and good citizen. This implies management, but the kind that grows out of the disciplines of community and stewardship. Management in this case means knowing what's manageable and what's not and having the good sense to leave the latter to manage itself. It is more like child-proofing a day-care center than piloting Spaceship Earth. It may be that our impulse to manage that which cannot be managed while leaving unmanaged that which could be is one source of the problem. In other words, lacking the will to control our appetites, economies, technologies, and propensity to breed, we must redesign the earth — an unlikely enterprise.

Among those of us engaged in conservation education — which is increasingly bound up with the teaching of management — the question of who manages what and how deserves to be discussed. The causes of biotic impoverishment have to do with the ownership and control of land, human numbers, the scale of technology, the manner and extent to which local economies are linked to larger economies, and the cultivation of knowledge that allows people to live sustainably in a particular place. Each of these factors has to do with politics, including the politics of knowledge. How should those calling themselves "conservation biologists" deal with politics and the question of management in their research, writing, and teaching? Comments are welcome.

David W. Orr
Meadowcreek Project, Inc.
Fox, AR 72051, U.S.A.

Literature Cited

Berry, W. 1989. The futility of global thinking. Harpers, September: 16–22.

Clark, W. C. 1989. Managing planet Earth. Scientific American, September: 53–54.

Repetto, R. 1986. World enough and time. Yale University Press, New Haven, Connecticut, p 8.

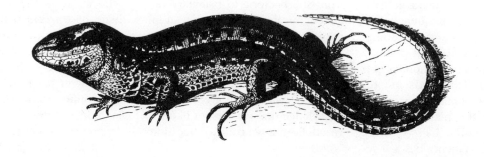

Techno-Arrogance and Halfway Technologies: Salmon Hatcheries on the Pacific Coast of North America

GARY K. MEFFE

Savannah River Ecology Laboratory
University of Georgia
Drawer E
Aiken, SC 29802, U.S.A.

Abstract: *Humankind has adopted an arrogant and ultimately self-defeating attitude toward nature that places technological mastery over nature at the forefront of our approach to many environmental problems. This "techno-arrogance" fails to recognize limitations on, and ramifications of, attempted control of nature. An example of techno-arrogance is the flawed attempt to recover Pacific salmonid fisheries through technological application in the form of hatcheries. Countless salmon stocks have declined precipitously over the last century as a result of overfishing and widespread habitat destruction. A central feature of recovery efforts has been to build many hatcheries to produce large quantities of fish to restock streams. This approach addresses the symptoms but not the causes of the declines (an example of a halfway technology), because the habitats remain largely unsuitable for salmon. There are at least six reasons why the hatchery approach will ultimately fail: (1) data demonstrate that hatcheries are not solving the problem—salmon continue to decline despite decades of hatchery production; (2) hatcheries are costly to run, and divert resources from other efforts, such as habitat restoration; (3) hatcheries are not sustainable in the long term, requiring continual input of money and energy; (4) hatcheries are a genetically unsound approach to management that can adversely affect wild populations; (5) hatchery production leads to increased harvest of declining wild populations of salmon; and (6) hatcheries conceal from the public the truth of real salmon decline. I recommend that salmonid management turn from the symptoms to the causes of decline. Overharvest and habitat destruction must be directly addressed in a major, landscape-level effort, on a scale comparable to the hatchery pro-*

Resumen: *La humanidad ha adoptado una actitud arrogante, y en última instancia destinada al fracaso, que pone a la maestría tecnológica por encima de la naturaleza en la vanguardia de nuestro ataque a muchos de los problemas ambientales. Esta arrogancia tecnológica falla en reconocer las limitaciones y las ramificaciones que tienen los intentos en controlar la naturaleza. Un ejemplo de arrogancia tecnológica es el intento fallido de recuperar las pesquerías salmoneras del Pacífico a través de aplicaciones tecnológicas en forma de criaderos. Innumerables stocks de salmones han decrecido precipitadamente en la última centuria como resultado de la sobrepesca y la destrucción masiva de hábitats. Una característica central de los esfuerzos de recuperación ha sido la construcción de numerosos criaderos a los efectos de producir grandes cantidades de peces para sembrar los rios. Esta estrategia esta dirigida a los síntomas pero no a las causas do la declinación (un ejemplo de tecnología a medio camino), puesto que los hábitats permanecen en su mayoria-no aptos para el salmón. Existen por lo menos seis razones por las cuales la estrategia de criaderos está, en última instancia, destinada al fracaso: (1) datos demuestran los criaderos no estan resolviendo el problema; el salmón continúa declinando a pesar de décadas de producción en los criaderos; (2) los criaderos son costosos en su operación, y desvían recursos que podrían ser destinados a otros esfuerzos tales como la restauración del hábitat; (3) los criaderos no son sostenibles en el largo plazo, requiriendo un flujo contínuo de dinero y energía; (4) los criaderos son una estrategia errónea desde un punto de vista genético y pueden afectar adversamente a las poblaciones silvestres; (5) la producción de los criaderos conlleva a un incremento en la cosecha de poblaciones silvestres de salmón; y (6) las estaciones de cría ocultan al público las verdaderas causas de la declinación de las poblaciones silvestres de salmón. Recomiendo que el manejo del salmón*

Paper submitted September 25, 1991; revised manuscript accepted April 9, 1992.

gram, if salmonid fisheries are to remain a part of the ecological, recreational, commercial, and asthetic arenas in the long term.

pase de los síntomas a las causas de la declinación. La sobrepesca y la destrucción del hábitat tienen que ser consideradas en un esfuerzo de grandes magnitudes a nivel paisajístico, y a una escala comparable a la de las campañas de criaderos, a los efectos de que las pesquerías de salmón continúen formando parte de las arenas ecológicas, recreativas y comerciales en el largo plazo.

In the preservation of biological diversity, the use of technology is a last resort. (Conway, 1986)

Perhaps unknowingly, and probably largely through historical momentum, humankind has adopted a short-sighted and ultimately self-defeating philosophy toward nature and our modifications of it. We seem to feel that we can solve any man-induced problem in the natural world, be it habitat destruction, the spread of exotic species, the dumping of toxicants, and even global climate change, through even further modifications using a concerted application of technology. The notion is that we can right virtually any wrong, given enough money, motivation, and innovation. We also seem to believe we can overcome most obstacles presented by nature through similar efforts. Thus, the attitude of many is, if it floods, channelize it; if it's a desert, irrigate it; if it grows too many mosquitoes, drain it or spray pesticides. And if any of those "solutions" cause unanticipated problems, simply apply more technology, perhaps calling up a different type of expertise.

I will call this approach to nature "techno-arrogance," borrowing from Ehrenfeld's (1981) work, *The Arrogance of Humanism*. In that book, he decries the arrogant and prevailing attitude of our species that we can control most of the important aspects of our lives and of nature through technology, irrespective of ultimate and perhaps devastating consequences. Such arrogance fails to recognize or accept limitations and ramifications of the attempted control of our human environment and of nature.

I wish to address a particular conservation problem that is the result of techno-arrogance and that is being "solved" through further application of technology. I refer to the precipitous loss of various salmonid fishes (salmon and anadromous trout) along the Pacific coast of North America. The problem is clear: numerous genetic stocks are being lost quite rapidly, largely through overharvest and a host of environmental effects, including hydropower development, clearcutting, siltation, channel manipulation, water diversion for agriculture, and pollution (Northwest Power Planning Council [NPPC] 1987; National Marine Fisheries Service [NMFS] 1991). For example, the annual return of anadromous salmon and trout to the Columbia River Basin has decreased from an estimated 12–16 million individuals in the 1880s to 2.5 million in the 1980s (NPPC 1987). Furthermore, Nehlson et al. (1991) identified 214 stocks of Pacific salmonids from California, Oregon, Idaho, and Washington that they considered to be of special concern, as they face a high or moderate risk of extinction.

A central feature of the mainstream solution to this debacle is technological: build hundreds of hatcheries to spawn thousands of fish and produce millions of eggs to stock back into the environment. There is a fundamental problem with this approach, however: much of the natural environment remains largely unsuitable for salmonid survival, reproduction, or migration, and continues to deteriorate. Millions of fish are being placed into degraded or even lethal environments and have little chance of survival to maturity and reproduction.

I maintain that a management strategy that has as a centerpiece artificial propagation and restocking of a species that has declined as the result of environmental degradation and overexploitation, without correcting the causes for decline, is not facing biological reality. Salmonid management based largely on hatchery production, with no overt and large-scale ecosystem-level recovery program, is doomed to failure. Not only does it fail to address the real causes of salmonid decline, but it may actually exacerbate the problem and accelerate the extinction process. There are at least six reasons why the current use of hatcheries in salmonid management is counter-productive and should be reconsidered:

First, the data demonstrate that hatcheries are not solving and likely will not solve the problem of salmon decline. Salmonids have continued to decline throughout the Pacific Northwest, despite decades of hatchery production and the expenditure of millions of dollars (see, Federal Register 1991; Hilborn 1991, 1992; Matthews & Waples 1991). It should be obvious that this is not a reasonable solution to the problem, as it clearly is not working. For example, as of this writing, the 1992 oceanic fishing season is in danger of being reduced or eliminated altogether, due to the now alarming decline of both natural and hatchery runs of fish.

Second, hatcheries are enormously costly to run. Severely limited state and federal monies spent on hatcheries could be redirected to local and ecosystem-level habitat restoration, or to prevention of further decline through land purchases. The latter would also benefit other species and maintain ecosystem services in the region (Ehrlich & Mooney 1983).

Third, hatcheries are not sustainable in any long-term sense. They require continual infusion of energy and money, and they are only a piecemeal, year-to-year approach to the problem. Will hatcheries continue to operate in fifty years? Five hundred years? Five thousand years? At some point, for economic or other reasons, hatcheries will cease to operate, and the system will collapse. A long-term, self-sustaining solution is needed.

Fourth, hatcheries are a biologically unsound approach to management that can result in negative genetic changes in natural populations (Allendorf & Ryman 1987, and references therein). The most basic concept in quantitative genetics is that an individual's phenotype reflects genotypic and environmental influences, plus interactions of these factors; hatcheries have never demonstrated the ability to properly manage either the genotype or the environment in any way that reasonably approximates nature. Although hatchery management practices have been changing to accommodate genetic concerns (see, Ryman & Utter 1987), most hatcheries have historically ignored basic principles of population genetics, such as genetically effective population size, and have purposely transferred stocks among subbasins and drainages, disregarding potential local adaptations and site fidelities. This has resulted in the genetic and ecological interaction of native and hatchery stocks, with repeated degradation or loss of native populations (Hindar et al. 1991, and many references therein).

Fifth, hatchery production leads to greater harvest of salmonids, including those from natural populations, resulting in decline of the very stocks being protected. Hilborn (1991) stated that "There is wide concern throughout the Northwest that we have allowed our fisheries harvest rates to match the potential productivity of hatchery stocks, causing wild stocks to be overfished." He continues with an example: "Just north of Puget Sound...harvest rates on Coho Salmon are as high as 95%, sustainable only by the most successful hatchery stocks. The net result of these high harvest rates is that as hatchery production has increased, wild stocks have declined. But the Canadians have no more Coho now than they did 15 years ago. They have swapped hatchery fish for wild fish." Successful hatchery production seems to provide a psychological license to increase harvest rates, which reduces wild stocks, thus defeating the initial purpose of hatcheries.

Sixth, hatcheries are at best a palliative that conceals from the public the real problems and dangers facing a valued resource. This, I believe, is the most serious objection to the hatchery approach. By financially supporting hatchery production as a standard mitigation practice, the hydropower companies and other development projects that are largely responsible for environmental degradation can "buy out" of their moral responsibilities for salmonid losses and habitat destruction by demonstrating their concern for and dedication to the declining resource. They, along with the fishing industry, have created a popular mythology, foisted on managers and the public, that hatcheries are a viable solution to environmental destruction and loss of salmon. This is an insidious deception of the public trust, and this particular mythology must be challenged. The taxpayer and voter is deceived (whether by commission or omission) into believing that technological advances can simultaneously allow environmental degradation and sustained production of an economically, aesthetically, and recreationally valuable resource. The public is also led to believe that their native salmonids are in reasonable condition and in good hands. Consequently, the public is insulated from the reality that their rivers and terrestrial ecosystems are rapidly degrading, and that native fishes, including the salmon they like to catch and eat, are disappearing.

The hatchery approach to salmon conservation is a good example of what Lewis Thomas (1974) has called "halfway technology," a reference to medical practices that treat symptoms rather than eliminate causes of disease. To quote Frazer's (1992) essay on sea turtle conservation,

> Thomas defined halfway technology as "the kinds of things that must be done after the fact, in efforts to compensate for the incapacitating effects of certain diseases whose course one is unable to do very much about. It is a technology designed to make up for disease, or to postpone death." In short, halfway technology does little or nothing to address the cause or the cure of disease. It's what we use to treat a disease when we don't really understand it.

Essentially, halfway technology in salmonid management recognizes the symptom (fewer fish) and treats that symptom (grow more fish) without making a concerted effort to identify and eliminate the underlying causes (environmental destruction and overexploitation). Hatchery rearing of millions of fish does nothing to address the causes of declining populations of fish but simply tries to make more fish available. A medical analogy would be to save the life of a bleeding patient by continual blood transfusions rather than by identifying and stopping the source of bleeding.

Again borrowing from Frazer's work on sea turtle conservation:

> In short, my point is simple. If the cause of the problem (disorientation of adult or hatchling sea turtles) is lighting on the beaches, the solution should address lighting on the beaches. If the cause of mortality is incidental capture in shrimp trawl nets, the solution should address the capture of turtles in nets. Unfortunately, at

present, the problem is too often defined simply as there being too few turtles, and the solution is likely to be viewed as anything that increases the numbers of turtles. This is short-sighted and cannot serve to ensure the cohabitation of the planet by humans and sea turtles in the long run. We would do well to concentrate our efforts on reducing our own negative effects on sea turtle populations instead of attempting to tip the balance between the turtles and their nonhuman predators.

Change "sea turtles" to "salmonids," and "lighting on the beaches" and "shrimp trawl nets" to "hydropower dams" and "siltation from logging," and Frazer has nicely described the Pacific salmonid situation. Halfway technology for salmonids ignores the many causes of decline, focuses on reduced numbers, and invents technological methodologies to increase those numbers. It ignores the fact that, no matter how many millions of eggs or fry or juveniles are "headstarted" in hatcheries, most are doomed in their riverine and oceanic environments: they will not grow and return to the ocean; if they happen to grow and reach the ocean, they will be harvested before their return migration for reproduction; if they should escape harvest and attempt to return, they will not be able to pass the many dams and reservoirs in their way; if they happen to pass them, they cannot reproduce in their natal streams due to siltation, pollution, or other habitat change; if they happen to reproduce, their offspring probably will not make it back to the marine system. Such is the fate of the contemporary Pacific salmonid.

What then do I suggest for the management of salmonid fisheries? We must re-orient recovery efforts from the *symptoms* of decline to the *causes* of decline. The present hatchery-led approach deludes the public (and probably the managers themselves) into thinking we are really doing something beneficial toward restoration of native salmon fisheries in the Pacific Northwest. In fact, those fisheries continue to decline (Matthews & Waples 1991) and many are headed toward extinction (Nehlson et al. 1991), even after and perhaps partly because of decades of intense hatchery production.

Running a multitude of expensive hatcheries while ineffectively dealing with turbines and dams, diversion of water for irrigation, dumping of mine tailings, sedimentation from road building and logging, overgrazing of watersheds, overharvesting, and genetic homogenization of populations, is halfway technology at its worst. Valuable and limited resources are being invested in a dead-end technology, while the causes of the problem continue unabated and even increase. Hatcheries may placate some individuals in the short-term, may please politicians, and may even sustain some fish populations for the present, but they will not rejuvenate a dying system without a great deal of effort being put into the fish's environment. This requires complete reevaluation of our basic philosophies of nature, technology, and resource use. This line of reasoning with respect to salmonids was beautifully developed by Scarnecchia (1988) and will not be further pursued here, other than to borrow a quote: "Rational salmon management is not just a search for technologies: it is a search for values." I contend that hatchery-centered management is based on misguided values.

The only sensible basis for management of salmonid fisheries (or any species in nature) is a clear understanding and acceptance of the evolutionary history of the species and adoption of measures that work within the constraints of that history. We know that anadromous salmonids must have both healthy riverine and marine systems to complete their life cycles. We know that free passage for adults returning upstream and juveniles migrating downstream is essential. We know that high mortality on the open seas will result in fewer adults available for spawning. We know that spawning site fidelity is high, and that changes in river odors may disrupt navigational abilities. These and many other life history facts are the result of thousands to millions of generations of evolutionary history and cannot be easily molded to the needs of man without seriously disrupting the system.

We as a society have adopted a techno-arrogant and self-defeating approach in trying to alter evolutionary history to the short-term benefit of humans. It seems more biologically reasonable to use our intellectual powers to recognize the limitations and liabilities of this approach and instead modify our own behavior through cultural change, rather than to manipulate the natural history of other species. We will not change the nature of salmonid life history quickly enough to allow the fish to respond successfully to polluted, dammed, and overexploited waterways. If we want these species to continue to exist and to be a usable resource, we need to make our polluted, dammed, and over-exploited waterways more compatible with salmonid life history.

I do not wish to imply that hatcheries are all bad; they may in fact be able to play a valid role in recovery of some salmonid populations. However, their purpose and operational philosophy needs to change from production to genetic conservation. Hatcheries can potentially play a critical role in genetic rehabilitation of depleted or genetically degraded stocks if they adopt strict genetic operational guidelines (Meffe 1986; Allendorf & Ryman 1987; Kapuscinski & Jacobson 1987; Kapuscinski et al., in review). Designation of genetic resource reserves for salmonid stocks (Currens et al., in review), along with ambitious genetic rehabilitation of decimated populations in hatcheries, would go a long way toward restoring a once remarkable system. However, no reasonable progress can be made without a concerted effort toward environmental restoration.

I also do not wish to imply that all salmonid managers are enamored with the hatchery response. Many, in fact, are quite opposed to this approach and feel that hatcheries are not the answer and actually contribute to the problem (Hilborn 1992). An active and healthy debate is ongoing in the Pacific Northwest between the pro- and anti-hatchery management groups (Goodman 1990; Martin et al. 1992), the outcome of which should be of interest to all conservationists.

The problems discussed here are by no means restricted to Pacific salmonids. A nearly identical situation occurs with the various species of endangered sea turtles, mentioned above. A litany of endangered vertebrates, plants, mollusks, insects, and other life forms throughout the world speaks volumes to the degraded environments that techno-arrogance has created. Such species can be maintained through halfway technology only for limited periods; money or available space will eventually run out, if genetic decline does not destroy the species first. Reasonable habitat appropriate to the species of concern must be the central goal of any recovery program, as technology can take us only partway, and often down the wrong path at that.

The ultimate outcome of our techno-arrogance is the increasingly intensive and essentially perpetual management of a multitude of species in a world unfit for their natural existence. Besides being prohibitively expensive, it represents techno-arrogance to the point of absurdity. We would do well to remember the centuries-old admonition of Francis Bacon: "Nature is only to be commanded by obeying her." We seldom have obeyed, or even considered, nature with respect to salmonid fishes, and our techno-arrogance has gotten us and salmonids into quite a mess; a large dose of humility is in order to help get us out.

Acknowledgments

Support was provided by contract DE-AC09-76SR00-819 between the U.S. Department of Energy and the University of Georgia. I thank Fred Allendorf, Nat Frazer, Willa Nehlson, Jay Nicholas, Robin Waples, and two anonymous reviewers for their comments, which were not always in agreement with all of the views herein.

Literature Cited

Allendorf, F. W., and N. Ryman. 1987. Genetic management of hatchery stocks. Pages 141–159 in N. Ryman and F. Utter, editors. Population genetics and fishery management. University of Washington Press, Seattle, Washington.

Conway, W. 1986. Can technology aid species preservation? Pages 263–268 in E. O. Wilson, editor. Biodiversity. National Academy Press, Washington, D.C.

Currens, K. P., C. A. Busack, G. K. Meffe, D. P. Philipp, E. P. Pister, F. M. Utter, and S. Yundt. A hierarchical approach to conservation genetics and production of anadromous salmonids in the Columbia River basin. Fishery Bulletin. In review.

Ehrenfeld, D. 1981. The arrogance of humanism. Oxford University Press, New York.

Ehrlich, P. R., and H. A. Mooney. 1983. Extinction, subdivision, and ecosystem services. BioScience 33:248–254.

Federal Register. 1991. Endangered and threatened species; proposed threatened status for Snake River spring, summer, and fall Chinook salmon; proposed rules. 56(124):29542–29554.

Frazer, N. 1992. Sea turtle conservation and halfway technology. Conservation Biology 6:179–184.

Goodman, M. L. 1990. Preserving the genetic diversity of salmonid stocks: a call for federal regulation of hatchery programs. Environmental Law 20:111–166.

Hilborn, R. 1991. Hatcheries and the future of salmon and steelhead in the Northwest. The Osprey 11:5–8.

Hilborn, R. 1992. Hatcheries and the future of salmon in the Northwest. Fisheries 17:5–8.

Hindar, K., N. Ryman, and F. Utter. 1991. Genetic effects of cultured fish on natural fish populations. Canadian Journal of Fisheries and Aquatic Sciences 48:945–957.

Kapuscinski, A. R., and L. D. Jacobson. 1987. Genetic guidelines for fisheries management. Minnesota Sea Grant, University of Minnesota, Duluth, Minnesota.

Kapuscinski, A. R., C. R. Steward, M. L. Goodman, C. C. Krueger, J. H. Williamson, E. Bowles, and R. Carmichael. Genetic conservation guidelines for salmon and steelhead supplementation. Fishery Bulletin. In review.

Martin, J., J. Webster, and G. Edwards. 1992. Hatcheries and wild stocks: are they compatible? Fisheries 17:4.

Matthews, G. M., and R. S. Waples. 1991. Status review for Snake River spring and summer Chinook salmon. National Oceanic and Atmospheric Administration Technical Memorandum NMFS F/NWC-200.

Meffe, G. K. 1986. Conservation genetics and the management of endangered fishes. Fisheries 11(1):14–23.

National Marine Fisheries Service. 1991. Factors for decline. A supplement to the notice of determination for Snake River spring/summer Chinook salmon under the Endangered Species Act. Environmental and Technical Services Division, Portland, Oregon.

Nehlson, W., J. E. Williams, and J. A. Lichatowich. 1991. Pacific salmon at the crossroads; stocks at risk from California, Oregon, Idaho, and Washington. Fisheries 16(2):4–21.

Northwest Power Planning Council. 1987. Columbia River Basin and Wildlife Program. Portland, Oregon.

Ryman, N., and F. Utter, editors. 1987. Population genetics and fishery management. University of Washington Press, Seattle, Washington.

Scarnecchia, D. L. 1988. Salmon management and the search for values. Canadian Journal of Fisheries and Aquatic Sciences 45:2042–2050.

Thomas, L. 1974. The lives of a cell. Notes of a biology watcher. Viking Press, New York.

Notes

"Costs" and Short-Term Survivorship of Hornless Black Rhinos

JOEL BERGER
CAROL CUNNINGHAM

Program in Ecology, Evolution, and Conservation Biology
University of Nevada
Reno, Nevada 89512, U.S.A.

A. ARCHIE GAWUSEB

P.O. Box 220
Khorixas, Kunene Province, Namibia

MALAN LINDEQUE

Etosha Ecological Institute
Okaukuejo via Outjo, Namibia

The devastation of Africa's black rhinos (*Diceros bicornis*) by poaching (Western 1987) highlights the challenges found at the interface of biology and economics (Leader-Williams & Albon 1988). One of the most controversial conservation actions to date has been the removal of horns from black and white rhinos (*Ceratotherium simum*), tactics adopted in both Namibia (Lindeque 1990) and Zimbabwe (Kock 1991). While acts such as these have been debated for more than a decade, the efficacy of "dehorning" remains unknown (Western 1982; Leader-Williams 1989), in part due to an absence of data on (1) rates of horn regrowth and the subsequent monetary worth of the horns; and (2) whether the vulnerability of dehorned rhinos or their neonates is altered with respect to poachers and predators. Here we explore economic and biological implications of horn removal using data gathered in Namibia in 1991, 1992, and 1993. Specifically, we make three points. First, horn regrowth is rapid, averaging nearly 9 cm of total horn per animal per year, a finding that suggests new horns on an average animal are worth $1775–7750 one year after dehorning. Second, because poachers fail to discriminate between large- and small-horned rhinos, recently-dehorned animals may not be immune from poaching. However, neither horned nor hornless rhinos differed in their vulnerability to poachers more than four years after the initial dehorning. Third, for mothers that varied naturally in horn length, calf age and not horn size affected responsiveness to dangerous predators such as lions (*Panthera leo*) and spotted hyenas (*Crocuta crocuta*).

Horns were removed from a total of 20 black rhinos in 1989 and 1991 by personnel of the Namibian Ministry of Wildlife, Conservation, and Tourism in the Kunene Province. This hyperarid region in the northern Namib Desert has lower densities (about $0.002/km^2$) than those reported for any species of rhinoceros (Hitchins & Anderson 1983; Conway & Goodman 1989; Owen-Smith 1989; Dinerstein & Price 1991). We as-

Paper submitted September 22, 1992; revised manuscript accepted March 29, 1993.

sessed possible consequences of dehorning by establishing four study populations: (1) two populations in the Kunene Province, one with horned and one with mostly hornless animals, at the *Namib* and *Escarpment* areas, respectively, which are up to 150 km apart and isolated by a country-length Veterinary cordon fence; and (2) two high-density populations *Pan* and *Vlei*, both with horned animals, about 225 km apart at Etosha National Park. Variation in horn size and growth rates (changes over time, calculated here on a per annum basis) were assessed at night with a Mitutoyo (500 Series) Photogrammetric Digimatic Caliper attached to a 300-mm Nikon lens. This equipment has been used to estimate the size of flukes of killer whales and horns of bison with about 98% accuracy when used within 37 m (Berger & Cunningham, in press). During the day, when rhinos are more dangerous to approach on foot, a Lietz rangefinder with a 500-mm Nikon lens was used at less than 70 m, a distance where the average error of morphological measures is less than 5% (Berger, unpublished).

Horn regrowth is now known for 80% ($n = 16$) of the dehorned black rhinos. Anterior horns grew more rapidly than posterior ones (6.9 cm/yr [SE = 0.7] versus 3.3 (0.4), $t = 5.96, p < 0.0001$). The shape is somewhat tubular and, four years after the 1989 dehorning, all of the horns are still blunt rather than pointed. Regrown adult male anterior ($\overline{X} = 5.3 \pm 0.3$ cm/yr) and posterior (2.3 ± 0.3 cm/yr) horns did not differ from those for females ($6.4 \pm 1.0, 3.0 \pm 0.5; t = 0.71, 0.61$, respectively), but juvenile horns grew faster ($8.9 \pm 1.2, 4.4 \pm 0.5; t = 2.14, 2.88, p < 0.05, p < 0.01$, respectively). On average then, adults and juveniles supported about 8.7 and 13.3 cm of total horn per year, which means that adults carried about 26 cm of horn three years after initial pruning.

Our measures of horn growth represent minimum values because rubbing may obscure maximum horn growth, but the rates we report exceed those from elsewhere (Table 1). Although the extent to which factors such as habitat or population density affect growth rates remains unknown, it is possible that the intrinsic growth of Kunene Province animals is compensatory; that is, dermal proliferation may be stimulated by horn removal.

To estimate the possible monetary worth of regrown horns, we first assessed relationships between horn size and horn mass based on measures of 104 confiscated horns. More than 83% of the variance in mass (Y) was explained by the product (X) of horn length (m along outside curve) and basal circumference (m) with either power ($14.32X^{1.05}$) or linear ($15.49X-.21$) regressions ($p < 0.0001$). Independently, horn height and basal circumference explained 73% and 65% of the variance. Adults had greater basal circumferences than juveniles ($t = 5.81; p < 0.001, n = 27$). The above relationships allow prediction of the regrown horn mass per animal yr^{-1} (anterior and posterior summed) which, on average is 0.54 and 0.33 kg per adult and juvenile black rhino. Since rates of intrinsic horn growth are similar for horned black and white rhinos (Table 1), we must assume that regrowth is also approximately similar for both species.

To convert mass measures to monetary worth, we relied on the 1990 market value of African rhino horns (per kg), which in Asia varies considerably: $3737 in Taiwan (Martin & Martin 1991), $15,205 in Thailand (Vigne & Martin 1991), and $16,240 in Hong Kong (Milliken et al. 1991; for South Korea in 1988 it was $4410). Using the minimum and maximum values, average regrown horn mass per black rhino is worth $1780–7750 by the end of the first year. If unpruned for four years, the value of regrown horn in a small population of 20 adults, assuming no change in market price, would be $142,000–$620,000. Given the minimum existing dollar value per kg, regrown horns are worth more than the cost of removing them in less than 10 months. The inevitable conclusion is that rhinos can be completely devalued only by annual pruning. Nevertheless, other factors are involved, including the costs absorbed by the host countries and nongovernmental organizations.

The estimated total expenses (government and nongovernment combined) of the two Namibian dehorning operations were $49,000, or $2450 per rhino. When the costs of relocating younger animals are subtracted, however, the price of dehorning the 20 animals drops to $1400 per animal. By contrast, the cost of removing the horns of 59 Zimbabwean white rhinos living at high

Table 1. Comparison of mean horn growth rates per yr^{-1} (cm) in intact and dehorned rhinos.

Location	Adults			Juveniles			Comment
	Black Rhinos						
Kaokoveld, Namibia	6.0,	2.7	(11)	8.9,	4.4	(5)	dehorned
Etosha Park, Namibia	0.7,	0.3	(5,3)	—	—		horned; 3 animals estimated at 20+ yrs
Kruger Park, South Africa	5.3	—	(4)	—	—		horned; radio-implant in horn
	White Rhinos						
Kruger Park, South Africa	4.9	—	(6)	—	—		horned; radio-implant in horn

Juveniles as used here are 3–5 years old (values refer to anterior and posterior horns, respectively; sample sizes are in parentheses). Namibian data are from this study (horn growth in Etosha animals was determined by calculating the change in distance in the location of a groove in the horn); Kruger data are from Pienaar et al. (1991).

densities was estimated at $425 per animal (Kock 1991). By including the estimated salaries of participating personnel (which were included in the Namibian but not the Zimbabwean sample), the mean cost becomes $704 per white rhino. Using the Namibian value and assuming a future (and conservative) annual inflation rate of 15%, a scenario can be developed representing extreme horn pruning schedules (one time only and annual) in relation to existing market prices (Fig. 1).

Although the frequency of dehorning exercises will undoubtedly be affected by monetary costs, to evaluate dehorning as a conservation tactic it is also critical to know whether poachers prefer rhinos with larger horns. If poachers discriminate, then rhinos with smaller or regrowing horns would be less at risk, an assumption that alters the economics of how to render rhinos valueless. Although one of us was convicted of felony rhino poaching and feels that preferences for horn size trophies do not exist, it seems prudent to evaluate the issue of poachers' choices statistically. Hence, we used data from our four study populations, in which more than 95% of the horn sizes of adults and subadults were known; the frequency distribution was then compared to that for horns matched by animal and confiscated by enforcement agencies (Fig. 2). Since rhinos carry two unequal-sized horns, a poacher might select for cumulative value rather than for the length of only one horn.

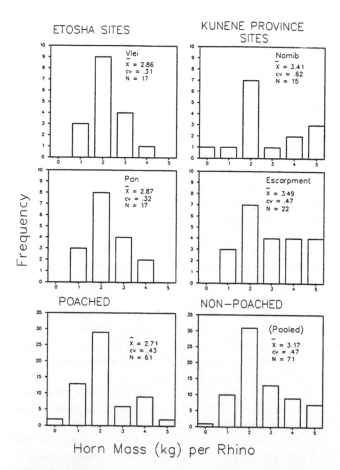

Figure 2. Frequency distribution of horn mass (kg per adult and subadult) in four living Namibian black rhino populations, and that (nonpoached) sample pooled for comparison with a confiscated (poached) sample. Horn mass estimated by method described in the text.

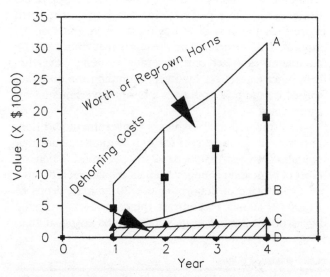

Figure 1. Comparison of 1990 Asian market value (mean $$ per kg per rhino)—dark squares—and range—vertical lines through squares (A—maximum, B—minimum) with two horn-pruning schedules—cross-hatching ((C) annual and (D) one-time only) based on mean regrowth rate of anterior and posterior black rhino horns. Dehorning costs assume annual inflation rate of 15%. Year 0 is the initial year of dehorning at $1400 per rhino.

Hence, the mass of each horn (derived previously; see above) was added so that a frequency distribution of mass per animal could be contrasted between the live and confiscated samples. It was necessary to check first whether differences in the frequency distribution of horn mass varied among live populations; if it did, we could not safely assume that the confiscated horns came from the same population. However, because interpopulation differences were not evident (Heterogeneity Chi Square Analysis using four horn mass categories; $X^2 = 13.47$, df = 9, NS), the data were pooled and contrasted with those from the poached sample (Fig. 2). The lack of differences in mean horn mass ($t = 1.41$, df = 130, NS) or frequency distributions (Kolmogorov-Smirnoff Test; $D = 0.0444$, NS) supported our suspicion that poachers did not discriminate among rhino horn sizes. This information suggests that even animals with small regrown horns are as likely to be killed as animals who have not been dehorned.

Economic points aside, the critical *biological* issue is

whether demographic viability is enhanced by horn pruning? Assessment is necessary at two levels. First, we have not seen interactions between dehorned mothers and dangerous predators. If hornless animals were more vulnerable to predators than were intact rhinos, however, then dehorned animals in the process of regrowing their horns should also be at a greater risk, since their horns would be smaller than those of intact animals. Therefore, we inferred whether hornless rhinos might react differentially to predators by contrasting the behavior of mothers that varied naturally in horn size during 27 witnessed interactions with lions and spotted hyenas in Etosha. Those ($n = 15$) with anterior horns below the mean (31.5 cm) were no more likely to run from predators than were larger-horned mothers (18% versus 19%, respectively; $p = 0.68$, Fisher's Exact Test). What affected maternal responsiveness was calf size; those with neonates younger than about 18 months fled more than those with older calves (50 versus 10%) ($p = 0.056$), though the sample is still too small to remove the possibility that horn size and calf size do not exert independent effects.

Second, the claim that demographic viability improves with dehorning is untenable for at least three reasons: (1) although no dehorned black rhinos have been poached, neither have any horned ones elsewhere in the Kunene Province, even though they outnumber hornless conspecifics by about 4–5:1; (2) all ($n = 3$) calves of hornless mothers that were sympatric with hyenas have died whereas no calves of hornless mothers in predator-free areas or of horned mothers who were sympatric with hyenas and transient lions have perished (Berger & Cunningham 1993); and (3) a country-length veterinary cordon fence isolates the dehorned animals so that a high degree of uncertainty about their long-term population viability exists.

With respect to dehorning as a conservation tactic, it is worthwhile to recall that such radical actions were *not* intended as a long-term biological solution. Therefore, they cannot be judged solely on biological criteria; economic and political environments will always mediate the success of any long-term conservation plan. While poaching of Kunene Province animals has not occurred since 1989, both black and white rhinos are still killed in other regions of northern Namibia, areas where poaching had not previously occurred. Perhaps the overly sensational, massive, and often inaccurate publicity about dehorning, both local and international, may have caused poachers to shift actions away from the Kunene Province to areas where only horned rhinos occur. Ostensibly, this may even have been the purpose of the media campaign. Nevertheless, without adherence to a time frame in which the demographic response of target species can be evaluated and without attention to monies allocated for resource protection (Leader-Williams & Albon 1988; Milner & Leader-Williams 1992), it is imprudent to attempt persuasive arguments about the soundness of dehorning, particularly given the lack of precision in predicting the timing and intensity of poaching epidemics. Finally, political events in 1990 unrelated to the media but associated with Namibian independence may have prompted an overall decline in poaching. Although these may relate in some combination to the cessation of hostilities involving South Africa, Angola, Namibia, and the removal of South African Defense Forces, it is also likely that the development of auxiliary game guard systems (Owen-Smith 1986; Owen-Smith & Jacobsohn 1989) has facilitated rhino conservation. Whatever the cause or causes, it is clear that biological solutions to regional conservation problems will be effective only as far as geography, socio-economics, and cultural influences on behavior are considered.

Acknowledgments

Our research has been graciously sponsored by the American Philosophical Society, the Frankfurt Zoological Society, the Hasselblad Foundation, the National Geographic Society, the Conservation and Research Center of the Smithsonian Institution, Rhino Rescue Ltd., Save the Rhino Trust, the University of Nevada, Wildlife Conservation International (New York Zoological Society), the National Science Foundation, the Office of Forestry, Environment, and Natural Resources of the Bureau of Science and Technology of the U.S. Agency for International Development (under NSF Grant BSR–9025018), and the World Wildlife Fund–USA. We thank Namibia's Ministry of Wildlife, Conservation, and Tourism, W. Gasaway, E. Joubert, M. Jacobsohn, M. Kock, R. and B. Loutit, G. Owen-Smith, J. Rachlow, and an anonymous reviewer for discussions or comments on a prior draft.

Literature Cited

Berger, J., and C. Cunningham. Mating, conservation, and bison. Columbia University Press, New York. In press.

Conway, A. J., and P. S. Goodman. 1989. Population characteristics and management of black rhinoceros *Diceros bicornis minor* and white rhinoceros *Ceratotherium simum simum* in Ndumu Game Reserve, South Africa. Biological Conservation 47:109–122.

Dinerstein, E., and L. Price. 1991. Demography and habitat use by greater one-horned rhinoceros in Nepal. Journal of Wildlife Management 55:401–411.

Hitchins, P. M., and J. L. Anderson. 1983. Reproduction, population characteristics and management of the black rhinoceros *Diceros bicornis minor* in the Hluhluwe/Corridor/Umfolozi Game Reserve Complex. South African Journal of Wildlife Research 13:78–85.

Kock, M. D. 1991. Report on experimental dehorning operation on white rhinoceros (*Ceratotherium simum*) in Hwange National Park, Zimbabwe. Final report. Department of National Parks and Wildlife Management, Harare, Zimbabwe.

Leader-Williams, N. 1989. Desert rhinos dehorned. Nature **340:**599–600.

Leader-Williams, N., and S. D. Albon. 1988. Allocation of scarce resources for conservation. Nature **336:**533–536.

Lindeque, M. 1990. The case for dehorning the black rhinoceros in Namibia. South African Journal of Science **86:**226–227.

Martin, C. B., and E. B. Martin. 1991. Profligate spending exploits wildlife trade in Taiwan. Oryx **25:**18–20.

Milliken, T., E. B. Martin, and K. Nowell. 1991. Rhino horn trade controls in East Asia. Traffic Bulletin **12:**17–21.

Milner-Gulland, E. J., and N. Leader-Williams. 1992. Illegal exploitation of wildlife. Pages 195–213 in T. M. Swanson and E. B. Barbier, editors. Economics for the wild. Earthscan Publications, London, England.

Owen-Smith, G. 1986. The Kaokoveld. African Wildlife **40:**104–115.

Owen-Smith, G., and M. Jacobsohn. 1989. Involving a local community in wildlife conservation. A pilot project at Purros, South-western Kaokoland, SWA/Namibia. Quagga **27:**21–28.

Owen-Smith, N. 1989. Megaherbivores. Cambridge University Press, Cambridge, England.

Pienaar, D. J., A. J. Hall-Martin, and P. A. Hitchins. 1991. Horn growth rates of free-ranging white and black rhinoceros. Koedoe **34:**97–105.

Vigne, L., and E. B. Martin. 1991. Pachyderm **14:**39–41.

Western, D. 1982. Dehorn or not dehorn? Swara **5:**22–23.

Western, D. 1987. Africa's elephants and rhinos: Flagships in crisis. Trends in Ecology and Evolution **2:**343–346.

Lion–Human Conflict in the Gir Forest, India

VASANT K. SABERWAL*
Wildlife Institute of India
Post Box #18
Dehra Dun, 248 001, India

JAMES P. GIBBS
Yale School of Forestry and Environmental Studies
205 Prospect Street
New Haven, CT 06511, U.S.A.

RAVI CHELLAM
Wildlife Institute of India
Post Box #18
Dehra Dun, 248 001, India

A. J. T. JOHNSINGH
Wildlife Institute of India
Post Box #18
Dehra Dun, 248 001, India

Abstract: *Asiatic lions (*Panthera leo persica*) now occur in the wild only as a small population (about 250 animals) within a single reserve, the Gir forest in Gujarat state in western India. Persistent attacks by lions on humans hinder support among local peoples for lion conservation. We analyzed 193 attacks by lions on humans and conducted interviews with 73 villagers to identify the spatial, temporal, and social factors associated with lion-human conflict in the region. An average of 14.8 attacks by lions and 2.2 lion-caused deaths occurred annually between 1978 and 1991, and most attacks (82%) occurred on private lands outside the forest reserve. A drought in 1987–1988 precipitated an increase in rates of conflicts (from 7.3 to 40.0 attacks/year) and in the proportion of attacks that occurred outside the reserve (from 75% to 87%). The spatial pattern of lion attacks could not be distinguished from random before the drought, whereas attacks were clustered after the drought in village subdistricts with a higher ratio of revenue land to*

---

* Current address: Yale School of Forestry and Environmental Studies, 205 Prospect St., New Haven, CT 06511. Address correspondence to J. P. Gibbs.
Paper submitted February 17, 1993; revised manuscript accepted July 30, 1993.

Conflicto entre leones y humanos en el bosque de Gir, India

Resumen: *Los leones Asiáticos (*Panthera leo persica*) se encuentran en estado salvaje sólo como una pequeña población (ca. 251 animales) dentro de una única reserva, en el bosque de Gir en el estado de Gujarat, al oeste de India. Los persistentes ataques de leones a humanos obstaculizan el apoyo de la población local para la conservación de leones. Nosotros analizamos 193 ataques de leones a humanos y conducimos encuestas a 73 pobladores locales para identificar los factores espaciales, temporales y sociales asociados con los conflictos entre leones y humanos en la región. Un promedio de 14.8 ataques por leones y 2.2 muertes causadas por leones ocurrieron anualmente entre 1978–91, la mayoría de los ataques (82%) ocurrieron en tierras privadas fuera de la reserva del bosque. Una sequía en 1987–91 precipitó un incremento en la tasa de los conflictos (de 7.3 a 40.0 ataques/año) y en la proporción de ataques ocurridos fuera de la reserva (de 75% a 87%). El patrón espacial de ataque de los leones no pudo ser distinguido del aleatorio antes de la sequía, mientras que los ataques después de la sequía pudieron ser agrupados en sub-distritos poblacionales con una relación tierras explotables: bordes de foresta más alta y aquellos más cercanos a sitios dónde los leones*

forest edge and those closer to sites where lions were formerly baited for tourist shows. Subadult lions were involved in conflicts in disproportion to their relative abundance. A majority of villagers interviewed expressed hostile attitudes toward lions owing to the threat of personal injury and economic hardship (mainly livestock damage) posed by lions. The escalation in lion-human conflict following the drought probably resulted from a combination of increased aggressiveness in lions and a tendency for villagers to bring their surviving livestock into their dwellings. Dissatisfaction with the government's compensation system for lion-depredated livestock was reported widely. The current strategy for coping with problem lions—that is, returning them to areas in the Gir forest already saturated with lions—is inadequate, as indicated by the sharp increase in lion-human conflict since 1988. Prohibiting lion baiting for tourist shows, consolidation of reserve boundaries, and implementation of a more equitable and simpler system for compensating villagers for livestock destroyed by lions could provide short-term alleviation of lion-human conflict in the region. Long-term alleviation may entail reducing the lion population by relocating or culling lions.

eran cebados para espectáculos turísticos. Los leones subadultos se vieron envueltos en conflictos en forma desproporcionada a su abundancia relativa. La mayoría de los pobladores encuestados expresaron su hostilidad hacia los leones debido al peligro de daños personales y pérdidas económicas (principalmente daños en el ganado) causados por los leones. La escalda de conflictos entre leones y humanos que siguió a la sequía, es probablemente el resultado de la combinación del incremento en la agresividad de los leones y la tendencia de los pobladores a llevar el ganado sobreviviente a sus moradas. Se reportó ampliamente una desatisfacción con respecto a la compensación que el gobierno brinda por la predación por leones del ganado. La presente estrategia para enfrentar los leones problemáticos, consistente en retornarlos a las áreas en el bosque de Gir que ya están saturadas con leones, es inadecuada, tal como lo indica el marcado incremento en el aumento de conflictos entre leones y humanos desde 1988. El prohibir que los leones sean cebados para los espectaculos turísticos, la consolidación de los límites de la reserva y la implementación de un sistema más equitativo y simple para compensar a los pobladores por el ganado perdido a causa de los leones puede mitigar en el corto plazo el conflicto entre leones y humanos en la región. Una mitigación a largo plazo podría implicar la reducción de la población de leones a través de la relocalización o eliminación de los leones menos aptos.

Introduction

Conflicts between wildlife and humans in India are escalating owing to increasing human population, extensive loss of natural habitats, and, in some regions, increasing wildlife populations due to successful conservation programs (Rodgers 1989; Gadgil 1992). Conflicts are most acute when the species involved is critically imperiled while its presence in an area poses a significant threat to human welfare. Such is the case in Gujarat state in western India, where about 250 Asiatic lions (*Panthera leo persica*), members of a subspecies once distributed widely between northern Greece and western Bengal (Joslin 1973), are now confined to a single 1400-km^2 reserve that comprises the Gir Forest. Cultivated lands extend to the edge of the protected area, and more than 400,000 people inhabit the subdistricts that adjoin the reserve (Anonymous 1983). Although protection of the reserve lands is relatively secure, prospects are uncertain for the survival of the Gir lions given the multitude of increasing threats now facing the population, including habitat alteration, occasional poaching, and loss of genetic variation through prolonged inbreeding (Oza 1974; Joslin 1984; O'Brien et al. 1987*b*; Wildt et al. 1987; Chellam & Johnsingh 1993). Additionally, the recent increase in attacks by lions on humans has been alarming (Chellam & Johnsingh 1993). Continued lion attacks could lead to political repercussions from the affected human population that may greatly complicate efforts to conserve Asiatic lions in the wild.

We examined attacks by lions on humans that occurred between 1978 and 1991 in an attempt to identify spatial and temporal factors associated with lion-human conflict. We describe rates of conflict, distances from the reserve at which conflicts occurred, characteristics of problem lions, and recent drought-related changes in rates and patterns of lion attacks. We also describe the attitudes and perceptions of villagers who lived near the Gir forest toward the lions and their management. Finally, we suggest means of alleviating lion-human conflict in the region.

Methods

Study Area

The Gir forest is located in the Junagadh district of Saurashtra, Gujarat State, in northwestern India (20–23° N latitude, 70–72° E longitude). Gir lies in a semiarid region with a highly seasonal climate where 65–100 cm of precipitation is received annually, typically between June and September. Annual maximum and minimum temperatures recorded are 46° and 7° C. Vegetation is comprised mainly of tropical, mixed dry deciduous forest, dominated by teak (*Tectona grandis*) and grassy scrubland (Chellam & Johnsingh 1993).

The Gir forest is comprised of the Gir Wildlife Sanctuary (1153 km²), constituted in 1965 and expanded in 1974, and Gir National Park (259 km²), constituted in 1975—otherwise known together as the Gir protected area. Although the Gir forest has been reduced to about a third of its former size over the last century, the protected area now covers most of the extant forest. The Gir Wildlife Sanctuary supports about 2200 buffalo-herding pastoralists (the Maldharis), but it is otherwise unoccupied by permanent settlers.

The lion population was estimated at 284 animals in 1990 (Gujarat Forest Department, unpublished data), an approximately 16-fold increase from a low of about 18 animals in 1893 (Wynter-Blyth & Dharmakumarsinhji 1950). Densities of one lion per 5–7 km² now occur in the forest (Chellam & Johnsingh 1993), which is comparable to the highest densities reported for lions in East Africa (Rodgers 1974; Elliot & Cowan 1978). The lions feed primarily on wild ungulates (65% of 142 kills investigated, mainly Chital [*Axis axis*]) and livestock (35% of kills; Chellam & Johnsingh 1993). The lions have been afforded strong protection measures since the late 1800s, originally by local royalty (the Nawab of Junagadh) and presently by the Indian government (Gujarat Forest Department). Killing lions for any reason is, at present, strictly prohibited.

Analysis of Lion Attacks

Our analysis focused on records of lion attacks obtained from the Gujarat Forest Department for the 13-year period between April 1978 and March 1991. We considered a lion-human conflict to be any interaction between a lion and human that resulted in physical injury to a human. Information in the records available to us ($n = 193$) included the date and location of each attack, whether an attack involved injury or death, and, in cases of death, whether the corpse was fed upon by a lion.

To examine the spatial pattern of lion attacks, we superimposed a grid composed of square cells equivalent to 100 km² in land area over a map of the Gir reserve and portions of the surrounding landscape. We then tallied the number of lion attacks for each grid cell. Observed attack frequencies per cell were compared with frequencies expected from a Poisson distribution of the same number of attacks to infer whether the observed pattern of attacks deviated from one expected at random (Brower & Zar 1984:135–143). Grid cells 100 km² in area were used because they provided an average of one to two attacks per cell, which is recommended for analyses of dispersion (Brower & Zar 1984:135).

For the 12 subdistricts in which at least one attack occurred during the study period, we correlated the subdistrict-specific intensity of lion-human conflict (attacks/year/100 km²) with (1) human density in each subdistrict (individuals/km²; Anonymous 1983), (2) the ratio of the length of edge of the Gir forest boundary to the amount of revenue (private) land in a subdistrict (km forest edge/100 km² revenue land), and (3) proximity to former lion-baiting shows. Lion show proximity was estimated as the average distance of points spaced at 5-km intervals throughout each subdistrict to the nearest historical lion-bating site, as recorded by Spillett (1968). All relationships among variables were assessed with Spearman rank correlation (Zar 1984).

Information on the age (subadult versus adult) and sex of lions involved in attacks on humans was obtained during captures that we witnessed of 22 problem lions.

Data on drought incidence in the region were obtained from monthly rainfall records collected over a 98-year period (1892–1990) at Veraval, located about 25 km from the study area.

Survey of Attitudes and Perceptions of Local Villagers

We visited 56 villages in subdistricts abutting the Gir forest between March 1 and April 25, 1990, and interviewed 73 villagers concerning their attitudes about and perceptions of lions. All villagers interviewed were male, and most were the headmen (sarpanch) of particular villages. Individuals were asked to respond to a standardized set of questions concerning (1) the frequency of lion sightings on their lands, (2) problems villagers encountered owing to the presence of lions in their villages, (3) villager's perceptions of the Gir forest generally and lions specifically, and (4) villager's opinions on the government's compensation system for livestock depredated by lions.

Results

Temporal Patterns of Lion-Human Conflict

A total of 193 attacks by lions on humans (14.8/year) and 28 lion-caused human mortalities (2.2/year) occurred between 1977 and 1991 (Table 1). Most attacks involved maulings; on average, 14.5% of attacks resulted in human mortality. A drought in 1987–1988, which our analysis of rainfall records indicated was the driest year in the region since 1918, profoundly altered the frequency of lion-human conflict. Prior to the drought, lion attacks averaged 7.3/year, but they increased sharply to 40.0/year following the drought. Although rates of lion-caused mortality also increased following the drought (from 0.8 to 6.7 deaths/year), the proportion of attacks resulting in human mortality was not different before versus after the drought ($\chi^2 = 1.6$, 1 df, $p = 0.20$). Following the drought, lions began to feed off corpses of humans they had killed (in 7 of 20 mortalities recorded), a behavior not observed over the 10-year period prior to the drought during which eight lion-caused mortalities were recorded.

A marked seasonality was observed in the occurrence of lion attacks during two of three years following the

Table 1. Human injuries and deaths caused by lions in the vicinity of the Gir forest, Gujarat State, India, 1978–1991.

Year**	Injured	Killed	Eaten
1978–1979	1	1	0
1979–1980	2	1	0
1980–1981	8	0	0
1981–1982	11	2	0
1982–1983	6	0	0
1983–1984	4	1	0
1984–1985	14	3	0
1985–1986	9	0	0
1986–1987	4	0	0
1987–1988	6	0	0
1988–1989	38	6	4
1989–1990	33	7	2
1990–1991	29	7	1
Total	165	28	7

*A severe drought began in March 1986, intensified in 1987, and ended in June 1988.
Source: Gujarat Forest Department.
**April 1 to March 31.

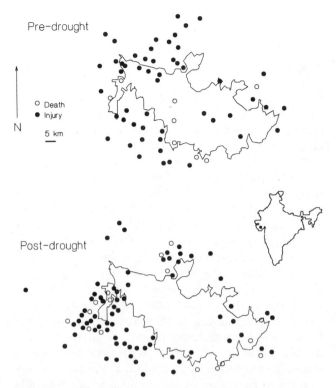

Figure 1. Lion-human conflict in the vicinity of the Gir forest, Gujarat State, India, 1978–1991 (source: Gujarat Forest Department). Pre-drought data are from 1978–1988, post-drought data are from 1988–1991. Only those locations of lion-human conflict that are exactly known are mapped. The location of the Gir forest is indicated by a dot on the map of India.

drought. A disproportionate number of attacks occurred during the monsoon months (June-September) in 1988 (56.8% of 44 attacks, $p = 0.032$) and 1989 (52.5% of 40 attacks, $p = 0.070$) but not in 1990 (38.9% of 36 attacks, $p = 0.62$) in comparison to that expected by a random distribution of attacks across months (33.3% during the fourth-month monsoon period).

Spatial Patterns of Lion-Human Conflict

Most lion attacks (82%) occurred outside the boundaries of the Gir protected area (Fig. 1). Lions ranged long distances from the reserve prior to attacking humans; 50% of attacks occurred within 9.5 km of the reserve boundary, 75% within 18.5 km, 95% within 32 km, and 100% within 73 km (median distance = 9.7, $n = 138$). There was no difference in the median distance from the reserve at which lion attacks occurred before versus after the drought (Kruskal-Wallis Statistic = 0.67, $p = 0.41$). However, a larger proportion of attacks occurred outside the reserve after the drought (87%) than before (75%; $\chi^2 = 3.41$, 1 df, $p = 0.065$).

Prior to the drought, the spatial distribution of lion attacks (injuries and deaths combined) could not be distinguished from a random distribution of incidents in the vicinity of Gir ($\chi^2 = 4.89$, $p = 0.18$), whereas attacks following the drought were highly clustered ($\chi^2 = 27.9$, $p < 0.001$). Attack clusters occurred primarily in the subdistricts of Talala, Maliya, and Visavadar. Among the 12 subdistricts in which at least one lion attack occurred during the study period, subdistricts with a higher ratio of revenue land to forest edge ($r_s = 0.73$, $p < 0.01$) and those closer to locations of former lion baiting sites ($r_s = -0.50$, $0.05 < p < 0.1$) had an increased incidence of lion-human conflict. These factors contributed independently to variation among subdistricts in attack intensity, because ratio of revenue land area to forest edge and proximity of lion-baiting sites were uncorrelated ($r_s = -0.26$, $p > 0.10$). Subdistrict-specific attack rates were unrelated ($p > 0.10$) to the density of the rural human population ($r_s = -0.056$).

Characteristics of Problem Lions

We observed three subadult males, six subadult females, one adult male, and four adult female lions involved in lion-human conflicts. In comparison to the demographic composition of the Gir lion population in 1979, the most recent census in which lions were classified by age and sex (Gujarat Forest Department, unpublished data), subadult lions were represented disproportionately in conflicts with humans ($\chi^2 = 15.53$, 1 df, $p = 0.006$). Among adults, however, males and females were involved in conflicts in proportion ($\chi^2 = 1.28$, 1 df, $p = 0.25$) to their abundance in the 1979 population.

Attitudes and Perceptions of Local Villagers

Of the 73 villagers interviewed, 61% expressed hostility toward the presence of lions near their villages owing to the threat of injury and of economic loss (mainly livestock damage) posed by lions. An annual loss of at least five livestock animals to lions from each village was reported by most villagers (92%). Also, many villagers (61%) agreed that the presence of lions compelled the villagers to curtail their activity after sunset, which reportedly imposed hardship on the villagers because electricity, diverted to factories during daytime, was available to power tube wells for irrigating fields only at night. Villagers also reported having to move about in groups of four or five to protect themselves against lions while irrigating their fields at night, and they considered this an inefficient use of their time. The majority of villagers (78%) kept their livestock within compound walls or in thorn enclosures to protect the animals from lions, and 22% reportedly underwent the discomfort of sharing their houses with livestock at night.

Many respondents (62%) reported that poorer villagers, who typically left their animals unprotected outside at night, lost substantially more livestock to starvation and lion depredation during the drought than did wealthier villagers, who could afford to build sheds and other structures to protect their animals. An aggressiveness in lions following the drought that was not noticeable prior to the drought was widely reported (76%). Reports of aggressiveness included lions entering villages with increased frequency, jumping compound walls to gain access to livestock, and trying to enter villagers' houses in search of livestock.

Dissatisfaction with the system used by the government to compensate villagers for depredated livestock was widely reported among villagers. Most respondents (86%) complained that compensation levels for livestock kills were too low in comparison to the purchase price of replacement animals. Furthermore, 81% stated that they did not file for compensation upon losing livestock to lions because of the procedural problems associated with filing such claims. Problems cited included excessive travel to report losses, the likelihood that an official would not be available to register a report of livestock depredation within the mandatory reporting period (less than 24 hours after the livestock death), subjective assessments by officials of the worth of depredated livestock, and difficulties associated with receiving payments for settlements.

Discussion

This study highlights problems associated with the conservation of large, carnivorous mammals through protection of nature reserves of limited area in regions densely inhabited by humans. A significant but unknown portion of the Gir lion population roams or may even reside outside the reserve owing to the shortage of unoccupied habitat within the reserve. For example, a radio-collared male has been located regularly at distances up to 40 km from the Gir forest over a two-year period (Ravi Chellam, unpublished data), and problem lions captured in agricultural areas have been known to return quickly to areas outside the reserve following their release in the Gir forest (V. K. Saberwal, personal observation). Lands outside the protected area are intensively cultivated and lack populations of native ungulates. Hence, livestock are the only available prey for lions living in these areas, a situation that readily brings lions into direct conflict with the human population also occupying these areas. The distances from the reserve at which lion-human conflicts were regularly observed (up to 30 km; Fig. 1) and distances from the reserve that marked lions habitually roam (up to 40 km; R. Chellam, unpublished data) suggest that efforts to manage the lion population should focus as much on agricultural areas abutting the reserve as on the reserve itself.

The recent spate of lion attacks mirrors an earlier episode in 1901–1904 that followed the drought of 1899–1900 and that involved the deaths of at least 66 people (Wynter-Blyth & Dharmakumarsinhji 1950). It remains unclear what drought-related mechanism precipitates sharp increases in lion-human conflict in the region. Peripheral settlements of poorer villagers probably absorbed the majority of lion depredations prior to the drought. Differential loss of their livestock during the drought, however, likely forced lions to penetrate farther into villages in search of prey. The few livestock that remained in villages following the drought were assiduously protected within compounds and houses. Encounters in these confined spaces between humans and lions could have lead to the increased incidence of conflicts.

The elevated incidence of lion attacks on humans during the monsoon can be partly explained by increased overlap in the daily activity patterns of lions and humans. During the monsoon, skies are overcast and temperatures remain cool (about 32°C). These conditions permit lions, typically most active at night, to be active throughout the day when humans are also most active. Denser vegetation during this season may also increase the likelihood of accidental encounters between humans and lions and increase the opportunity for lions to ambush humans and livestock.

Prevalence of forest edge in a subdistrict was probably an important factor underlying the spatial pattern of attacks, because the likelihood that lions encounter livestock and humans would be greatest along fringes of the protected area. Projections of the Gir protected area, particularly those at Visavadar and Devalia (Fig. 1), may even act as funnels that direct wandering lions out of the forest and onto revenue lands.

Habituation of lions to humans at baiting sites may also have affected the spatial pattern of lion-human conflicts. Baiting involved attracting lions on a regular basis to sites where buffalo calves were tethered, thereby facilitating viewing opportunities for large groups of tourists. Those lions habituated to large numbers of tourists would likely have been less intimidated by attempts by villagers to drive them off livestock kills, and hence they would have been more likely to have been involved in conflicts with humans over livestock in areas near former baiting sites. Baiting lions for tourists was halted in 1987, coincident with the occurrence of the drought and the sharp rise in lion-human conflict.

The disproportionate tendency for subadult lions to be involved in lion-human conflicts may reflect social dynamics occurring in the lion population within Gir. The subadult lions involved in attacks are likely those displaced from or dispersing from their natal territories (Schaller 1972; Bertram 1973; Rudnai 1979; Hanby & Bygott 1987). Wandering subadults would be unlikely to obtain territories within Gir owing to the unusually high density of lions that occurs in the protected area, and they would be displaced to poorer quality habitats along the fringes or outside of the protected area, where there are no wild ungulates to prey upon and where the likelihood of interaction with livestock and humans is much higher. Similarly, many livestock-raiding lions in Namibia (Stander 1990) and man-eating tigers in Nepal (McDougal 1987) and India (Seidensticker et al. 1976) are subadult individuals displaced from their former home-ranges.

The recent escalation in lion-human conflict, the advent of direct predation by lions on humans, and the pervasiveness of hostility among local villagers toward lions represents an alarming shift in the intensity and dimensions of lion-human conflict in Gujarat State, which could greatly complicate efforts to conserve the remaining wild population of Asiatic lions. Furthermore, the disproportionate participation of subadult lions in lion-human conflicts and the high density of the lion population within the Gir forest may be undermining the primary strategy used at present to resolve lion-human conflicts—that is, capturing problem animals outside the park and subsequently releasing them within the park. Subadults likely are leaving the protected area because resources available to them there are insufficient. Returning displaced animals to unsuitable habitats within the forest may only perpetuate rather than resolve lion-human conflict in the region.

Our study indicates that the following management actions could alleviate lion-human conflict in the region over the short-term. (1) A permanent prohibition should be placed on lion-baiting activities because frequency of lion-human conflict and proximity to former baiting sites appear to be related. (2) Park boundaries could be consolidated to minimize forest edge to revenue land ratios, which were associated with elevated rates of lion attacks in certain subdistricts. For example, many attacks have occurred in the vicinity of the Visavadar and Devalia projections of Gir, lands that could potentially be exchanged for revenue lands that intrude into the protected area if original vegetation could be effectively restored. (3) Local support for lion conservation, currently at an alarmingly low level, could be boosted by creation of a more equitable and simpler system for compensating villagers for livestock lost to lion depredation.

The following management actions should be considered to reduce lion-human conflict in the region over the long-term. (1) Translocation of lions to other areas of suitable habitat within the former range of the Asiatic lion should be examined critically (see Anderson 1981; Panwar & Rogers 1982), a process that could lead to establishment of a second population of Asiatic lions and thereby increase the prospects for the Asiatic lion's survival in the wild. (2) Problem lions could be sold internationally to zoos to replace current captive stocks of Asiatic lions, the majority of which appear to be hybrids between African and Asiatic subspecies rather than genetically pure representatives of the Asiatic lion (O'Brien et al. 1987a). (3) Problem lions could be culled. Options for culling include selling the rights to shoot problem lions outside the protected area to sport-hunters, a program that if structured properly could potentially generate valuable revenue for local peoples and lion management programs. Overt shooting of problem lions by state officials also could help rally support for lion conservation among local peoples (Mishra et al. 1987). Implementation of any culling program must be contingent upon the collection of detailed information on the age and sex structure and dynamics of the Gir lion population; such information is currently lacking and would facilitate the application of models to assess the potential effects of a culling program on the demography and genetic structure of the lion population. (4) A system should be developed to redistribute to local peoples a significant portion of revenues generated by sale of the lions to zoos, shooting permits bought by sport-hunters and ecotourism. (5) To motivate people to live in an area along with lions, a program could be enacted in which financial incentives are provided by the government to all residents of subdistricts that abut the Gir forest, possibly in proportion to lion densities or lion-associated damages occurring in a given subdistrict. This program should complement, not replace, the existing program for compensating individuals harmed by lions. (6) Lastly, continued examination of the perceptions and opinions of villagers living near the Gir forest will be necessary for perpetuating an effective and socially acceptable strategy for conservation of the Asiatic lion.

Acknowledgments

We gratefully acknowledge the Gujarat Forest Department for its support and for providing access to its data.

Funding for the study was provided by the Wildlife Institute of India, the United Nations Education and Science Organization, and the INLAKS Foundation. H. S. Panwar and W. A. Rodgers were instrumental in initiating the study. Dhanna Lakshman, Mohammad Juma, and Karsan Nanji provided valuable field assistance, and Rajesh Thapa, Manoj Aggarwal, and Lek Nath Sharma helped prepare the manuscript. M. L. Hunter and M. Gadgil reviewed the manuscript.

Literature Cited

Anderson, J. L. 1981. The re-establishment and management of a lion *Panthera leo* population in Zululand, South Africa. Biological Conservation **19:**107–117.

Anonymous. 1983. Census of India, 1981, series 5: Gujarat. Government Central Press, Delhi, India.

Bertram, B. C. R. 1973. Lion population regulation. East African Wildlife Journal **11:**215–225.

Brower, J. E., and J. H. Zar. 1984. Field and laboratory methods for general ecology. 2nd edition. Wm. C. Brown, Dubuque, Iowa.

Elliot, J. P., and I. M. Cowan. 1978. Territoriality, density and prey of the lion in Ngorongoro Crater, Tanzania. Canadian Journal of Zoology **56:**1726–1734.

Gadgil, M. V. 1992. Conserving biodiversity as if people matter: A case study from India. Ambio **21:**266–270.

Hanby, J. P., and J. D. Bygott. 1987. Emigration of subadult lions. Animal Behavior **35:**161–169.

Joslin, P. 1973. The Asiatic lion: a study of ecology and behavior. Ph.D. Thesis. University of Edinburgh, Edinburgh, U.K. 249 pp.

Joslin, P. 1984. The environmental limitations and future of the Asiatic lion. Journal of the Bombay Natural History Society **81:**648–664.

McDougal, C. 1987. The man-eating tiger in geographical and historical perspective. Pages 435–448 in R. L. Tilson and U. S. Seal, editors. Tigers of the world. Noyes Publications, Park City, New Jersey.

Mishra, H. R., C. Wemmer, and J. L. D. Smith. 1987. Tigers in Nepal: Management conflicts with human interests. Pages 449–463 in R. L. Tilson, and U. S. Seal, editors. Tigers of the world. Noyes Publications, Park City, New Jersey.

O'Brien, S. J., P. Joslin, G. L. Smith, R. Wolfe, N. Schaffer, E. Heath, J. Ott-Joslin, P. P. Rawal, K. K. Bhattacharjee, and J. S. Martenson. 1987a. Evidence for African origins of founders of the Asiatic lion Species Survival Plan. Zoo Biology **6:**99–116.

O'Brien, S. J., J. S. Martenson, C. Packer, L. Herbst, V. de Vos, P. Joslin, J. Ott-Joslin, D. E. Wildt, and M. Bush. 1987b. Biochemical genetic variation in geographic isolates of African and Asiatic lions. National Geographic Research **3:**114–124.

Oza, G. M. 1974. Conservation of the Asiatic lion: Now limited to Gujarat State. Biological Conservation **6:**225–227.

Panwar, H. S., and W. A. Rodgers 1982. The re-introduction of large cats into wildlife protected areas. Indian Forester **112:**939–944.

Ravi Chellam and A. J. T. Johnsingh. 1994. Management of Asiatic lions in the Gir forest, India. Symposia of the Zoological Society of London 65.

Rodgers, W. A. 1974. The lion (*Panthera leo,* Linn.) population of the eastern Selous Game Reserve. East African Wildlife Journal **12:**313–317.

Rodgers, W. A. 1989. Policy issues in wildlife conservation. Indian Journal of Public Administration **35:**461–468.

Rudnai, J. 1979. Ecology of lions in Nairobi National Park and adjoining Kitespela Conservation Unit in Kenya. African Journal of Ecology **17:**85–95.

Seidensticker, J., R. K. Lahiri, K. C. Das, and A. Wright. 1976. Problem tiger in the Sunderbans. Oryx **13:**267–273.

Schaller, G. B. 1972. The Serengeti lion. The University of Chicago Press, Chicago, Illinois.

Spillett, J. J. 1968. A report on wild life surveys in south and west India, November-December 1966. Journal of the Bombay Natural History Society **65:**1–46.

Stander, P. E. 1990. A suggested management strategy for stock-raiding lions in Namibia. South African Journal of Wildlife Research **20:**37–43.

Wildt, D. E., M. Bush, K. L. Goodrowe, C. Packer, A. E. Pusey, J. L. Brown, P. Joslin, and S. J. O'Brien. 1987. Reproductive and genetic consequences of founding isolated lion populations. Nature **329:**328–331.

Wynter-Blyth, M. A., and K. S. Dharmakumarsinhji. 1950. The Gir forest and its lions, Part II. Journal of the Bombay Natural History Society **49:**456–470.

Zar, J. H. 1984. Biostatistical analysis. 2nd edition. Prentice-Hall, Englewood Cliffs, New Jersey.

Ecotourism: New Partners, New Relationships

JOAN GIANNECCHINI
New York University*
New York, NY 10003, U.S.A.

"Ecotourism" is the buzzword being used to describe a new partnership among the travel industry, tourists, and the conservation community to promote and enhance environmental sensitivity through responsible travel (Ehrenfeld 1992). Each of these partners brings to the arena their own hopes and aspirations for ecotourism. The travel industry, aided by the upsurge in nature travel, expects soon to become the largest industry in the world. Tourists, with leisure time and money to spend, want to escape from their urban existence into the beauty, simplicity, and adventure of nature. The conservation community hopes to promote their programs for the protection and maintenance of our dwindling natural areas.

While each of these entities shares a concern for the environment, the stakes are particularly high for the conservation community. As shepherds of the world's parks and protected areas, the primary destination of most ecotravel, conservationists must be concerned about the impact of the burgeoning numbers of visitors to the wilderness. These visitors represent the potential to generate revenue for the support of protected areas, as well as to destroy them through overuse and abuse. It will be a major challenge to the conservation community to participate with the travel industry and the world's tourists to plan and control the growth and development of ecotourism in a manner that will conserve and protect the planet's natural assets.

The size and strength of the travel industry insures its position as a powerful and influential partner in determining the direction of ecotourism. By any measure, the growth of this industry is staggering. At the Ecotourism Management Workshop held at George Washington University in June 1991, Dr. Donald Hawkins, Director of the George Washington University World Tourism Organization, stated that the Travel Industry World Yearbook reported world tourism receipts in 1986 totalling $2 trillion. These receipts were expected to increase to nearly $3 trillion in 1996, making travel and tourism the largest single industry in the world.

At any given moment, 300 million people worldwide are in the process of traveling (Warpole 1991). Projections indicate that spending on international tourism will increase at a rate of 4.5–5% per year through 1999, and faster in developing countries (Economist 1989a). Dr. Hawkins estimated that 20–25% of leisure travel (leisure travel comprises 55–60% of the world travel market) could be defined as nature tourism, or broadly defined as ecotourism. He indicated that thus far it had not been possible to quantify that portion of the pleasure tourism expenditure which could be attributed to "ecotourism" per se, in part because a strict definition of ecotourism had not been agreed upon. However, available statistics indicated that, as a result of a growing interest in personal improvement and environmental awareness, it was the strongest growth sector in the industry overall.

A further analysis of tourist destinations confirms the increased interest in nature travel: 11.5 million U.S. citizens took trips with environmental themes in 1990 (Weiner 1991). World Wildlife Fund's recent poll of international travelers at Latin American airports found that 47% cited nature destinations as an important factor in vacation planning (Brooke 1991). A 1990–1991 survey by US Travel Data Center revealed that 77% of U.S. citizens surveyed considered recreation a priority in their lives, and 36% had taken part in an organized effort to improve their environment (New York Times 1991). Travel to exotic destinations has also risen dramatically.

---
*Address correspondence to the author as follows: 310 Crestview Drive, Ukiah, California 95482, U.S.A.
Paper submitted May 4, 1992; revised manuscript accepted September 30, 1992.

During 1988, arrivals increased by 24% in Hong Kong, 22% in Thailand, 35% in Malaysia, and 31% in Australia (Economist 1989b).

It is evident from these statistics that protected and pristine areas, in which moderately funded and staffed conservation organizations have traditionally operated, will be increasingly scrutinized for development by this new, profit-driven, international partner, one capable of becoming a dangerous competitor for any resource, particularly those which are scarce or fragile.

Both the conservation community and the tourist must understand that, for the travel industry, ecotourism is not merely a new form of travel "simpatico" with the environment. It is a powerful marketing device currently being employed to develop and sell an aspect of specialty travel. Conservation ideals, including sustainable use of resources and development, are shared only in part by the tour industry. Their customary goal of quick optimum profits is in direct conflict with long-range goals of protection and conservation. This does not mean that the only, or even the primary, relationship between the tourist industry and conservationists must be adversarial. But it does mean that whatever laudable, environmentally sound policies and goals the industry articulates, they will remain subsidiary to the demand for profit. Therefore, if the tourist industry becomes the principal force in the development of ecotourism, it will almost certainly be detrimental to long-range environmental concerns.

The traditional functions of the conservationist in relation to business and government have been research, advice, and education, and these are the functions that the tourist industry anticipates conservation organizations will adopt regarding ecotourism. A recent case in point was the Ecotourism Management Conference at George Washington University, which I attended. This conference included representatives of the tourist industry (mainly operators), government functionaries, and conservation organizations. Interestingly enough, it appeared tacitly understood among these groups that policies and regulations concerning ecotourism, from carrying capacities to sustainable development, would be determined between the tourist industry and government. Conservationists were cast in the role of altruistic and cost-free shepherds of the resources that would insure ecotour profits. More noteworthy, perhaps, was that the conservationists seemed to accept this role willingly. They continue to perceive themselves, and be perceived by other professionals, as consultants.

However, it is likely that the limits of this role have been reached. The arena in which conservation organizations have traditionally held sway is expanding and coming under pressure as a result of the advent of broadly marketed, internationally financed ecotourism. Therefore, the conservation community will increasingly be required to adopt a more active, aggressive posture vis-a-vis the tourism industry and to seek to participate in development projects at earlier stages. All conservation organizations involved in the protection of natural areas, whether they like it or not, have become competitors, not merely with other nongovernmental organizations for the limited funds available but with a $3-trillion service industry, for some of the world's most fragile ecosystems.

To develop an effective ecotourism strategy, the conservation community must pool its institutional resources and lobby for its own goals. This could best be accomplished through the development of an umbrella organization incorporating the many groups that support conservation as foremost in policy development. It should be the responsibility of such an organization to formulate a position on ecotourism, to maintain a presence at conferences and meetings concerning these issues, to lobby for its acknowledged goals, to monitor compliance in the field, and to publicize the results to the prospective customer—the tourist—thus rewarding ecologically responsible businesses.

This monitoring function can be particularly effective in the case of ecotourism, as a result of the unique relationship between the tourist as consumer and the tourist as partner in ecotourism. While it may be difficult to persuade a prospective buyer in Europe or America to forgo the purchase of an electrical appliance because of the manner in which the copper was mined in Zambia, a negative report exposing a tour operator's—or government's—abuse of the environment has a potentially more immediate and greater impact. This could lead banks and other financial institutions to seek input from conservation organizations at the planning stage of a project as a way of nurturing and protecting a potential investment. Moreover, it would provide the third partner in the ecotourism triad, the tourist, with an effective means to influence the direction of ecotourism.

The importance of the tourist in this partnership cannot be overemphasized. The tourist dollar drives the travel industry. The conservation community must develop the tourist as an educated and sophisticated ally in global conservation efforts. To insure that ecotourism develops in an environmentally sensitive and sustainable manner will require that the tourists' collective consciousness be influenced to garner support both for the immediate concerns of parks and protected area programs and for the broader goals of environmental sustainability.

Today's tourists travel for the same reason tourists have always traveled: to escape from the mundane. Many seek to enhance their experiences through nature. Ordinary vacationers have been emboldened to seek out alien destinations and exotic settings, spurred on by a renewed interest in physical fitness, ubiquitous environmental news and wildlife films, and a growing number of tour operators willing to guide them through the most perilous of adventures. Many of these people consider themselves ecotourists; in the broad sense of Donald Hawkins' definition, they are.

Yet, a strict definition of "ecotourism" has not been agreed upon, although several exist. Karen Ziffer's definition, contained in *Ecotourism, an Uneasy Alliance,* prepared for Conservation International, provides an insight into the expectations surrounding this new type of travel, as well as the rigors of the concept as perceived by the conservation community:

> *Ecotourism:* a form of tourism inspired primarily by the natural history of an area, including its indigenous cultures. The ecotourist visits relatively undeveloped areas in the spirit of appreciation, participation and sensitivity. The ecotourist practices a non-consumptive use of wildlife and natural resources and contributes to the visited area through labor or financial means aimed at directly benefiting the conservation of the site and the economic well-being of the local residents.... (Ziffer 1989)

Despite the lack of reliable statistics, it is evident that the majority of tourists who include an aspect of nature travel in their itinerary would be unable to meet Ziffer's laudable if somewhat stringent criteria, and that a great disparity exists between her ecotourist and the millions of leisure travelers broadly defined as ecotourists by Dr. Hawkins. Sheer volume alone dictates that it will be Hawkins' ecotourists, rather than Ziffer's, who will generate the greatest impact on their chosen destinations.

An itemized list of possible negative effects brought about through increased tourism is long and imposing: pollution, disruption of wildlife habitat and migration, increased cost of living for locals, overuse of scarce resources such as water and firewood, and the breakdown of indigenous cultures—to name just a few. In addition to the hardships imposed upon indigenous populations and ecosystems, exposure to these interactive complexities will inevitably subject the broadly defined ecotourist to frustration, disappointment, and even resentment when a perceived vacation in exotic and pristine surroundings is transformed into a difficult lesson in world politics. Lack of satisfaction with such an experience is almost guaranteed. The resulting negative reaction may mean the loss of an ambassador for park programs. More important, it could lead to a lack of cooperation and support on the part of ecotourists for a broad range of conservation initiatives.

The advent of ecotourism provides the conservation community with an opportunity to demonstrate firsthand to a burgeoning audience the connections between the health and preservation of protected ecosystems and that of the global environment. Knowledge of these operative systems and the issues surrounding their support is crucial to their survival, but research indicates that the public's understanding of ecology is superficial (Kellert 1986). David Orr refers to the ability to distinguish between health and disease in natural systems, and their relationship to health and disease in human ones, as "ecological literacy," a literacy that is most effective when acquired out of doors where it can be driven by a sense of wonder, what E. O. Wilson refers to as "biophilia." Orr believes that ecological literacy will become increasingly difficult to achieve, not because there are fewer books about nature (there are more), but because people are becoming further removed from experiencing nature first-hand, with the result that the perception of nature is becoming more abstract (Orr 1989).

Tourists and parks hold the potential to form a perfect partnership to nurture ecological literacy: a captive, self-declared audience in an outdoor, natural setting. However, there are dangers inherent in this approach. As Richard Bangs, President of Sobek Travel, remarked after his company launched a series of disappointing environmental trips to the Brazilian rain forest, "People don't want to spend their hard-earned money being lectured" (Weiner 1991). This would seem to suggest that while ecotourists, particularly ecotourists as defined broadly, constitute a potentially receptive audience, they are still on vacation, and a delicate blending of form and content will have to be employed in attempting to recruit this constituency for environmental goals.

An educated traveler is also a necessary component of the conservation organizations' monitoring of ecotourism. They must know what to expect from their tour operators in the field and be able to recognize the difference between exploitive and protective behavior when traveling in environmentally and culturally sensitive areas. As the end-users of ecotourism, it will be the multitudes of tourist/consumers who reward environmentally responsible tour operators with their business, and who boycott those who are not.

The ecotourism trend is consistent with the world economy's unflagging search for new commercial enterprises. Ecotourism must be considered unique, however, because its very existence depends on the exploitation of natural areas that have been specifically set aside for conservation and protection. From a market point of view, there is little difference between a macaw or an indigenous people or a fragile river system, and a lode of iron ore. Those to whom these precious resources have been entrusted are responsible for ensuring that they do not become tourism's raw materials. In order to meet this challenge, the conservation community must become a stronger, more aggressive player in the ecotourism arena. To paraphrase Gede Ardika, head of the Bali office of Indonesia's Tourism Industry; the partnership conservation and tourism is not a marriage of love but an arranged marriage that must be managed with great care (Carey 1991). Conservation organizations must carefully weigh their strengths and weaknesses, identify the areas in which they can make influential contributions, and form alliances with both tourists and the travel industry to guarantee that ecotourism develops in a manner that will support rather than defeat sustainable conservation of the world's protected areas and the world itself.

Literature Cited

Brooke, J. 1991. Brazil's forests in the balance. New York Times, Travel Section. May 19.

Carey, S. 1991. Tourist spots developing "green" images. Wall Street Journal. May 10.

Economist. 1989a. Third-world tourism: visitors are good for you. **310(7593)**:19–22.

Economist. 1989b. Hither and thither. **311(7598)**:73–74.

Ehrenfeld, D. 1992. The business of conservation. Conservation Biology **6(1)**:1–3.

Kellert, S. R. 1986. Public Understanding and Appreciation of the Biosphere Reserve Concept. Environmental Conservation. **13(2)**:101–106.

New York Times. 1991. Americans are happy campers, survey shows. Travel Section, May 19.

Orr, D. W. 1989. Ecological literacy. Conservation Biology **3(4)**:334–335.

Warpole, K. 1991. Travel in mind. New Statesman & Society **3(133)**:12–15.

Weiner, E. 1991. Ecotourism: can it protect the planet? New York Times, Travel Section. May 19.

Ziffer, K. 1989. Ecotourism: the uneasy alliance. Working paper no. 1. Conservation International, Washington, D.C.

Essays

The Limits to Caring: Sustainable Living and the Loss of Biodiversity

JOHN G. ROBINSON

NYZS The Wildlife Conservation Society
Bronx, NY 10460, U.S.A.

Abstract: Caring for the Earth *represents current, middle-of-the-road thinking on the relationship between conservation and development. This IUCN/UNEP/WWF document has embraced a purely utilitarian perspective: it considers the conservation and development of natural resources to be the same process. In this analysis, I argue that the goal of creating the sustainable society, as defined in* Caring for the Earth, *is an unattainable utopia, and that the mechanisms proposed to attain this goal will lead irrevocably to the loss of biological diversity. I consider the history of the concept of sustainable development, and then document the constraints on sustainable use of natural resources. Sustainable use only occurs when both human needs are met and the losses of biodiversity and environmental degradation are acceptable. These conditions are not always met when natural resources are used, and I consider the fundamental contradictions between resource potential and human needs. I conclude by emphasizing that while sustainable use is a powerful approach to conservation, it is not the only one, and the conservation of many species and biological communities also requires a preservationist approach.*

Los limites de cuidari: vida sostenible y la pérdida de biodiversidad

Resumen: Cuidar la Tierra *representa una forma moderada de pensar en relación a los conceptos de conservación y desarrollo. Este documento de la IUCN/UNEP/WWF abarca una perspectiva totalmente utilitarista. En éste se considera que la conservación es lo mismo que el desarrollo de los recursos naturales. En este análisis, yo argumento que la meta de crear una sociedad sostenible como está definida en* Cuidar la Tierra, *es una utopía que no se puede alcanzar, y que los mecanismos que se proponen para obtener esta meta llevarán a la irrevocable pérdida de la diversidad biológica. Se examina la historia del concepto de desarrollo sostenible, y después se documentan las limitaciones del uso sostenible de los recursos naturales. El uso sostenible solamente ocurre cuando las necesidades humanas están satisfechas y cuando las pérdidas de la biodiversidad y la degradación del ambiente son aceptables. Estas condiciones no siempre son satisfechas cuando los recursos naturales son utilizados, y examino las contradicciones fundamentales entre el potencial de los recursos y las necesidades humanas. Concluyo enfatizando que mientras el uso sostenible es un enfoque eficaz para la conservación, no es el único, y que la conservación de muchas de las especies y comunidades biológicas también requiere de un enfoque preservacionista.*

Introduction

Caring for the Earth: A Strategy for Sustainable Living, was launched in October 1991. It is an important manifesto for two reasons. First, it is the explicit successor to the *World Conservation Strategy,* the original global conservation blueprint published in 1980. Second, like the *World Conservation Strategy,* *Caring for the Earth* is authored by and bears the imprimatur of some of the world's most respected conservation organizations: the World Conservation Union (IUCN), the United Nations Environmental Programme (UNEP), and the World Wide Fund for Nature (WWF). The *World Conservation Strategy* was the first influential conservation document aimed at government officials and development practicioners, as well as at conservationists. It legitimized the involvement of government and development agencies in conservation. In the world envisioned by the *World Conservation Strategy,* parks and reserves were seen

Paper submitted July 20, 1992; revised manuscript accepted December 4, 1992.

not only as bastions for wildlife but also as integral components in national strategies of development. National Conservation Strategies, following the format developed in the *World Conservation Strategy*, were prepared in over 50 countries. The World Bank, the U.S. Agency for International Development, and the European Community poured millions of dollars and ecus into conservation projects. *Caring for the Earth* seeks to assume the mantle of this tradition. This document will be perceived as the middle-of-the-road, pragmatic, responsible thinking on conservation issues by those deciding governmental and developmental policy throughout the world. This explicit objective is stated in the introduction; "*Caring for the Earth* is intended to be used by those who shape policy and make decisions that affect the course of development and the condition of our environment" (p. 3). And it will be.

Yet *Caring for the Earth* represents a significant departure from previous thinking on the conservation of natural resources and of biodiversity. (1) The sustainable society as defined in *Caring for the Earth* is an unattainable utopia because it states goals and principles incompatible with one another. By not acknowledging the conflicts and contradictions inherent in conservation and development, the analysis is simplistically optimistic. (2) The goals of sustainable use and sustainable development, as defined in *Caring for the Earth*, will lead irrevocably to the loss of biological diversity. This biological diversity is not simply numbers of species, many of which are supported in human-managed ecosystems (Pimentel et al. 1992), but it concludes "the variety and variability among living organisms and the ecological complexes in which they occur" (OTA 1987), a definition that encompasses ecosystem, species, and genetic diversity.

Caring for the Earth can best be understood by first examining how the emphasis on development and conservation has changed when compared with the earlier document. In 1980, the *World Conservation Strategy* stated that "the object of development is to provide for social and economic welfare, [and] the object of conservation is to ensure Earth's capacity to sustain development and to support all life." The significant concept popularized in the *World Conservation Strategy* was that conservation and development were not necessarily mutually exclusive. By incorporating the conservation approach into global policy, development can be made sustainable and natural systems, together with their biological diversity, can be conserved. The *World Conservation Strategy* popularized the term "sustainable development" as a process in which conservation and development were mutually dependent. The specific objectives of the *World Conservation Strategy* were to maintain ecological processes and life-support systems, to preserve biological diversity, and to ensure that use of natural resources was sustainable. The focus therefore is on the natural world and human dependence on our environment. The *World Conservation Strategy* promulgated the concept of *conservation through sustainable development.*

The goals of *Caring for the Earth* are superficially similar, but there is a different emphasis. The goal is "to help improve the condition of the world's people" through two actions: development of a sustainable society and integration of conservation and development. *Caring for the Earth* argues that in a sustainable society, conservation and development are totally compatible with one another (not, as envisioned in the *World Conservation Strategy,* as two separate but mutually dependent activities). *Caring for the Earth* states that "conservation and development . . . are essential parts of *one* indispensable process" (p. 8, my italics). In a sustainable society (by definition), people will improve the quality of their lives, while conserving the Earth's vitality and diversity and keeping within the Earth's carrying capacity. In other words, *Caring for the Earth* promulgates the concept of *sustainable development:* conservation will be an inevitable consequence of such development.

Caring for the Earth includes a long list of general principles that define the sustainable society. This list includes the specific objectives of the *World Conservation Strategy:* maintaining ecological processes and life support systems, preserving genetic diversity, and ensuring the sustainable utilization of species and ecosystems. Additional principles include respecting and caring for the community of life, improving the quality of human life, minimizing the depletion of nonrenewable resources, keeping human numbers and life-styles within the earth's carrying capacity, changing personal attitudes and practices, enabling human communities to care for their own environments, providing a national framework for integrating development and conservation, and creating a global alliance for sustainability. These principles are emotionally appealing, and the recognition that humans need to redefine their relationship with the natural world is a necessary one.

Caring for the Earth within its specified limits of interest and within its own definitions is an admirable document and sets out basic tenets of living that we all must follow if we are to survive on this planet. Its failure is that it does not acknowledge that the goals of development are different from the goals of conservation, and it offers no general principles by which we might resolve conflicts and balance contradictory demands. *Caring for the Earth* does not recognize that while improving the quality of human life, we will inevitably decrease the diversity of life. If we do not acknowledge the contradictions, we will smugly preside over the demise of biological diversity while waving the banner of conservation.

Sustainable Development

The goal of national development in the 1950s and 1960s was to increase the Gross National Product (GNP) of countries, especially in the South. The mechanism was technological progress. The result was supposed to be an increase in the consumption of natural resources and their use by people, an increase in national export and import of goods, and an increase in the human standard of living. The problem was that by the 1970s it was clear that the process was not working in much of the world (Redclift 1987; Sachs 1991). Consumption of natural resources was up, but the disparity between and the economic dependence of the South on the North had increased. In many countries, per capita incomes were down, deforestation, overgrazing, and overcultivation were up. Environmental degradation in the countries of the South was becoming increasingly evident, and the economic costs of this were becoming appreciated. The loss of wildlands and the disappearance of species were provoking alarm (Ehrlich & Ehrlich 1981).

Popular concerns with the consequences of national development crystalized in the publication *The Limits to Growth* (The Club of Rome 1972), which examined the long-term trends in world population, resource use, food production, and industrialization. In the same year, the United National Conference on the Human Environment was held in Stockholm, and led to the establishment of the U.N. Environmental Program (UNEP). These initiatives were followed by the *World Conservation Strategy* in 1980.

The *World Conservation Strategy* retained the traditional concept of development, which it defined as activities that "satisfy human needs and improve the quality of human life." Its innovation was that it acknowledged that alone this approach was not sufficient. The *World Conservation Strategy* advocated the approach of *sustainable* development, which incorporated social and ecological considerations for *long-term* as well as short-term advantages. Conservation was then linked to development, by defining it as activities that "yield the greatest sustainable development to present generations while maintaining its potential to meet the needs and aspirations of future generations." This concept goes back, at least, to the utilitarian philosophy of Gifford Pinchot, the father of North American forestry and the person who defined conservation in this 1947 autobiography as "the greatest good for the greatest number for the longest time."

The *World Conservation Strategy* vision of conservation was not purely utilitarian, embracing as it does "preservation, maintenance, sustainable utilization, restoration, and enhancement of the natural environment." Sustainable development requires conservation, but it is not the same process as conservation. The *World Conservation Strategy* recognized that conservation action can limit development (e.g., "Protected areas and other conservation measures, however, may restrict access to fuel, food, forage, and other products"), and that development can decrease biodiversity (e.g., "Most changes of use, other than protection, involve a loss of equilibrium in the ecosystem concerned, sometimes a radical loss"), and it devoted an entire section to the methods of balancing competing land uses.

In the 1987 report of the World Commission on Environment and Development (otherwise known as the Brundtland Commission), there was a significant redefinition of sustainable development, which was defined as development that "seeks to meet the needs and aspirations of the present without compromising the ability to meet those of the future." This definition is virtually identical to the *World Conservation Strategy*'s definition of "conservation." This definitional shift followed from the Commission's focus on the failure of development and from the environmental consequences of that failure. The Brundtland Commission focused on the environmental problems associated with development—not on the conservation of the natural environment. The concern is with "the impact of ecological stress—degradation of soils, water regimes, atmosphere, and forests—*upon our economic prospects*" (my italics). The Commission was able to appropriate the language of conservation in its definition of sustainable development because it adopted an exclusively utilitarian approach—not considering the need to conserve any life that was not explicitly useful to human beings.

In *Caring for the Earth,* the goal is to build a sustainable society. This requires sustainable development, which is defined as "improving the quality of human life while living within the carrying capacity of supporting ecosystems" (p. 10). Whatever its intent, this definition emphasizes traditional development at the expense of the conservation of natural resources and biodiversity.

This definition, while differing from that of the Brundtland Commission, follows its lead by ignoring the dichotomy between conservation and development. However, unless one adopts a purely utilitarian approach, conservation and development are *not* the same process. Frequently the two processes are compatible (a realization that was at the heart of the *World Conservation Strategy*), but when they are not, human development can lead to species extinction, and conservation can limit development. *Caring for the Earth* does not recognize this incompatibility.

This definition of "sustainable development" reverts to a definition that approximates the *World Conservation Strategy*'s definition of "development." The goal is to improve the quality of human life (defined broadly in terms of social, cultural, and economic welfare). *Caring for the Earth* states that "it is important to remember that we are seeking not just survival but a sustainable improvement in the quality of life of several billion people" (p. 43). Unfortunately, stating that development should be sustainable does not make it so. Sustainable development ultimately requires that renewable resource consumption be stationary, that the product of number of people and the resource amount which each consumes not increase (Daly 1980). This requires: (a) technological advances that would allow our planet to support more people at a higher quality of life (Simon 1980, but see Redclift 1987); (b) a dramatic reduction in the world's human population; and/or (c) a dramatic reduction in the level of consumption in affluent countries to allow an improvement in the quality of life in poorer countries. None of these possibilities seems likely unless there are significant changes in the process of development. Yet *Caring for the Earth* provides no analysis of how to achieve these goals. Indeed, *Caring for the Earth* does not appear to deviate from the traditional development formulae of the 1950s and 1960s, and a continuation of such policies will surely decrease biological diversity.

This definition of sustainable development requires that it take place "within the carrying capacity of supporting ecosystems" (p. 10). However, carrying capacity is not an ecosystem characteristic, but is defined for the population of a given species (Begon et al. 1986). In the case of *Caring for the Earth*, the species of interest is the human being. The carrying capacity of earth for humans depends on a complex interaction of environmental potential, lifestyle aspirations, technologies, and sociopolitical and economic organization (see Daily & Ehrlich 1992). Yet as a general rule, human beings are more able to use ecosystems at young successional stages, which tend to be more productive. Accordingly, a general characteristic of human development is that we tend to maximize productivity by creating and maintaining ecosystems at such stages. This requires energy input, in forms such as irrigation, insecticides, fertilizers, mechanical alterations of the environment, etc. In contrast, undisturbed ecosystems, not subject to such inputs, become mature, and tend to be less productive, but more biologically diverse. In other words, the goal of maximizing the carrying capacity of human beings will encourage intensive agriculture at the expense of natural systems, pine plantations in place of hickory-oak forests, and maize fields instead of tropical savannas.

This definition of sustainable development reverts to a popular misunderstanding of natural systems—that they exist at equilibrium, and that there is a "balance of nature" (see Brussard 1991; Pimm 1991). While some natural systems are relatively stable and maintaining them at stasis will tend to preserve species biodiversity, other systems require disturbance. Many biological communities are not structured by intraspecific competition alone (Andrewartha & Birch 1954) and require nonequilibrium conditions to maintain biological diversity (e.g., Connell 1978; Hubbell 1979).

Sustainable Use

Caring for the Earth formally links sustainable use to sustainable development. To keep within the carrying capacity of an ecosystem requires that resources be used sustainably. Sustainable use requires that resources are used "at rates within their capacity for renewal" (p. 10). Defined in these very general terms, sustainable use is noncontroversial and generally supportable by all thoughtful people.

Two ideas lie at the heart of the sustainable use concept. One is that the resources are renewable. All living resources are renewable by definition, but some resources are more renewable than others. In the language of economics, the interest rate varies. The other idea is that people can balance their consumption with resource production. Spend the interest, not the principal. "Humanity must take no more from nature than nature can replenish. This in turn means adopting lifestyles and development paths that respect and work within nature's limits." This view optimistically generalizes the concept of the noble savage to all of humankind—"Living sustainably depends on accepting a duty to seek harmony with other people and with nature" (p. 8). Whether such enlightened resource use has ever been a characteristic of human groups is debatable (see for instance Redford 1990). Nevertheless, in the following discussion, I will consider the following question: If natural resources are used sustainably, in a manner acceptable to *Caring for the Earth*, what will be the effect on biodiversity?

To understand sustainable use, one needs to consider three interdependent questions. What will be the impact of human use on the *environment* or the *biological resource*? This considers the ecological sustainability of human activities. What are the *needs* and *aspirations* of resource users? This is a consideration of economic sustainability. Finally, what are the *rights* of different user groups to the resource? This is a social and political consideration.

Ecologically Sustainable Use—Species

Ecologically sustainable activities are defined as those that do not degrade the natural resource. Consider the sustainable harvest of a species. The only requirement for ecological sustainability is that harvest from the population must not exceed the potential yield. Yield is total production subtracting natural mortality. There are therefore many population levels at which any species can be sustainably harvested. Very small populations, for instance, will have a very small potential yield, but as long as harvest does not exceed that yield, it will be sustainable. There is a population level at which yield is maximized—termed the maximum sustainable use (MSU), maximum sustainable yield (MSY), or maximum sustainable cut (MSC)—and managing populations to this level is the goal of many resource managers.

The extent to which a species can be harvested or used by humans depends in large part on whether it exhibits density compensation: As the density falls, does the species population "compensate" by increasing its rate of increase or rate of growth? If it does, then down to a certain level an exploited population is more productive than an unexploited one; the overall yield is higher. The extent to which populations exhibit density compensation, whether populations grow sigmoidally, and the point at which yield maxima are achieved, remain under examination (Reimers 1975; Caughley 1985; Hall 1988). Nevertheless, some general trends among species are discernible. Species that exhibit strong density compensation tend to have MSY points significantly lower than carrying capacity (Fredin 1984). Species that show weak density compensation, for which densities are determined by interactions with other species, tend to have MSY points close to their carrying capacity (Eberhardt 1977, 1981; Fowler 1981; Smith 1984). As a population will go to local extinction if the number of individuals harvested exceeds the MSY, then any significant harvest of species in this latter category will tend to drive density progressively lower than that which produces MSY, and such use cannot be sustainable.

Use is therefore much more feasible when people are exploiting species that show strong density compensation and have high rates of renewability. Species with these characteristics are much more likely to be found in ecosystems at younger successional stages. On the other hand, it will generally be more difficult to exploit species in complex, mature, high diversity ecosystems because density compensation is minimal in most species. For instance, significant harvests of white-tailed deer in perpetuity is a theoretical possibility, but it is likely that significant harvests of most tropical forest mammal species will systematically reduce their populations. Robinson and Redford (1991a), for instance, point out for the neotropics that "highly seasonal ecosystems [such as the llanos and the altiplano], with low species diversity, are more likely to contain large-bodied [wildlife] species with high densities and intrinsic rates of population increase. These are the species that have traditionally been exploited commercially." In contrast, "the more species-diverse habitats, such as tropical forests, do not appear to contain single species with high enough densities and rates of population increase to be commercially exploited." The potential harvest of many species is therefore minimal, and the possibility of human use of such species is limited. For these species, any significant harvest will drive populations to local extinction.

The human use of a species also has a more pervasive effect on overall biological diversity. Species occur as parts of biological communities, and harvest will have ramifications throughout the community. Larkin (1977), in an influential critique of MSY thinking in modern fisheries, argued that the unpredictability of fish stocks derived in part from managers ignoring the role of the target species in fish communities. Redford (1992) has pointed out that human hunters in tropical forests take the large bird and mammal frugivores, a preference that must affect seed dispersal and predation of the tropical trees, as well as removing the prey base of the large forest predators. Any exploitation of a species will remove a part of a biological community, with concomitant effects on community dynamics and ecosystem functioning.

Species therefore can be exploited sustainably, but requiring sustainability does not prescribe the intensity of exploitation. One must also specify a minimum population level that is acceptable, and this presumably is defined by the requirements of population viability and by the importance of the species as part of its biological community. Any use of a species however is likely to encourage the overall loss of biological diversity.

Ecologically Sustainable Use—Biological Communities

Ecologically sustainable activities at the level of the biological community are defined as those that do not degrade the capacity of that community to sustain human beings. One must recognize however, that any use of a biological community will ultimately involve a loss of biological diversity. While one can naively imagine human beings in the modern world living as part of a natural community, taking, as envisioned in *Caring for the Earth,* "no more from nature than nature can replenish," humans will always encourage desirable species and remove competitor species, and all human groups have the capacity to change their environments radically.

This point can be illustrated by considering again a tropical forest. The forest itself could be managed, it could be used for swidden agriculture, or it could be converted for intensive agriculture. These represent increasing human manipulation of the community, increasing external inputs, and increasing the carrying capacity of humans. All are conceivably sustainable. But each has a different effect on biological diversity.

Extraction of forest products, practiced throughout the tropics, represents the lowest level of forest manipulation. The net result of such activities is a loss of forest species (e.g., Anderson 1990), but humans can sometimes make a living from the forest, although usually at only subsistence levels (Fearnside 1989). Increased human manipulation of the forest is represented by swidden (or slash-and-burn) agriculture. Over the long-term, such agriculture can have a significant impact on species composition in the forest (Gómez-Pompa & Kaus 1990), but these activities certainly can be sustainable and support more people over the long term (Flowers et al. 1982). Finally, intensive agricultural production in tropical forest ecosystems (Vasey 1979) represents the most intrusive use. Pedro Sanchez and his colleagues (e.g., Sanchez et al. 1982) have been major proponents of the position that the technology is available, and the economics amenable to developing intensive and sustainable agriculture on tropical forest soils. The method requires removal of tropical forest, burning, and significant application of fertilizer and pesticides. These activities could also constitute sustainable use, and as such would be totally compatible with the *Caring for the Earth*'s goals, and should be promoted in a sustainable society. But a tropical forest has a much higher biological diversity than a maize field, even though the latter can support more people with a higher quality of life (as defined in *Caring for the Earth*).

The point therefore is not that a tropical forest cannot be sustainably used—any intensity of use is potentially ecologically sustainable. The more intense the human use of a forest, however, the greater will be the loss of biological diversity. This statement is not at variance with the observations that biological diversity is frequently higher in communities subject to "intermediate disturbance" (Connell 1978)—the range of disturbances resulting from human activities is on an altogether greater scale.

Socioeconomically Sustainable Activities

Economically sustainable activities are defined as those that meet the economic needs and aspirations of the human users. The level of use of a species or a biological community, however, also varies with the identity of the human users. Different human users require different levels of use to meet their economic needs. Accordingly, any plan of sustainable resource use needs to specify the human consumers. Who will use the natural resources? Is it the indigenous groups, local communities, local landowners, regions, an entire nation, or foreign countries? The decision to allocate rights to the resource is usually based on social, political, and perhaps ethical criteria. And it is a decision to empower specified social groups or classes. Socioeconomically sustainable activities therefore are those that meet the economic needs and aspirations of those users who have been allocated rights to those resources.

Identifying the human consumers of natural resources also requires considering other socioeconomic or political groups. The needs and interests of these groups frequently contradict those of the direct users. The socioeconomic sustainability of resource use therefore will depend on the conflicts of interests and/or synergistic effects *among* local communities, businesses, and national institutions (Robinson 1990). Most discussions of sustainable use assume that it is in the interests of all social groups that resource use be sustainable. This is frequently not the case. For instance, the decision, at the national level, to create rubber "extractive reserves" in western Brazil—areas managed by local communities and reserved for specific resource extraction—is commonly trumpeted as a successful approach to sustainable resource use (e.g., McNeely 1988). But the applicability of this approach depends on the political power of the rubber tappers union, the interests of local cattle ranchers, the market demand for their products, the ability of local communities to get their products to the market, to name a few considerations (see Browder 1992). The international political and economic structures will also have an impact on the long-term viability of such extractive endeavors (Redclift 1987). Until these influences are understood, it is unclear whether a resource use will be socioeconomically sustainable. In isolation, a local community might be able to meet their socioeconomic needs, but when national or international politics or markets are considered, they might be unable to do so.

Ecologically Sustainable Socioeconomic Activities

Ecologically sustainable socioeconomic activities are those that are both ecologically and socioeconomically sustainable. Both types of sustainability must be considered because each alone does not specify an intensity of use. In the case of ecological sustainability, one must also define an acceptable loss of biodiversity or environmental degradation. For socioeconomic sustainability, one must also define the consumer group that is the beneficiary of the resource use and consider the interaction with other human interest groups. *Sustainable use only occurs when the rights of different user groups are specified, when human needs are met, and when the losses in biodiversity and environmental degradation are acceptable.*

The history of natural resource use in modern times bears witness to the frequency that resource potential and human needs are incompatible. Even systems like marine fisheries, which are highly productive and heavily managed, have been consistently exploited, and stocks of many economically important species are today highly precarious (How to fish 1988). Tropical forest management may be even more difficult. The failure of management efforts like the Tropical Forestry Action

Plan (TFAP) and government fora like the International Tropical Timber Organization (ITTO) have been notably unsuccessful at stopping tropical forest loss (Empire of the Chainsaws 1991; Vincent 1992). In most cases, the economic constraints on the user groups force this overexploitation (Clark 1973a, 1973b; Larkin 1977). In other words, catches and cuts that are ecologically sustainable do not meet the socioeconomic needs of the users.

With many types of resource extraction under many different socioeconomic conditions, sustainable use will be impossible. This conclusion contrasts with *Caring for the Earth*, which believes as an article of faith that sustainable use is always possible. The failure to use resources sustainably is viewed by *Caring for the Earth* as being a consequence of poor planning, inefficient bureaucracies, inappropriate institutions, and the unthinking waste of human and financial resources. Once people are enlightened, and with appropriate intellectual input, then any resource can be used sustainably (see Chapter 6). *Caring for the Earth* promotes a utopian vision in which belief in sustainable use promotes sustainable use. Unfortunately, there are real contradictions underlying the frequent failure to use resources sustainably.

My discussion has focused on the theoretical constraints of the sustainable use concept. I have not addressed the practical and a management issue inherent in the sustainable use concept. It remains an open question whether we have the management capacities or can develop appropriate economic incentives (McNeely 1988) to ensure that use is sustainable, even in those cases in which it is possible (see Robinson & Redford 1991c). I have also not discussed the danger that advocating sustainable use will give a green light to nonsustainable exploitative use. Resource extraction schemes are proliferating everywhere, and most advertise themselves—without justification—as sustainable. Instead I have restricted my discussion to limitations inherent in the concept itself.

Conclusion

Caring for the Earth presents us with a simplistic vision of development and conservation. Critical concepts, such as human "quality of life" and "carrying capacity" are loosely used. There is confusion between the concept of sustainability and that of ecological equilibria. There is a failure to distinguish ecological from socioeconomic sustainability, and the necessary relationship between these two types of sustainability—a relationship that defines sustainable use—is never explored. *Caring for the Earth* places no limits on the loss of biodiversity that is acceptable, neither does it acknowledge that different human consumers have different interests and needs. The result is that the process of socioeconomic development is never examined critically, and sustainable development appears to be easily attainable. While positive thinking is praiseworthy, there must also be a reality check.

Caring for the Earth is also limited in its vision, focusing as it does almost exclusively on human beings. It is concerned with improving the quality of life of people. This is a worthy goal, but it is not the same goal as conserving the full spectrum of biological diversity. Because of this anthropocentric orientation, *Caring for the Earth* emphasizes sustainable use of natural resources as the only approach to conserving natural systems. Sustainable use is a powerful approach to conservation (see Robinson & Redford 1991c), but it is not the only one. Many species and biological communities will be lost unless they are protected and managed with the express goal of their conservation. Sustainable use is very appropriate in certain circumstances, but it is not appropriate in all. It will almost always lower biological diversity, whether one considers individual species or entire biological communities, and if sustainable use is our only goal, our world will be the poorer for it.

Acknowledgments

This essay benefitted from discussions and comments by many colleagues, including Elizabeth Bennett, Dorene Bolze, Peter Brussard, William Conway, Susan Jacobson, John Hart, Terese Hart, Douglas Murray, Alan Rabinowitz, Kent Redford, Stuart Strahl, Simon Stuart, and reviewers of an earlier draft. Not all agreed with my analysis. To these people I offer my sincere thanks.

Literature Cited

Anderson, A. B. 1990. Extraction and forest management by rural inhabitants in the Amazon estuary. Pages 65–85 in A. B. Anderson, editor. Alternatives to deforestation: steps towards sustainable use of the Amazon rain forest. Columbia University Press, New York.

Andrewartha, H. G., and L. C. Birch. 1954. The distribution and abundance of animals. University of Chicago Press, Chicago.

Begon, M., J. L. Harper, and C. R. Townsend. 1986. Ecology. Sinauer Press, Sunderland, Massachusetts.

Browder, J. O. 1992. The limits of extractivism. BioScience 42:174–182.

Brussard, P. F. 1991. Nature in myth and reality. Review of Discordant Harmonies, by Botkin, D. B. 1990. Oxford University Press. Conservation Biology 5:571–572.

Caughley, G. 1985. Harvesting of wildlife: past, present, and future. Pages 3–14 in S. L. Beasom and S. F. Roberson, editors. Game harvest management. Caesar Kleberg Wildlife Research Institute, Kingsville, Texas.

Clark, C. W. 1973a. Profit maximization and the extinction of animal species. Journal of Political Economics 81:950–961.

Clark, C. W. 1973b. The economics of overexploitation. Science 181:630–634.

Club of Rome. 1972. The limits to growth. Universe Books, New York.

Connell, J. H. 1978. Diversity of tropical rainforests and coral reefs. Science 199:1302–1310.

Daily, G. C., and P. R. Ehrlich. 1992. Population, sustainability, and the earth's carrying capacity. BioScience 42:761–771.

Daly, H. E. 1980. The steady-state economy: toward a political economy of biophysical equilibrium and moral growth. Pages 325–356 in H. E. Daly, editor. Economy, ecology, and ethics. Essays toward a steady-state economy. W. H. Freeman, San Francisco.

Eberhardt, L. L. 1977. Optimal policies for conservation of large mammals, with special reference to marine ecosystems. Environmental Conservation 4:205–212.

Eberhardt, L. L. 1981. Population dynamics of Pribilof fur seals. Pages 197–200 in C. W. Fowler and T. D. Smith, editors. Dynamics of large mammal populations. Wiley-Interscience, New York.

Ehrlich, P. R., and A. Ehrlich. 1981. Extinction: the causes and consequences of the disappearance of species. Random House, New York.

Empire of the chainsaws. 1991. The Economist, August 10:36.

Fearnside, P. M. 1989. Extractive reserves in Brazilian Amazonia. BioScience 39:387–393.

Flowers, N. M., D. R. Gross, M. L. Ritter, and D. W. Werner. 1982. Variation in swidden practices in four central Brazilian societies. Human Ecology 10:203–217.

Fowler, C. W. 1981. Comparative population dynamics in large mammals. Pages 437–456 in C. W. Fowler and T. D. Smith, editors. Dynamics of large mammal populations. Wiley-Interscience, New York.

Fredin, R. A. 1984. Levels of maximum net productivity in populations of large terrestrial mammals. Pages 381–387 in W. F. Perrin, R. L. Brownell, Jr., and D. P. DeMaster, editors. Reproduction in whales, dolphins, and porpoises. Reports of the International Whaling Commission, Special Issue 6.

Gómez-Pompa, A., and A. Kaus. 1990. Traditional management of tropical forests in Mexico. Pages 45–64 in A. B. Anderson, editor. Alternatives to deforestation: steps towards sustainable use of the Amazon rain forest. Columbia University Press, New York.

Hall, C. A. S. 1988. An assessment of several of the historically most influential theoretical models used in ecology and of the data provided in their support. Ecological Modeling 43:5–31.

How to fish. 1988. The Economist, December 10:93–96.

Hubbell, S. P. 1979. Tree dispersion, abundance, and diversity in a tropical dry forest. Science 203:1299–1309.

IUCN/UNEP/WWF. 1980. World Conservation Strategy. Living resource conservation for sustainable development. Gland, Switzerland.

IUCN/UNEP/WWF. 1991. Caring for the Earth: a strategy for sustainable living. Gland, Switzerland.

Larkin, P. A. 1977. An epitaph to the concept of maximum sustained yield. Transcripts of the American Fisheries Society 106:1–11.

McNeely, J. A. 1988. Economics and biological diversity. Developing and using economic incentives to conserve biological resources. IUCN, Gland, Switzerland.

OTA (Office of Technological Assessment, U.S. Congress). 1987. Technologies to maintain biological diversity. U.S. Government Printing Office, Washington, D.C.

Pimentel, D., U. Stachow, D. A. Takacs, H. W. Brubaker, A. R. Dumas, J. J. Meaney, J. A. S. O'Neil, D. E. Onsi, and D. B. Corzilus. 1992. Conserving biological diversity in agricultural/forestry systems. BioScience 42:354–362.

Pimm, S. L. 1991. The balance of nature. Ecological issues in the conservation of species and communities. University of Chicago Press, Chicago.

Pinchot, G. 1947. Breaking new ground. Harcourt, Brace & Co., New York.

Redclift, M. 1987. Sustainable development. Exploring the contradictions. Methuen, London.

Redford, K. H. 1990. The ecologically noble savage. Orion 9:24–29.

Redford, K. H. 1992. The empty forest. BioScience 42:412–422.

Reimers, E. 1975. Age and sex structure in a hunted population of reindeer in Norway. University of Alaska Biological Population Special Report 1:181–188.

Robinson, J. G. 1990. Economic approaches to conservation. Ecology 71:410–411.

Robinson, J. G., and K. H. Redford, 1991a. The use and conservation of wildlife. Pages in J. G. Robinson and K. H. Redford, editors. Neotropical wildlife use and conservation. University of Chicago Press, Chicago.

Robinson, J. G., and K. H. Redford 1991b. Sustainable harvest of neotropical forest animals. Pages 415–429 in J. G. Robinson and K. H. Redford, editors. Neotropical wildlife use and conservation. University of Chicago Press, Chicago.

Robinson, J. G., and K. H. Redford (eds.). 1991c. Neotropical wildlife use and conservation. University of Chicago Press, Chicago.

Sachs, W. 1991. Environment and development: the story of a dangerous liaison. The Ecologist 21:252–257.

Sanchez, P. A., D. E. Bandy, J. H. Vallachica, and J. J. Nicholaides. 1982. Amazon basin soils: management for continuous crop production. Science **216**:821–827.

Simon, J. L. 1980. Resources, population, environment: an oversupply of false bad news. Science **208**:1431–1437.

Smith, T. D. 1984. Estimating the dolphin population size yielding maximum net production. Pages 187–190 in W. F. Perrin, R. L. Brownell, Jr., and D. P. DeMaster, editors. Reproduction in whales, dolphins, and porpoises. Reports of the International Whaling Commission, Special Issue 6.

World Commission on Environment and Development. 1987. Our common future. Oxford University Press, Oxford.

Vasey, D. E. 1979. Population and agricultural intensity in the humid tropics. Human Ecology **7**:269–283.

Vincent, J. R. 1992. The tropical timber trade and sustainable development. Science **256**:1651–1655.

Comments

Limits to Caring: A Response

MARTIN HOLDGATE

Director General
IUCN–The World Conservation Union
Rue Mauverney 28
1196 Gland, Switzerland

DAVID A. MUNRO

2513 Amherst Avenue
Sidney, British Columbia
V8L 2H3 Canada

We are grateful to John Robinson for a thoughtful essay on the relationship between conservation and development. While awareness of that vital relationship is spreading, many people who wield political and economic power in the world remain ignorant of it, and there are few if any who understand all its ramifications in the varying circumstances of the human environment. It is clear that many years of debate will be required to fully clarify and widely disseminate the concept of sustainable development, and every positive contribution to that process will be welcome.

Robinson criticizes *Caring for the Earth* (International Union for the Conservation of Nature and Natural Resources/United Nations Environmental Program/World Wildlife Fund 1991) on the basis that "the goal of creating the sustainable society, as defined in *Caring for the Earth,* is an unattainable utopia, and that the mechanisms proposed to attain this goal will lead irrevocably to the loss of biological diversity." He also takes issue with its treatment of conservation and development as part of one process, saying that it thereby ignores the cases where they can conflict. We hope that our response to his criticisms will help to promote a broader appreciation of this very important topic.

Unattainable utopia or not, we make no apology for describing a state of the human environment toward which we should strive in the belief that progress even part way would be preferable to the state in which we would surely languish if we remained unaware of our predicament and made no effort to escape it. We agree with many of the arguments that Robinson uses to support his contentions, but we question whether those contentions are helpful in considering how to achieve sustainability or maintain biodiversity. More fundamentally, we question whether they are helpful in trying to improve the condition of the world's people, which is the avowed aim of *Caring for the Earth.*

We would like to comment on these and other points against a background in which two points are preeminent. The first is that *Caring for the Earth* was written for a political purpose. It was, as Robinson recognizes in quoting from it "intended to be used by those who shape policy and make decisions that affect the course of development and the condition of our environment," yet Robinson seems to forget this point as he proceeds with his critique. Its scope and style necessarily reflect its political purpose. The second point is that social, economic, and environmental change are continuous. It would indeed be unattainably utopian to believe that we could halt all change, not least in the dimensions of biodiversity.

A document with a political purpose must seek to relate the primary objectives that it espouses to the aspirations of the widest possible audience. The main theme and principal arguments of *Caring for the Earth* are, as Robinson noted, derived from utilitarian conservation; but they are strongly influenced by what is termed an "ethic for living sustainably," which emphasizes the duty of each person to care for others now and in the future and for the other forms of life with which we share the earth. It is not enough, of course, that *Caring for the Earth* should appeal to the conservationist and humanitarian constituencies in which it has roots: it must reach out to a much larger public. Indeed, success in maintaining biodiversity and in eventually achieving sustainability is utterly dependent upon the collaboration and good will of the majority of people in all countries. It rests most especially on those who are at once terribly disadvantaged economically and de facto custodians of biodiversity. It is to support the efforts of governments in the species-rich tropical countries who are trying to improve the desperate conditions of hundreds of millions of their citizens that *Caring for the Earth* emphasizes sustainable use of natural resources. But since sustainable use embraces conservation, *Caring for the Earth* most certainly does not exclude protection and management with the express goal of conservation.

Indeed, conserving the Earth's vitality and diversity is what the 14 priority actions in Chapter 4 are all about. It is to that end that Chapter 4 says "biological diversity should be preserved as a matter of principle, because all species deserve respect regardless of their use to humanity and because they are all components of our life support system"; "prudence dictates that we keep as much variety as possible"; and "biological diversity should be conserved by a combination of measures to safeguard species and genetic stocks, the establishment and maintenance of protected areas, and wider strategies that combine economic activities and conservation over entire regions." The action on protected areas (4.9), species conservation (4.10), biological research and monitoring (4.11), and *in situ* and *ex situ* conservation (4.12) all call for more to be done to achieve "traditional conservation" than any state is now doing.

A document with a political purpose must also rest on arguments that are broadly understandable. Points that do not reflect an appreciation of human welfare or that are based on academic distinctions are unlikely to be politically compelling. Robinson complains that the concepts of human quality of life and carrying capacity are loosely used. In a document of global scope and limited length they could hardly be otherwise. A detailed analysis of all the factors that must be considered in establishing the parameters of carrying capacity and quality of life under different circumstances would hardly help to promote the broader understanding of those and other concepts that *Caring for the Earth* tries to bring about. Each of these terms must be interpreted in the light of the particular circumstances in which it is being applied, but each also has a general relevance that is quite clear.

We should, however, digress to respond more specifically to Robinson's comments on the use of "carrying capacity" (p.10), where sustainable development is defined as taking place "within the carrying capacity of ecosystems." It would be correct to say that carrying capacity is not *usually* considered to be an ecosystem characteristic but is defined for the population of a given species. In *Caring for the Earth,* it is quite clear that it is the human species upon which the definition of carrying capacity depends. Beyond that, however, *Caring for the Earth* attributes a new and broader meaning to carrying capacity, considering it to be the "capacity of an ecosystem to support healthy organisms while maintaining its productivity, adaptability, and capability of renewal" (Glossary, p. 210). This expanded meaning is important because it comprehends the notion that carrying capacity for humans must be determined within the context of the health and productivity of other species. This idea is central to *Caring for the Earth,* as is the desirability of conserving the diversity of life on Earth, to which a good part of Chapter 4 is devoted and which is widely reflected in other parts of the text. To attribute a meaning to carrying capacity that takes account of the desirability of maintaining the adaptability of ecosystems is particularly important because it recognizes that ecosystems are dynamic, although Robinson believes that *Caring for the Earth*'s definition of sustainable development reflects the popular misconception of "a balance of nature."

We would perhaps not need to emphasize the continuity of all kinds of change as a background to *Caring for the Earth* were it not that the criterion by which Robinson judges it seems to be the preservation of biological diversity. We have already mentioned the passages in *Caring for the Earth* which emphasize the need to maintain biological diversity, but we do find it necessary to remind readers that, with or without drastic human intervention, change in the form of extinction always has occurred and always will. *Caring for the Earth* is aimed at decelerating that process to the extent allowed by a scale of development necessary to achieve a tolerable level of human welfare throughout the world.

Caring for the Earth recognizes that there are conflicts between conservation and development (why otherwise would it call for the reduction of levels of consumption as well as population, for integrated resource management, environmental impact assessment, and so forth?), but it assumes that they can be resolved only by adequate knowledge, understanding, and good will. We have not yet reached the point at which those conditions prevail: *Caring for the Earth* is aimed at moving us toward that point. At a certain level, conservation and development may well be distinct activities, but they should not be separate; they should have a common context of concern for the whole community of life. They are essential parts of one indispensable process, which is the achievement through sustainable development of a decent future for humanity within the biosphere. The alternative is a continuation of the present profligate and destructive impact of people on the biosphere. Unless development is sustainable—based on conservation principles—there will be a massive breakdown in the biosphere, with grave loss of biodiversity and untold human strife and misery.

We should perhaps draw attention to a number of particular points in *Limits to Caring* with which we disagree, or where we feel that *Caring for the Earth* has been misinterpreted.

Page 22: Contrary to Robinson's view, sustainable development as described in *Caring for the Earth* both requires and *includes* conservation. It is possible, for example, to manage a forest for sustainable development by providing simultaneously—in adjacent, judiciously chosen, and carefully defined blocks—for both sustainable harvest and complete preservation of all species.

Page 23: *Caring for the Earth* is said to provide no analysis of how to change the process of development. We would say it is full of relevant suggestions: Chapter 11, for example, is devoted to practical suggestions of how business, industry, and commerce can become sustainable; Chapter 13 does the same for agriculture.

Page 24: We doubt that "*any* use of a biological community will ultimately involve a loss of biological diversity." We accept that that has often happened, but we don't believe it to be inevitable. Indeed, the prescriptions of *Caring for the Earth* are intended to prevent it. What is important is that all use of biological resources requires an appropriate balance between sustainable use and strict protection.

Page 26: We would consider it foolish to place limits on the loss of biodiversity that would be acceptable. It is realistic to accept that there will be losses but to define what would be acceptable would be tantamount to setting targets.

We should also draw attention to a number of points in *Limits to Caring* with which we fully agree. The discussions of ecologically sustainable use of species and biological communities are comprehensive and generally helpful. We particularly approve of the statement that "sustainable use only occurs when the rights of different user groups are specified, when human needs are met, and when the losses in biodiversity and environmental degradation are acceptable."

Limits to Caring characterizes *Caring for the Earth* as simplistically optimistic. So be it. The complexity of a dynamic biosphere no doubt exceeds our understanding, and human greed and frailty may well set limits to caring. We are not naive about this. But if a sustainable society, which means a society that minimizes losses of biological diversity, is worth striving for, we see as yet no alternatives to the policies and actions recommended in *Caring for the Earth*.

Certainly no such alternatives are proposed in *Limits to Caring*. We would have been deeply grateful if there had been.

Literature Cited

International Union for Conservation of Nature and Natural Resources/United Nations Environmental Program/The World Wide Fund for Nature. 1991. Caring for the Earth: A strategy for sustainable living. Gland, Switzerland.

Robinson, J. G. 1993. The limits to caring: Sustainable living and the loss of biodiversity. Conservation Biology 7:20–28.

"Believing What You Know Ain't So": Response to Holdgate and Munro

JOHN G. ROBINSON

NYZS The Wildlife Conservation Society
Bronx, NY 10460, U.S.A.

Martin Holdgate and David Munro have cogently summarized the intent and strategy followed by those who framed *Caring for the Earth* (International Union for the Conservation of Nature and Natural Resources/United Nations Environmental Program/World Wildlife Fund 1991). The pragmatic approach that they advocate is "to relate the primary objectives that [*Caring for the Earth*] espouses to the aspirations of the widest possible audience." Holdgate and Munro make no apologies for the fact that *Caring for the Earth* might be an "unattainable utopia" or "simplistically optimistic." Instead they argue, persuasively as always, that political documents must be broadly understandable and acceptable to the people that wield political and economic power in the world. In the words of the American author Mark Twain, the pragmatic approach is to go on "believing what you know ain't so."

The widespread acceptance of *Caring for the Earth* by those in government agencies and international development organizations is testimony to the success of this approach. But the ease with which the document has been accepted demonstrates that *Caring for the Earth* does not advocate a sea change in our patterns of resource exploitation. And such a transformation is necessary if we are to live sustainably. Consider the chapter on business, industry, and commerce, which is mentioned by Holdgate and Munro as "full of relevant suggestions." The chapter advocates resource and energy efficiency, technological innovation, waste reduction, and pollution prevention. These are all actions that can and do win support from industry, because they frequently increase profits. But there are no specific suggestions on how to make business operations more sustainable (beyond identifying sustainability as a need).

There is no discussion of the need for greater geographic equity in the distribution of industrial production, or of a more realistic valuation of the environmental costs of doing business. There is nothing to give a CEO pause for thought. Business as usual, though more palatable to world leaders, will not promote sustainable living.

Holdgate and Munro scold those that make "academic distinctions," and they suggest that understanding would not be promoted by "a detailed analysis of all the factors that must be considered." I disagree. Conservation policy must be rooted in the most complete understanding of the ecological, social, and economic contexts that we have available. We must examine the consequences of development and conservation initiatives on biodiversity, on the lives of rural peoples, and on the balance of power among different interest groups. And we must adjust our strategies so as to achieve the optimal balance between conflicting and frequently contradictory aims. To do otherwise relegates conservation to empty generalities and politically correct phrases. The need to incorporate scientific understanding is illustrated by Holdgate and Munro's response to *Limits to Caring* (Robinson 1993) in which they suggest that it is possible to manage "adjacent, judiciously chosen and carefully defined" blocks of forest for both sustainable harvest and complete preservation of all species. A huge literature on the effect of forest fragmentation on biodiversity argues otherwise. Realistic recommendations must derive from the applied research of many disciplines, and from the experience of those who have implemented conservation and development projects.

None of this denigrates the "avowed aim of *Caring for*

the Earth" which is "to improve the condition of the world's people." Neither does it belittle the useful and constructive suggestions with which *Caring for the Earth* is packed. But it does suggest that the information is available to develop more appropriate conservation policy, even though that policy may not be acceptable to all. Unless human society confronts its failure to use resources sustainably, then sustainable living will prove illusory.

Limits to Caring discusses the effect of sustainable living, as defined by *Caring for the Earth,* on biodiversity. A central theme of my critique is that when human beings live in an area and use natural resources, they degrade biodiversity. I argue that ecological theory and field data suggest that this loss is inevitable. While Holdgate and Munro do not concede this point, they do recognize that biodiversity is often lost. This is an important recognition, because it differentiates the goals of biodiversity conservation from those of sustainable development. Once one recognizes that conservation and development are mutually interdependent but separate processes, then one can examine the tradeoffs inherent in any development or conservation initiative. If one is interested in biodiversity conservation, then one can examine the loss associated with development actions and seek ways to minimize that loss. Certainly it is worth striving for the sustainable society in which human needs and aspirations are met. But there will be costs, and one of those costs will be a loss of biodiversity.

If there are losses, then, as argued in *Limits to Caring,* we need to define the "acceptable loss of biological diversity or environmental degradation." While Holdgate and Munro shy away from "targets," it is imperative that we define when the loss is too much. Are there species and ecosystems that we cannot afford to lose? What level of species loss will shackle human endeavour, destroying the raw material on which technological innovation depends? How degraded does a forest have to be before it is no longer a forest? What loss of ecosystem functioning will degrade the environment to such an extent that the carrying capacity for human beings is decreased? Answers to these questions will help to decide what needs to be preserved and what can be used.

Limits to Caring does not advocate that we embrace a single goal—that of protecting biodiversity. It does not vilify the goal of improving the condition of the world's people. It does not argue for stasis versus change, preservation versus use, "traditional conservation" versus sustainable development. It does advocate a more realistic appraisal of the tradeoffs between conservation and development. Not all areas or all resources can be used sustainably. Biodiversity cannot be preserved everywhere. Instead of a single universally applied goal, we need multiple goals sensitive to significance, to place, and to time. Instead of the single goal of sustainable development, we need the goals of preservation, sustainable use, sustainable development, and socioeconomic growth. Rather than seeking to make everything everywhere sustainable, we should aspire to a "sustainable landscape"—a landscape made up of a mosaic of different land uses, not all of which would be either productive or sustainable, but taken as a whole able to preserve biodiversity and allow sustainable living.

Finally, I would like to address the issue raised by Holdgate and Munro's final paragraph. Is it inappropriate to publicly examine important conservation documents before coming up with fully formed alternatives? The argument is frequently made that public discourse on goals and approaches will undermine the unity of conservation efforts. I believe that while *Caring for the Earth* is a praiseworthy document in many respects, it also promotes approaches that are ill-informed and incomplete. As such, its underlying concepts, and the approaches derived therefrom, need to be deliberately dissected. The field of conservation is mature enough to tolerate such analysis, and the resulting changes in its approaches will further the goals of conservation and sustainable living.

Acknowledgments

I would like to thank Bill Conway for his comments and his knowledge of Mark Twain, and Bill Weber, Linda Cox, and Dorene Bolze for a constructive reading of the manuscript.

Literature Cited

International Union for the Conservation of Nature and Natural Resources/United Nations Environmental Program/World Wildlife Fund. 1991. Caring for the Earth: A strategy for sustainable living. Gland, Switzerland.

Robinson, J. G. 1993. The limits to caring: Sustainable living and the loss of biodiversity. Conservation Biology 7:20–28.

Natural Capital and Sustainable Development

ROBERT COSTANZA

Director, Maryland International Institute for Ecological Economics
Center for Environmental and Estuarine Studies
University of Maryland
Box 38, Solomons, MD 20688, U.S.A.

HERMAN E. DALY*

Environment Department
The World Bank
1818 H. Street, NW
Washington, D.C. 20433, U.S.A.

Abstract: *A minimum necessary condition for sustainability is the maintenance of the total natural capital stock at or above the current level. While a lower stock of natural capital may be sustainable, society can allow no further decline in natural capital given the large uncertainty and the dire consequences of guessing wrong. This "constancy of total natural capital" rule can thus be seen as a prudent minimum condition for assuring sustainability, to be relaxed only when solid evidence can be offered that it is safe to do so.*

We discuss methodological issues concerning the degree of substitutability of manufactured for natural capital, quantifying ecosystem services and natural capital, and the role of the discount rate in valuing natural capital. We differentiate the concepts of growth (material increase in size) and development (improvement in organization without size change). Given these definitions, growth cannot be sustainable indefinitely on a finite planet. Development may be sustainable, but even this aspect of change may have some limits. One problem is that current measures of economic well-being at the macro level (i.e., the Gross National Product) measure mainly growth, or at best conflate growth and development. This urgently requires revision.

Finally, we suggest some principles of sustainable development and describe why maintaining natural capital stocks is a prudent and achievable policy for insuring sustainable development. There is disagreement between technological optimists (who see technical progress as eliminat-

Resumen: *Una condición mínima para el crecimiento sostenido es el mantenimiento del stock del capital natural total al presente nivel o por encima del mismo. Si bien un stock de capital natural menor podría ser sostenible, la sociedad no permite mayores declinaciones en el mismo debido a la gran incertidumbre y a las consecuencias lamentables que podría tener el adivinar erradamente. Esta regla "de constancia del capital natural total" puede por lo tanto ser considerada una prudente condicion mínima para asegurar sostenibilidad económica, que solo podría ser relajada cuando se den sólidas evidencias en contrario.*

Discutimos temas metodológicos que conciernen el grado de sostenibilidad económica de capital manufacturado por capital natural, cuantificación de los servicios del ecosistema y capital natural, y el rol de la tasa de descuento en la valoración de capital natural. Diferenciamos entre los conceptos de crecimiento (crecimiento material en tamaño) y desarrollo (mejoramiento en la organización sin cambio en tamaño). Dadas estas definiciones, el crecimiento no puede ser mantenido indefinidamente en un planeta limitado. El desarrollo puede ser sostenido, pero incluso este aspecto del cambio puede tener limites. Uno de los problemas es que las variables corrientemente usadas para medir el bienestar a nivel global (es decir el Producto Nacional Bruto) miden principalmente crecimiento, o como máximo relacionan entre si crecimiento y desarrollo. Esto requiere una revisión en forma urgente.

Finalmente, proponemos algunos principios de desarrollo

Paper submitted December 7, 1990; revised manuscript accepted August 5, 1991.
* *The views presented here are those of the author and should in no way be attributed to the World Bank.*

ing all resource constraints to growth and development) and technological skeptics (who do not see as much scope for this approach and fear irreversible use of resources and damage to natural capital). By maintaining natural capital stocks (preferably by using a natural capital depletion tax), we can satisfy both the skeptics (since resources will be conserved for future generations) and the optimists (since this will raise the price of natural capital depletion and more rapidly induce the technical change they predict).

sostenible y describimos porque el mantenimiento del stock de capital natural representa una política prudente y posible para asegurar un desarrollo sostenido. Existe un desacuerdo entre optimistas tecnológicos (que ven el progreso tecnológico como eliminando todos los límites, en cuanto a recursos, sobre el crecimiento y desarrollo) y escépticos tecnológicos (que no ven espacio suficiente para esta posibilidad y temen un uso irreversible de los recursos y un daño al capital natural). Manteniendo los stocks de capital natural (preferentemente usando un impuesto al uso exhaustivo de capital natural) podemos satisfacer tanto a los escépticos (dado que la recursos van a ser conservados para generaciones futuras) como a los optimistas (dado que esto va a incrementar el precio del uso exhaustivo de capital natural e inducirá más rápidamente los cambios técnicos que ellos predicen).

What Is Natural Capital?

Since "capital" is traditionally defined as produced (manufactured) means of production, the term "natural capital" needs explanation. It is based on a more functional definition of capital as "a stock that yields a flow of valuable goods or services into the future." What is functionally important is the relation of a stock yielding a flow—whether the stock is manufactured or natural is in this view a distinction between kinds of capital and not a defining characteristic of capital itself. For example, a stock or population of trees or fish provides a flow or annual yield of new trees or fish, a flow that can be sustainable year after year. The sustainable flow is "natural income"; the stock that yields the sustainable flow is "natural capital." Natural capital may also provide services such as recycling waste materials, or water catchment and erosion control, which are also counted as natural income. Since the flow of services from ecosystems requires that they function as whole systems, the structure and diversity of the system is an important component in natural capital.

We also need to differentiate between natural capital and income and natural resources. There are at least two possibilities here: (1) natural capital and natural income are simply the stock and flow components, respectively, of natural resources, and (2) natural capital and natural income are aggregates of natural resources in their separate stock and flow dimensions, and forming these aggregates requires some relative valuation of the different types of natural resource stocks and flows. Capital and income, in this view, have distinct evaluative connotations relative to the more physical connotations of the term "resources." We prefer the latter definition because it emphasizes the aggregate nature of terms such as "capital" and "income" while acknowledging that this aggregation is both a strength and a weakness.

We can differentiate two broad types of natural capital: (1) renewable or active natural capital, and (2) nonrenewable or inactive natural capital. Renewable natural capital is active and self-maintaining using solar energy. Ecosystems are renewable natural capital. They can be harvested to yield ecosystem goods (such as wood) but they also yield a flow of ecosystem services when left in place (such as erosion control and recreation). Nonrenewable natural capital is more passive. Fossil fuel and mineral deposits are the best examples. They generally yield no services until extracted. Renewable natural capital is analogous to machines and is subject to entropic depreciation; nonrenewable natural capital is analogous to inventories and is subject to liquidation (El Serafy 1989).

In addition, we can differentiate two broad types of human-made capital. One is the factories, buildings, tools, and other physical artifacts usually associated with the term "capital." A second is the stock of education, skills, culture, and knowledge stored in human beings themselves. The latter type is usually referred to as "human capital" while the former we will call simply "manufactured capital." Thus we have three broad types of capital: natural, human, and manufactured, corresponding roughly to the traditional economic factors of production of land, labor, and capital. In addition, we have the important distinction between renewable and nonrenewable natural capital, and for some purposes we can lump both human and manufactured capital together as "human-made capital."

Figure 1 elaborates these concepts and their interconnections. Manufactured capital (MC), human capital (HC), and renewable natural capital (RNC) decay at significant rates by the second law of thermodynamics and must constantly be maintained. Nonrenewable natural capital (NNC) also decays, but the rate is so slow relative to MC and RNC that this can be ignored. NNC

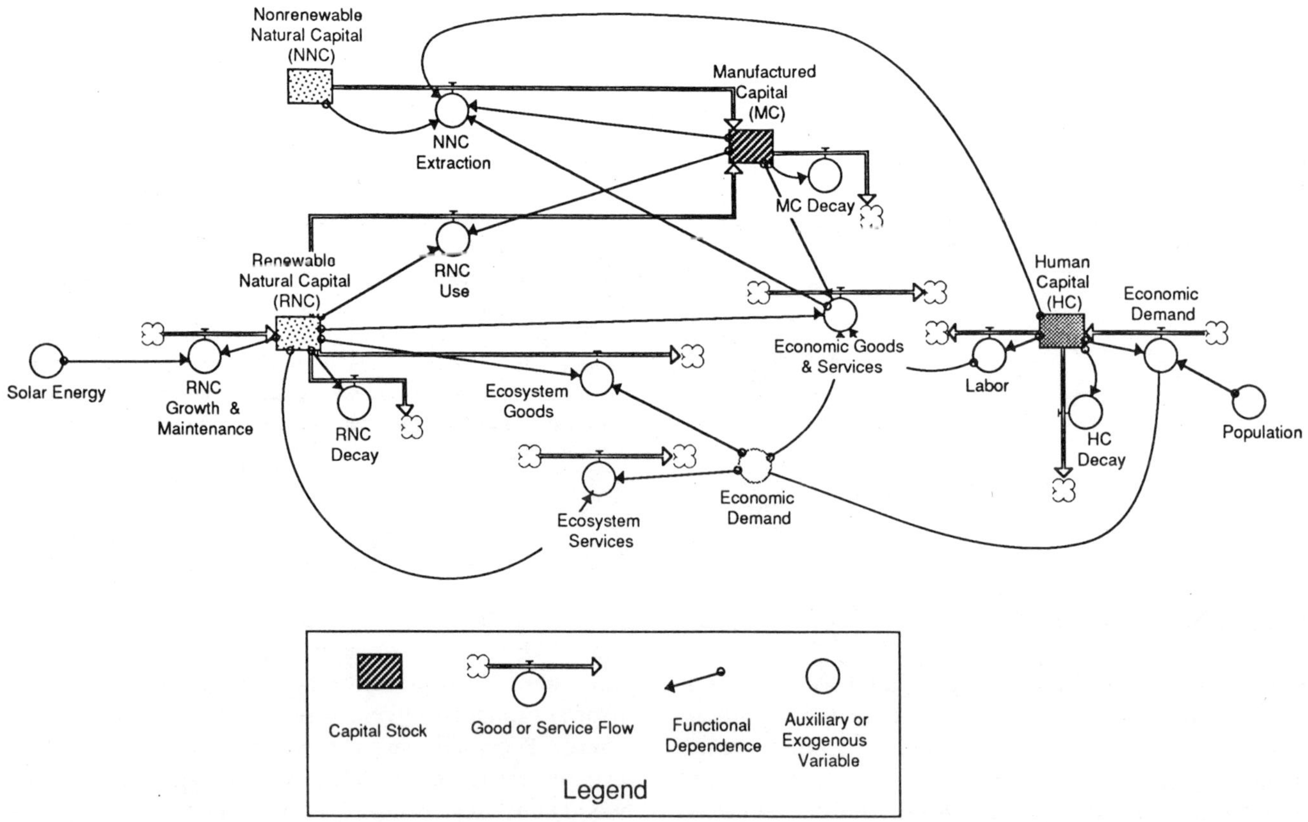

Figure 1. Types of natural and human-made capital stocks, good and service flows, and their interdependence.

can be viewed as a long-term inventory that will sit quietly until extracted and used, but once it is used it is gone. RNC produces both ecosystem goods (portions of the RNC itself) and ecosystem services, and renews itself using its own capital stock and solar energy. Excessive harvest of ecosystem goods can reduce RNC's ability to produce services and to maintain itself. MC, RNC, ecosystem services, and NNC interact with HC and economic demand to determine the level of "economic" (marketed) goods and services production. The form of this interaction is very important to sustainability, and it is not well understood (more on this later). Total income in the context of Figure 1 is a combination of traditional marketed economic goods and services, and nonmarketed ecosystem goods and services.

The concept of sustainability is implicit in the definition of income (following Hicks), so natural income must be sustainable; that is, any consumption that requires the running down of natural capital cannot be counted as income. This should at least be true for RNC. Since NNC must run down with use, a logical way to maintain constant income is to maintain as constant the total natural capital (TNC = RNC + NNC), which implies some reinvestment of the NNC consumed into RNC (as has been suggested by El Serafy [1989] for national income accounting [more on this later]).

Hence constancy of total natural capital (TNC) is the key idea in sustainability of development. It is important for operational purposes to define sustainable development in terms of constant or nondeclining TNC, rather than in terms of nondeclining utility (e.g., Pezzey 1989). While there are admittedly problems in measuring TNC, utility is beyond all hope of measurement. Aggregated, discounted future utility is what is really needed to operationalize the utility-based definition of sustainability, and that is even more of a will-o'-the-wisp. Also, an important motivation behind the sustainable development discussion is that of a just bequest to future generations. Utility cannot be bequeathed, but natural capital can be. Whether future generations use the natural capital we bequeath to them in ways that lead to happiness or to misery is beyond our control. We are not responsible for their happiness or utility—only for conserving for them the natural capital that can provide happiness if used wisely.

In the past, only manufactured stocks were considered as capital because natural capital was superabundant in that mankind's activities operated at too small a scale relative to natural processes to interfere with the free provision of natural goods and services. Expansion of manufactured and human capital entailed no opportunity cost in terms of the sacrifice of services of natural capital. Manufactured and human capital were the limiting factors in economic development. Natural capital was a free good. We are now entering an era, thanks to the enormous increase of the human scale, in which natural capital is becoming the limiting factor. Human economic activities can significantly reduce the capacity of natural capital to yield the flow of ecosystem goods and services and NNC upon which the very productivity of human-made capital depends.

Of course the classical economists (Smith, Malthus, Ricardo) emphasized the constraints of natural resources on economic growth, and several more recent economists, especially environmental and ecological economists, have explicitly recognized natural resources as an important form of capital that produces major contributions to human well-being (cf., Scott 1955; Daly 1968, 1973, 1977; Page 1977; Randall 1987; Pearce & Turner 1989). But environmental economics has, until now, been a tiny subfield far from the mainstream of neoclassical economics, and the role of natural resources within the mainstream has been de-emphasized almost to the point of oblivion. We believe that, if we are to achieve sustainability, the economy must be viewed in its proper perspective, as a subsystem of the larger ecological system of which it is a part, and that environmental and ecological economics need to become much more pervasive approaches to the problem (Costanza et al. 1991).

Why Is Accounting for Natural Capital So Important?

Natural capital produces a significant portion of the real goods and services of the ecological economic system, so failure to adequately account for it leads to major misperceptions about how well the economy is doing. This misperception is important at all levels of analysis, from the appraisal of individual projects to the health of the ecological economic system as a whole. Let us concentrate on the level of national income accounting, however, because of the importance of these measures to national planning and sustainability.

There has been much recent interest in improving national income and welfare measures to account for depletion of natural capital and other mismeasures of welfare (cf. Ahmad et al. 1989). Daly and Cobb (1989) have produced an index of sustainable economic welfare (ISEW) that attempts to account mainly for depletions of natural capital, pollution effects, and income distribution effects. Figure 2 shows two versions of their index compared to GNP over the 1950 to 1986 interval. What is strikingly clear from Figure 2 is that while GNP has been rising over this interval, ISEW has remained relatively unchanged since about 1970. When depletions of natural capital, pollution costs, and income distribution effects are accounted for, the economy is seen to be not improving at all. If we continue to ignore

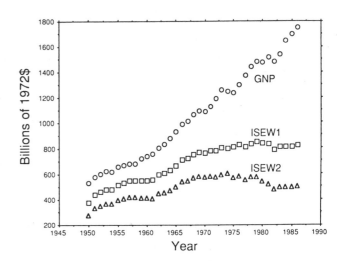

Figure 2. U.S. GNP compared with the Index of Sustainable Economic Welfare (ISEW, from Daly & Cobb 1989) for the interval 1950 to 1986. ISEW2 includes corrections for depletion of nonrenewable resources and long-term environmental damage; ISEW1 does not.

natural capital, we may well push welfare down while we think we are building it up.

Substitutability Between Natural and Man-made Capital

In addition to the former smallness of the human scale, a further reason for the neglect of natural capital has been the tenet of neoclassical economic theory that human-made capital is a near-perfect substitute for natural resources, and hence for the natural capital that generates the flow of natural resources. In the words of Nordhaus and Tobin (1972):

> The prevailing standard model of growth assumes that there are no limits on the feasibility of expanding the supplies of non-human agents of production. It is basically a two-factor model in which production depends only on labor and reproducible capital. Land and resources, the third member of the classical triad, have generally been dropped ... the tacit justification has been that reproducible capital is a near perfect substitute for land and other exhaustible resources.

The mathematical form assumed for the production function can also imply more substitutability than is there in reality. For example, even if natural capital is explicitly included in the production function, it makes little difference as long as the production function is a form (such as the Cobb-Douglas function) in which natural resources can approach zero with output remaining constant, and as long as reproducible (manufactured) capital or labor (human capital) are increased by a compensatory amount. In more technical terms, the elasticity of substitution of human-made for natural capital was assumed to be constant and high.

This assumption of near-perfect substitutability (high constant elasticity of substitution) has little support in logic or in fact. It was motivated more by mathematical convenience than anything else, except perhaps the hubris-driven technological dream of being independent of nature. Consider the following list of objections to the tenet of near-perfect substitutibility of human-made for natural capital:

1. If human-made capital were a perfect substitute for natural capital, then natural capital would also be a perfect substitute for human-made capital. But if the latter were the case there would be no reason to develop and accumulate human-made capital in the first place! Why does one need human-made capital if one already has an abundance of a near-perfect substitutes? Historically, we developed human-made capital as a complement to natural capital, not as a substitute. It should be obvious that the human-made capital of fishing nets, refineries, saw mills, and the human capital skill to run them does not substitute for, and would in fact be worthless without, the natural capital of fish populations, petroleum deposits, and forests.
2. Manufactured capital is itself made out of natural resources, with the help of human capital (which also consumes natural resources). Creation of the "substitute" requires more of the very thing that it is supposed to substitute for!
3. A physical analysis of "production" reveals that it is really a transformation process—a flow of natural resource inputs is transformed into a flow of product outputs by two agents of transformation, the stock of laborers (human capital) and the stock of manufactured capital at their disposal. Natural resources are that which is being transformed into a product (the material cause of production); manufactured and human capital are that which is effecting the transformation (the efficient cause of production). The relationship is overwhelmingly one of complementarity, not substitutability. The overwhelming reason for increasing the stock of human-made capital is to process a larger flow of natural capital, not to make possible a reduced flow. It is possible to reduce the waste of materials in process by investing capital in the recycling of prompt scrap, but this is marginal and limited.

The point is that the substitution of human-made physical capital for natural capital in the production of a given good is very limited, and that on the whole natural capital and human-made capital are complements in the production of any given good. There may remain considerable substitutability between human and manufactured capital (the two agents), or among various particular forms of natural capital (aluminum for copper, glass for aluminum), or even between NNC and RNC. That is not in dispute. Nor are we disputing the possibility of substituting a technically superior product that requires less energy and materials to render the same human service (e.g., cars that get more miles per gallon and light bulbs that give more lumens per watt). The latter is efficiency-increasing technical progress (development) as opposed to throughput-increasing technical progress (growth). But for any given product embodying any given level of technical knowledge, human-made capital and natural capital are, in general, complements, not substitutes.

Valuation of Natural Capital

The issue of valuation of natural capital is difficult but essential for many purposes, including aggregation and determining the optimal scale of human activities. The valuation of natural capital involves allocation of matter-energy across the boundary separating the economic subsystem from the ecosystem, and could be referred to as *macro-allocation*. By contrast *micro-allocation* is the allocation among competing uses of matter-energy that has already entered the economic subsystem—allocation proper. The logic defining the two optima is the same—the optimum is at the point where marginal costs equal marginal benefits. But the nature of the cost and benefit functions in the two cases is very different.

The cost and benefit functions relevant to the micro-allocation problem are those of individuals bent on maximizing their own private utility both as consumers and producers. The market coordinates and balances these individualistic maximizing efforts and in so doing determines a set of relative prices that measure opportunity cost. Individuals are allowed to appropriate matter-energy from the ecosystem as required for their individualistic purposes. Since the benefits of such expropriation are mostly private while the costs are largely social, there is a tendency to overexpand the scale of the economy—or to "allocate" too much of the matter-energy of the total ecosystem to the economic subsystem. Therefore the macro-allocation or scale problem should be viewed as a social or collective decision rather than an individualistic market decision. This means that the cost and benefit functions of macro-allocation are at the level of social preferences. A social preference function may give considerable weight to individual utility but is certainly not reducible to that alone. It has a community dimension. The value of community (with other people and other species, both present and future) must be counted in the cost and benefit functions associated with macro-allocation (Daly & Cobb 1989). These community costs and benefits are not captured in micro-allocation market prices.

How then are these nonmarket social costs and benefits measured? One approach is to imagine the valuation to be done by a different *Homo economicus* than the neoclassical pure individualist. This broader *Homo economicus* (call him *H-e* 2 to differentiate him from the neoclassical *H-e* 1) is a person in community rather than a pure individualist. *H-e* 2 is also fully informed about how the economy is related to the ecosystem and is constituted in his very identity by the relations of community with both future generations and other species with whom he shares a place in the sun. *H-e* 2 would value natural capital according to its relative long-term potential for supporting life and wealth in general. This long-term potential is closely associated with the low entropy matter-energy embodied in the natural capital. Therefore we offer as one hypothesis for investigation the idea that natural capital could be evaluated in proportion to its embodied energy (Costanza 1980; Cleveland et al. 1984). The willingness to pay of *H-e* 2 (person in community) is hypothesized to be in accordance with this long-run capacity to support life and wealth.

But it will be objected that this *H-e* 2 is not the "real" one. The "real" one (*H-e* 1) is generally ignorant of ecological relations, short-sighted, and individualistic. The "willingness to pay" of this more usual *H-e* 1 is the more common approach to the valuation of natural capital. Both concepts of *H-e* are abstractions from real people. For the micro-allocation problem we think people generally behave like the traditional individualistic *H-e* 1. But when confronted with the macro-allocation problem we think most people would behave more like *H-e* 2, the person in community. Therefore valuation of natural capital, we submit, should be done by individuals acting in an entirely different mode from that in which they operate in consumer markets. *H-e* 1 is different from *H-e* 2, but both are equally real as different aspects of real human beings relevant to different purposes. At any rate this is the interpretation we offer for the two methods of valuation we discuss here: the willingness-to-pay approach and the energy analysis approach.

Because natural capital is not captured in existing markets, special methods must be used to estimate its value. These range from attempts to mimic market behavior using surveys and questionnaires to elicit the preferences of current resource users (i.e., willingness-to-pay [WTP] to methods based on energy analysis [EA] of flows in natural ecosystems which do not depend on current human preferences at all). More complete discussions are given in Farber and Costanza (1987) and Costanza et al. (1989).

There are also problems common to valuing any kind of capital, including human-made capital. One can generally not value capital directly. The two options in use for MC are to value the net stream of services produced by the capital, or to value the cost of forming the capital.

With reference to Figure 1, for RNC this corresponds to estimating the present value of ecosystem goods and services production (with, for example, WTP) or to valuing the cost of RNC production (with, for example, EA). Table 1 summarizes results from a recent study of average wetland values in coastal Louisiana (a state containing 40% of the coastal wetlands in the United States) as an example. Details of the methods, especially their conceptual and empirical assumptions and uncertainties, are contained in Farber and Costanza (1987) and Costanza et al. (1989).

Discounting

Often the present-vs.-future issue is thought to be objectively decided by discounting. But discounting at best only reflects the subjective valuation of the future to presently existing individual members of human society. Discounting is simply a numerical way to operationalize the value judgment that (1) the near future is worth more than the distant future to the present generation of humans, and (2) beyond some point the worth of the future to the present generation of humans is negligible. Economists tend to treat discounting as rational, optimizing behavior based on people's inherent preferences for current over future consumption.

There is evidence, however, that discounting behavior may be symptomatic of a kind of semirational, suboptimizing behavior known as a "social trap." A social trap is any situation in which the short-run, local reinforcements guiding individual behavior are inconsistent with the long-run, global best interest of the individual or society (Platt 1973; Cross & Guyer 1980; Costanza 1987). We go through life making decisions about which path to take based largely on the "road signs," the short-run, local reinforcements that we perceive most directly. These short-run reinforcements can include monetary incentives, social acceptance or admonishment, and physical pleasure or pain. Problems arise, however, when the road signs are inaccurate or misleading. In these cases we can be trapped into following a path that is ultimately detrimental because of our reli-

Table 1. Summary of wetland Renewable Natural Capital (RNC) value for coastal wetlands in Louisiana. Estimates (1983 dollars).

	Per-acre present value at specified discount rate	
Method	8%	3%
WTP based		
Commercial fishery	$ 317	$ 846
Trapping	151	401
Recreation	46	181
Storm protection	1,915	7,549
Total	$2,429	$8,977
Option and existence values	?	?
EA based		
GPP conversion	$6,400–10,600	$17,000–28,200
"Best estimate"	$2,429–6,400	$8,977–17,000

ance on the road signs. Discounting may allow individuals to give too little weight to the future (or other species, other groups or classes of humans, etc.) and thus helps to set the trap. Economists, while recognizing that individual behavior may not always lead to optimal social behavior, generally assume that discounting the future is an appropriate thing to do. The psychological evidence indicates, however, that humans have problems responding to reinforcements that are not immediate (in time and space) and can be led into disastrous situations because they discount too much.

It can therefore be argued that the discount rate used by the government for public policy decisions (like valuing natural capital) should be significantly lower than the rate used by individuals for private investment decisions. The government should have greater interest in the future than individuals currently in the market because continued social existence, stability, and harmony are public goods for which the government is responsible, and for which current individuals may not be willing to fully pay (Arrow 1976).

Discounting future value by the rate of interest also provides a tight link between ecological destruction and macroeconomic policy. Any exploited species whose natural rate of population growth is less than the real rate of interest is under threat of extinction, even in the absence of common property problems. While Alan Greenspan and the Federal Reserve probably do not worry about the effect of U.S. interest rate policy on deforestation in the Amazon or destruction of Louisiana wetlands, such links really do exist, and they probably should be broken.

In terms of the natural capital valuation problem, all this merely increases the uncertainty concerning the total present value because the appropriate discount rate is uncertain and makes a big difference in the results. In the wetland valuation example mentioned above, estimates for a range of discount rates (3–8%) were given to demonstrate how much uncertainty is introduced by uncertainty in the discount rate. We've also given arguments for why a lower discount rate may be more appropriate for natural capital valuation decisions. Indeed there is a reasonable case to be made for a zero discount rate in decisions taken on behalf of society at large (Page 1977; Georgescu-Roegen 1981), since society, unlike the individual, is quasi-immortal. A zero discount rate gives infinite or very large values for any indefinitely sustainable stream of income. The wants of future generations will be just as immediate to them as ours are to us. And if the fears of many climatologists and ecologists prove correct, productivity growth will be negative in the long run, so that equity would even require discounting at a negative rate—that is, future resources should be valued more highly than present resources.

Another possibility (Hannon 1985) is that the appropriate discount rate for natural capital should be linked to the natural growth and decay rates (see Fig. 1). RNC will not produce a stream of benefits into the indefinite future unless it is constantly supplied with new energy to maintain it against entropic decay. If this energy were not put into the natural capital stock in question it could be used to maintain some other natural capital stock. The "natural" discount rate might therefore be tied to the average natural decay rate (probably somewhere on the order of 1–3% per year). This is an issue for further research.

Growth, Development, and Sustainability

Improvement in human welfare can come about by pushing more matter-energy through the economy or by squeezing more human want satisfaction out of each unit of matter-energy that passes through. These two processes are so different in their effect on the environment that we must stop conflating them. It is better to refer to throughput increase as *growth,* and efficiency increase as *development.** Growth is destructive of natural capital and beyond some point will cost us more than it is worth—that is, sacrificed natural capital will be worth more than the extra man-made capital whose production necessitated the sacrifice. At this point growth has become anti-economic, impoverishing rather than enriching. Development, that is qualitative improvement does not occur at the expense of natural capital. There are clear economic limits to growth, but not to development. This is not to assert that there are no limits to development, only that they are not so clear as the limits to growth, and consequently there is room for a wide range of opinion on how far we can go in increasing human welfare without increasing resource throughput. How far can development substitute for growth? This is the relevant question, not how far can human-made capital substitute for natural capital, the answer to which, as we have seen, is "hardly at all."

Some people believe that there are truly enormous possibilities for development without growth. Energy efficiency, they argue, can be vastly increased (Lovins 1977; Lovins & Lovins 1987); so can the efficiency of water use. Potential efficiency increases for other materials are not so clear. Others (Costanza 1980; Cleveland et al. 1984; Hall et al. 1986; Gever et al. 1986) believe that the coupling between growth and energy use is not so loose. This issue arises in the Brundtland Commission's Report (WCED 1987), which recognizes on the

---
**This distinction is explicit in the dictionary's first definition of each term. To grow means literally "to increase naturally in size by the addition of material through assimilation or accretion." To develop means "to expand or realize the potentialities of; bring gradually to a fuller, greater, or better state." (The American Heritage Dictionary of the English Language).*

one hand that the scale of the human economy is already unsustainable in the sense that it requires the consumption of natural capital, and on the other hand calls for further economic expansion by a factor of 5 to 10 to improve the lot of the poor without having to appeal too much to the "politically impossible" alternatives of serious population control and redistribution of wealth. The big question is, how much of this called-for expansion can come from development and how much must come from growth? This question is not addressed by the Commission. But statements from the leader of the WCED, Jim MacNeil (1990), that "The link between growth and its impact on the environment has also been severed" (p. 13), and that "the maxim for sustainable development is not 'limits to growth'; it is 'the growth of limits,'" indicate that WCED expects the lion's share of that factor of 5 to 10 to come from development, not growth. They confusingly use the word "growth" to refer to both cases, saying that future growth must be qualitatively very different from past growth. When things are qualitatively different it is best to call them by different names. Hence our distinction between growth and development. Our own view is that WCED is too optimistic—that a factor of 5 to 10 increase cannot come from development alone, and that if it comes mainly from growth it will be devastatingly unsustainable. Therefore the welfare of the poor, and indeed of the rich, depends much more on population control, consumption control, and redistribution than on the technical fix of a 5- to 10-fold increase in total factor productivity.

We acknowledge, however, that there is a vast uncertainty on this critical issue of the scope for economic development from increasing efficiency. We have therefore devised a policy that should be sustainable regardless of who is right in this debate. We save its description for the final section. First some general principles of sustainable development.

Toward Operational Principles of Sustainable Development

The concept of sustainable development has received much attention lately, but research into how the concept might be operationalized is only beginning (Pearce & Turner 1989; Daly 1990; Costanza 1991). Below we sketch out the broad outlines of some operational principles of sustainability, while acknowledging that we still have a long way to go (both scientifically and politically) to achieve them. All the more reason to get started.

Weak sustainability is the maintaining intact of the sum of human-made and total natural capital. Even that is not done currently. Strong sustainability is the maintaining intact of natural capital and man-made capital separately. Weak sustainability would require the pricing of natural capital, which as we have just argued itself requires a given scale, that is, the holding constant of natural capital at some level, which is to say strong sustainability. So we can concentrate on strong sustainability, maintaining total natural capital intact. What does this mean operationally?

(1) The main principle is to limit the human scale to a level which, if not optimal, is at least within the carrying capacity of the remaining natural capital and therefore sustainable. Once carrying capacity has been reached, the simultaneous choice of a population level and an average "standard of living" (level of per capita resource consumption) becomes necessary. Sustainable development must deal with sufficiency as well as efficiency and cannot avoid limiting physical scale.

(2) Technological progress for sustainable development should be efficiency-increasing rather than throughput-increasing. Limiting the scale of resource throughput by high resource taxes would induce this technological shift, as discussed further below.

(3) RNC, in both its source and sink functions, should be exploited on a profit-maximizing sustained-yield basis, and in general stocks, should not be driven to extinction since they will become ever more important as NNC runs out. Specifically this means that:
 (a) harvesting rates should not exceed regeneration rates; and
 (b) waste emissions should not exceed the renewable assimilative capacity of the environment.

(4) NNC should be exploited, but at a rate equal to the creation of renewable substitutes. Nonrenewable projects should be paired with renewable projects and their joint rate of return should be calculated on the basis of their income component only, since that is what is perpetually available for consumption in each future year. It has been shown (El Serafy 1989) how this division of receipts into capital to be reinvested and income available for current consumption depends on the discount rate (rate of growth of the renewable substitute) and the life expectancy of the NNC (reserves divided by annual depletion). The faster the growth of the renewable substitute and the longer the life expectancy of the NNC, the greater will be the income component and the less the capital set-aside. "Substitute" here should be interpreted broadly to include any systemic adaptation that allows the economy to adjust to the depletion of the nonrenewable resource in a way that maintains future income at present levels (e.g., recycling).

Specific application of principle (3) might, for example, involve such requirements as no net depletion of aquifers or of topsoil (on the input side) and no net increase in soil acidity, salinization, or toxification (on the waste output side). Principle (1), general respect for carrying capacity, can be straightforwardly applied in rangelands, but can also be extended to industrial projects by requiring that all natural capital used by the industry be maintained without depletion.

These principles move us some distance toward operationalizing the basic notion that we should satisfy the needs of the present without sacrificing the ability of future populations to meet their needs. But they clearly fall far short of an operational blueprint complete with measurements. However, as argued in the following section, the principles are operational enough to guide some important policy changes without precise measures of assimilative capacities and sustainable yields. Uncertainty itself is one of the critical factors that must be addressed in designing sustainable policies.

A Fail-Safe Policy Proposal to Achieve Sustainability

We end with a policy proposal that is simple in concept (though not in implementation) and that accomplishes much toward the end of sustainable development. In spite of the disagreement over how much to expect from development without growth, both sides should be able to agree on the following. Strive to hold throughput (consumption of TNC) constant at present levels (or lower truly sustainable levels) by taxing TNC consumption, especially energy, very heavily. Seek to raise most public revenue from such a natural capital depletion (NCD) tax, and compensate by reducing the income tax, especially on the lower end of the income distribution, perhaps even financing a negative income tax at the very low end. Technological optimists who believe that efficiency can increase by a factor of ten should welcome this policy, which raises natural resource prices considerably and would powerfully encourage just those technological advances in which they have so much faith. Skeptics who lack that technological faith will nevertheless be happy to see the throughput limited since that is their main imperative in order to conserve resources for the future. The skeptics are protected against their worst fears; the optimists are encouraged to pursue their fondest dreams. If the skeptics are proven wrong and the enormous increase in efficiency actually happens, then they will be even happier (unless they are total misanthropists). They got what they wanted, but it just cost less than they expected and were willing to pay. The optimists, for their part, can hardly object to a policy that not only allows but offers strong incentives for the very technical progress on which their optimism is based. If they are proved wrong at least they should be glad that the rate of environmental destruction has been slowed.

Implementation of this policy does not hinge upon the precise measurement of natural capital. The valuation issue remains relevant in the sense that our policy recommendation is based on the perception that we are at or beyond the optimal scale. The evidence for this perception consists of the greenhouse effect, ozone layer depletion, acid rain, and general decline in many dimensions of the quality of life. It would be helpful to have better quantitative measures of these perceived costs, just as it would be helpful to carry along an altimeter when we jump out of an airplane. But we would all prefer a parachute to an altimeter if we could take only one thing. The consequences of an unarrested free fall are clear enough without a precise measure of our speed and acceleration. But we would need at least a ballpark estimate of the value of natural capital depletion in order to determine the magnitude of the suggested NCD tax. This, we think, is possible, especially if uncertainty about the value of natural capital is incorporated in the tax itself, using, for example, the refundable assurance bonding system proposed by Costanza and Perrings (1990).

The political feasibility of this policy is an important and difficult question. It certainly represents a major shift in the way we view our relationship to natural capital and would have major social, economic, and political implications. But these implications are just the ones we need to expose and face squarely if we hope to achieve sustainability. Because of its logic, its conceptual simplicity, and its built-in market incentive structure leading to sustainability, the proposed NCD tax may be the most politically feasible of the possible alternatives to achieving sustainability.

We have not tried to work out all the details of how the NCD tax would be administered. In general, it could be administered like any other tax, but it would probably require international agreements or at least national ecological tariffs to prevent some countries from flooding markets with untaxed natural capital or products made with untaxed natural capital. By shifting most of the tax burden to the NCD tax and away from income taxes, the NCD tax could actually simplify the administration of the taxation system while providing the appropriate economic incentives to achieve sustainability.

Acknowledgments

This paper was originally prepared for a workshop on natural capital organized by Barry Sadler for the Canadian Environmental Assessment Research Council (CEARC) and held in Vancouver, British Columbia, Canada, March 15 and 16 1990. We thank the participants at the workshop for feedback on early drafts of the paper, as well as John Cumberland, Dennis King, Colin Clark, and one anonymous reviewer for detailed and helpful suggestions on a subsequent draft.

Literature Cited

Ahmad, Y. J., S. El Serafy, and E. Lutz. 1989. Environmental accounting for sustainable development. A UNEP—World Bank Symposium. The World Bank, Washington, D.C.

Arrow, K. J. 1976. The rate of discount for long-term public investment. In H. Ashley, R. L. Rudman, and C. Shipple, editors. Energy and the environment: a risk benefit approach. Pergamon Press, New York.

Cleveland, C. J., R. Costanza, C. A. S. Hall, and R. Kaufmann. 1984. Energy and the United States economy: a biophysical perspective. Science **255**:890–897.

Costanza, R. 1980. Embodied energy and economic valuation. Science **210**:1219–1224.

Costanza, R. 1987. Social traps and environmental policy. BioScience **37**:407–412.

Costanza, R. editor. 1991. Ecological economics: the science and management of sustainability. Columbia University Press, New York. 525 pp.

Costanza, R., H. E. Daly, and J. A. Bartholomew. 1991. Goals, agenda, and policy recommendations for ecological economics. Pages 1–21 in R. Costanza, editor. Ecological economics: the science and management of sustainability. Columbia University Press, New York. 525 pp.

Costanza, R., S. C. Farber, and J. Maxwell. 1989. The valuation and management of wetland ecosystems. Ecological Economics **1**:335–362.

Costanza, R., and C. Perrings. 1990. A flexible assurance bonding system for improved environmental management. Ecological Economics **2**:57–76.

Cross, J. G., and M. J. Guyer. 1980. Social traps. University of Michigan Press, Ann Arbor, Michigan.

Daly, H. E. 1968. On economics as a life science. Journal of Political Economy **76**:392–406.

Daly, H. E. 1973. Toward a steady-state economy. W.H. Freeman and Co., San Francisco, California.

Daly, H. 1977. Steady-state economics: the political economy of bio-physical equilibrium and moral growth. W.H. Freeman and Co., San Francisco, California.

Daly, H. E. 1990. Toward some operational principles of sustainable development. Ecological Economics **2**:1–6.

Daly, H. E., and J. B. Cobb, Jr. 1989. For the common good: redirecting the economy toward community, the environment, and a sustainable future. Beacon Press, Boston, Massachusetts. 482 pp.

El Serafy, S. 1989. The proper calculation of income form depletable natural resources. Pages 10–18 in Y. J. Ahmad, S. El Serafy, and E. Lutz, editors. Environmental accounting for sustainable development. A UNEP—World Bank Symposium. The World Bank, Washington, D.C.

Farber, S., and R. Costanza. 1987. The economic value of wetlands systems. Journal of Environmental Management **24**:41–51.

Georgescu-Roegen, N. 1981. Energy, matter, and economic valuation: where do we stand? Pages 43–79 in H. E. Daly and A. F. Umaña, editors. Energy, economics, and the environment: conflicting views of an essential interrelationship. AAAS Selected Symposium 64. Westview Press, Boulder, Colorado.

Gever, J., R. Kaufmann, D. Skole, and C. Vörösmarty. 1986. Beyond oil: the threat to food and fuel in the coming decades. Ballinger, Cambridge, Massachusetts. 304 pp.

Hall, C. A. S., C. J. Cleveland, and R. Kaufmann. 1986. Energy and resource quality: the ecology of the economic process. John Wiley & Sons, New York. 577 pp.

Hannon, B. 1985, Ecosystem, flow analysis. Canadian Journal of Fisheries and Aquatic Sciences **213**:97–118.

Lovins, A. B. 1977. Soft energy paths: toward a durable peace. Ballinger, Cambridge, Massachusetts.

Lovins, A. B., and L. H. Lovins. 1987. Energy: the avoidable oil crisis. The Atlantic (December): 22–30.

MacNeil, J. 1990. Sustainable development economics, and the growth imperative. Background Paper No. 3, Workshop on the Economics of Sustainable Development, January 23. Washington, D.C.

Nordhaus, W., and J. Tobin. 1972. Is growth obsolete? National Bureau of Economic Research, Columbia University Press, New York.

Page, T. 1977. Conservation and economic efficiency: an approach to materials policy. Resources for the Future, Washington, D.C. 266 pp.

Pearce, D. W., and R. K. Turner. 1989. Economics of natural resources and the environment. Wheatsheaf, Brighton, U.K.

Pezzey, J. 1989. Economic analysis of sustainable growth and sustainable development. Environment department working paper No. 15. The World Bank, Washington, D.C.

Platt, J. 1983. Social traps. American Psychologist **28**:642–651.

Randall, A. 1987. Resource economics. 2nd ed. John Wiley & Sons, New York.

Scott, A. D. 1955. Natural resources: the economics of conservation. University of Toronto Press, Toronto, Canada.

WCED. 1987. Our common future: report of the world commission on environment and development. Oxford University Press, Oxford, England.

Water, Endangered Fishes, and Development Perspectives in Arid Lands of Mexico

SALVADOR CONTRERAS-B.
M. LOURDES LOZANO-V.

Facultad de Ciencias Biologicas
Universidad Antonoma de Nuevo León
Apdo. Postal 504
San Nicolás de los Garza
Nueva León 66450, México

Abstract. Nearly half of Mexico is arid or semiarid, with scarce waters. At least 92 springs and 2500 km of river have dried in this area. Surface waters have diminished, and phreatic waters are sinking deeper, provoking intrusion of saline waters and salinization of agricultural wells in Sonora, reversing phreatic circulation in the Comarca Laguenera, allowing arrival of arsenic to agricultural waters, and threatening metropolitan Torreón. There are nearly 200 species of freshwater fishes in this region, 120 under some threat, 15 extinct through human impact. As of 1985, an average of 68% of species was eradicated in local fish faunas. Finally, salinization of the lower Rio Bravo del Norte has replaced 32 native fish of fresh or slightly brackish water with 54 mainly marine or highly salt-tolerant species; the salinization threatens all uses of water. Some marine fishes invade up to 400 km upstream. Pollution is strong, and fish kills have been reported. These low-quality and scarce waters comprise future resources for cities such as Monterrey, which, along with its border twin cities, is expected to double its human population by 2010. Redesign of regional development is urgently needed, in keeping with the real availability of water. All water use should be equal to or less than lower recharge averages; norms of integral basin management should rely on criteria of high-use efficiency, recuperation, recycling, and reutilization. Also necessary are reduced pollution and increased treatment of residual waters. Innovative environmental vision is especially essential in light of expectations for development through the North American Free Trade Agreement, the Border Integral Environmental Plan, industrial expansion, and the modernization

Paper submitted January 5, 1993; revised manuscript accepted September 7, 1993.

Aguas, peces en peligro, y perspectivas de desarrollo en las tierras áridas de México

Resumen: *Casi la mitad de México es árido o semiárido, con escasas aguas. Al menos, 92 manantiales y 2500 Km de ríos se han secado. Las aguas superficiales han disminuído y las aguas freáticas se encuentran a mayor profundidad, provocando la intrusión de aguas salinas y la salinización de pozos agrícolas en Sonora, revirtiendo la circulación de la capa freática en la Comarca Lagunera, provocando la llegada de arsénico a aguas agrícolas y amenazando el área metropolitana del Torreón. Existen apróximadamente 200 especies de peces de agua dulce en esta región, 120 están amenazadas y 15 extintas por impacto humano. En 1985 un promedio de 68 especies fueron erradicadas de las faunas de peces locales. Finalmente, la salinización del bajo Río Bravo del Norte ha causado un cambio de 32 especies nativas dulceacuícolas o salobres, a 54 especies principalmente marinas o de una alta tolerancia a la salinidad, poniendo en peligro todo uso del agua. Algunas especies marinas penetran hasta 400 Km aguas arriba. La contaminación es grave y ha sido reportada la mortandad de peces, sin embargo, estas aguas escasas y de baja calidad constituyen los recursos futuros de ciudades como Monterrey y las ciudades gemelas fronterizas, cuyas poblaciones se duplicarán para el año 2010. Urge replantear el desarrollo regional y medirlo en función de la disponibilidad real de agua. Todos los usos deben ser iguales o menores a la capacidad de recarga promedio, las normas de manejo integral de la cuenca deben basarse en un criterio de alta eficiencia de uso, recuperación y reciclaje. También, es necesario reducir la polución e incrementar el tratamiento residual de las aguas. En necesaria una visión ambiental innovativa, epecialmente debido a las expectativas de desarrollo a través del tratado de libre co-*

and internationalization of the northern Mexican border belt. These all conflict with the high priority recently decreed for species conservation by the Mexican Act, creating the National Committee on Knowledge and Use of Biodiversity.

mercio con Norte América, el Acuerdo de Protección Integral Ambiental Fronteriza, la expansión industrial, la modernización e internacionalización de la franja fronteriza norte. También debe considerarse la alta prioridad decretada recientemente por el Gobierno Mexicano para la Conservación de especies, creando la Comisión Nacional para el Conocimiento y Uso de la Biodiversidad.

Introduction

Arid and semiarid zones comprise from 40% to 60% of land in México, depending on the criteria applied. In this paper we (1) discuss consequences of economic development on diversity of endangered and threatened fishes, and problems of water availability, in the Mexican states of Baja California, Sonora, Chihuahua, Coahuila, Nuevo León, Tamaulipas, Durango, Zacatecas and San Luis Potosí (Figure 1); and (2) provide an overview of some ongoing environmental impacts that diminish biodiversity through abuse or misuse of water and display a lack of a broad environmental understanding. These points are particularly relevant to the North American Free Trade Agreement and the Border Environmental Plan, which do not take adequate account of biodiversity considerations (NAFTA 1992; SEDUE/EPA 1992).

This paper is not an exhaustive scientific review but rather a position paper summarizing personal research findings and other facts important for the future of the Méxican-U.S. border states.

In 1988, México promulgated legislation that made environmental impact evaluation a policy for all public or private works, mentioning, without defining or listing, endangered species and the legally recognized need of conservation. Until recently, there was no specific legal responsibility toward the knowledge, understanding, use, or protection of biodiversity as such. On March 16, 1992, the Mexican federal government promulgated the Decreto para el Uso y Conocimiento de la Biodiversidad, an act that assigned such responsibilities to all citizens and created the Comisión Nacional for said objectives. This act completes the legal background creating a program to protect Mexican flora and fauna, but it will be necessary to improve public understanding of

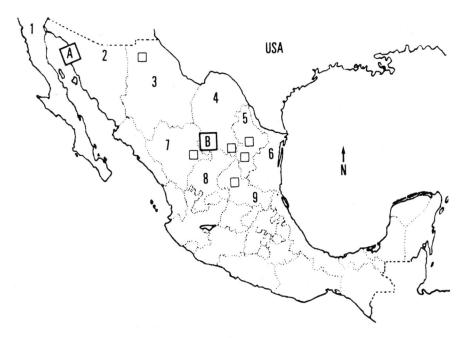

Figure 1. Northern Mexican States: (1) Baja California, (2) Sonora, (3) Chihuahua, (4) Coahuila, (5) Nuevo León, (6) Tamaulipas, (7) Durango, (8) Zacatecas, (9) San Luis Potosí. Large squares: (A) Sonora Coastal Plain, (B) Comarca Lagunera. Small squares; regions of drying springs referred to in text.

the survival of fishes, both as part of biodiversity and as indicators and sufferers of environmental degradation under global and regional views.

Water in Northern Mexico

Loss of Water Quantity

An early symptom of regional water deficiency is the drying of springs. Preliminary counts of dry springs in Northern México include the following: Nuevo León, 20 springs; Coahuila, more than 15 springs in the Saltillo Valley (a reference from the colonial period tells of 665 springs in this valley alone), and more than 10 springs each in the bolsones (playa lakes) of Viesca and Parras; Durango, 20 springs; Chihuahua, 5 springs; and San Luis Potosi, 12 springs (Fig. 1) (Contreras & Almada 1991). No data are available for Zacatecas or Sonora. The situation in Texas is not very different (Brune 1975, 1981). The data are from inspected localities only, and exploration is far from complete. Losses of springs have accelerated during the last 20 years, but only approximate dates of their dissappearance are known.

Another symptom of water deficiency is the lowering of riverine discharges, as in the following rivers: in Chihuahua, the lower Río del Carmen and Middle Rio Grande above Presidio; in Durango/Coahuila, Río Nazas and Bolson Mayrán, Río Aguanaval and Bolsón Viesca; in Coahuila, Río de Nadadores; in Coahuila/Nuevo León, Saltillo and Río Salinas; in Nuevo León, Río Sabinas and Río Santa Catarina; in San Luis Potosi, Río Ahualulco-Venado-Moctezuma region. These rivers have all become dry, intermittent, or occasional within the past two decades (Fig. 2) (Contreras & Lozano 1993).

Another indicator of water loss is lowering of the water table, which in Monterrey, Nuevo León, now lies at 200 m beneath the surface; in Comarca Lagunera, Coahuila, at 80 m, and in Sandia el Grande, Nuevo León, between 2 and 10 m, the last being close to the average for northern México. These are some of the same areas that contain dry springs. Some governmental offices still deny this collective information on loss of water resources.

In 1975, Plan Nacional Hidráulico (PNH 1975) issued a map of northern México with aquifer recharge, extraction, demand, and planned expansion sites. Using such

Figure 2. Rivers of Northern México that are becoming dry or intermittent: (1) Lower Río Carmen, (2) Middle Río Grande above Presidio, Texas (3) Lower Río Nazas and Laguna Mayrán, (4) Middle and Lower Río Aguanaval and Laguna Viesca, (5) Rio de Nadadores, (6) Rio Sabinas, (7) Valle de Saltillo and Río Salinas, (8) Río Santa Catarina, and (9) Río Ahualulco-Venado-Moctezuma region.

data, 21.1% of sites were considered underexploited, 26.3% were overexploited, 52.6% had unknown recharge rates, and the average overexploitation was 66.2%; the map included data on immediate demand and planned expansion that already represented an 80% increase in water use (PNH 1975; Contreras 1975). We submit that no planning for further development or determination of rates of exploitation can be done without knowledge of water use and recharge rates.

In contrast, a few years later, the Instituto Nacional de Estadística, Geografía e Informática's geographical syntheses (INEGI: Nuevo León 1981; Zacatecas 1981; Coahuila 1983; Tamaulipas 1983; Baja California 1984) showed that in the early 1980s, with most aquifers having unknown recharges, 14,445 out of 20,249 (72%) water wells were regarded as underexploited (Table 1), in spite of the overall decline of the aquifers and obvious rise in regional development and groundwater use. Public explanation is thus required for the criteria defining over- or underexploitation. How can it be defined in the

Table 1. Wells considered to be underexploited, overexploited, in equilibrium, or dry in northern Mexico.

	Total	Under Exploited	Over Exploited	In Equilibrium	Dry
Baja California	3,472	147	2,825	482	—
Coahuila	4,525	2,689	1,836	—	—
Nuevo Leon	11,149	10,623	—	526	—
Tamaulipas	986	986	—	—	9
Zacatecas	117	?	?	?	9
Totals	20,249	14,445	4,661	1,008	18

Source: INEGLI (1981; 1983).

absence of recharge data? We define overexploitation as water extraction above the lowest average recharge level, which causes a lowering of the water table and drying of springs. Rational exploitation would then be defined as regulating the maximum average extraction to below the minimal average recharge as a safe and permissible average (maximum sustainable) use. The Mexican federal government has received outdated information and urgently needs to reconsider actual water availability.

Degradation of Water Quality

There was an apparently accidental spill of sulphuric acid on the Río Salinas of Nuevo León in 1990 (Fig. 3), resulting in a fish kill that was roughly estimated as somewhat less than 2000 fish *per* linear meter of river (average width 20 m), or 96 dead fish per square meter. The upper 2 km of river were explored, with an estimated 4,000,000 fish killed. Visible impact of the spill extended 20 km more downstream. There were no estimates for the intermediate area, and hence total impact can only be estimated (S. Contreras-B, personal observation). Incidents of this type may be expected more frequently as the region becomes more densely populated and industrialized.

The lower Río Bravo del Norte (Rio Grande) and lower Río San Juan, into the latter of which the Río Salinas flows, comprise some of the best irrigation districts along the Méxican-U.S. border. Yet, as an indication of broad ecological change, marine and brackish-water fishes are invading this area. In the 1850s, only six marine species were recorded (Edwards & Contreras 1991); in 1953 they were seven; by 1976 their numbers had risen to 18 (Contreras et al. 1976) and by 1989 to 54 (Edwards & Contreras 1991). This penetration has extended up to 400 km upstream from the ocean (Contreras et al. 1976). One species, tidewater silversides (*Menidia beryllina*), had by 1978 reached within 20 km northeast of Monterrey, Nuevo León, at Laguna Montfort (Figure 3). This invasion was stopped by pollution from Monterrey municipal and industrial sewage. In contrast, some freshwater species have retreated from the lower 400 km of the river (Contreras 1975; Contreras et al. 1976). From a qualitative viewpoint, invasion of marine/brackish fishes and retreat of freshwater species indicate that salinization is increasing rapidly, represents a threat to agriculture, will corrode industrial machinery, and will involve costly water treatment.

The Comarca Lagunera (Torreon Region), Coahuila

The Comarca Lagunera and 30,000 of its human inhabitants are suffering from a complex problem. Maeda

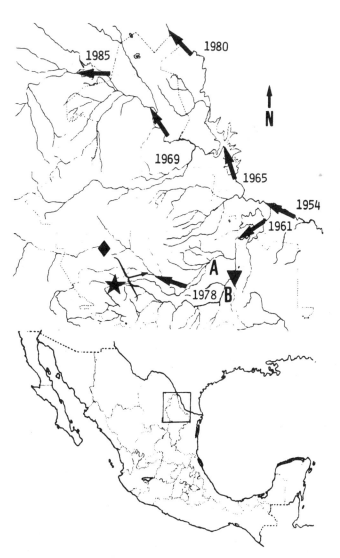

Figure 3. Map of Neuvo León, México. Arrows indicate progressive invasion of the Río Grande basin by the tidewater silversides, Menidia beryllina, *since 1954, in years shown. Bracket indicates area of residual water collectors and outflow planned to Río Pesquería (A) to bypass the future reservoir Presa El Cuchillo (inverted triangle) being built on Río San Juan (B) to provide tap water for Monterrey (star). The site of a major fish kill is marked with a diamond.*

(1990; 1992) reported numerous mild to severe cases of arsenic poisoning in resident farmers north and east of Torreón, resulting in lesions, pustules, finger losses, and even some deaths. There are several possible explanations for this arsenic poisoning, and the most plausible is as follows: Flores (1990) reported that overexploitation of ground water lowered the water table enough to cause a reversal of its flow; formerly flowing southwest to northeast and away from Torreón, ground-water now flows northeast to southwest toward the city. On its return, water flows through arsenic minerals, resulting in the local arsenicism. Questioned about the timing of the arrival of such poisoned waters to the metropolitan Torreón tap-water sources, Flores suggested 10 years if overexploitation rates continue. This overexploitation was first indicated by drying of springs and rivers in the area (Contreras & Maeda 1985).

The Coastal Plain of Sonora

The coastal plain of Sonora has been and is an agricultural emporium. Overexploitation has lowered the water table, however, resulting in replacement of fresh by brackish water. This brackish water may come from marine intrusion or from replacement by salinized agricultural runoff; studies are needed to identify the source of salt. Salinization has reached up to 100 km inland, as in Caborca. Newspapers have reported that some salinized water wells have been used for culturing shrimp and may become part of local governmental plans. Large-scale change from irrigated agriculture to shrimp culture does not seem to us environmentally or economically advantageous.

Monterrey Tap Water: "Plan 4"

Widely publicized through newspapers and local hearings, "Plan 4" is a program that relies on the flow of Río San Juan to provide future tap water for metropolitan Monterrey, Nuevo León. This river, a tributary of Río Bravo, is the largest stream in Nuevo León, draining most of the middle portion of the state. It has a large and nearly dry headwater in Coahuila, and a small run in Tamaulipas near its discharge to Río Bravo (Fig. 3). Its basin area is 33,830 km^2 (Tamayo & West 1964), with two major subbasins—Río Pesquería and Río San Juan proper—that join near China, Nuevo León. The Río San Juan has been termed, freely translated, as a "first-order problem requiring immediate attention," and a "public collector of residual waters, without aquatic life signs, offensive to sight and smell and noxious to the ecology of the area" (INEGI 1981).

The Río San Juan originates along the heavily populated axis from Monterrey to Montemorelos, an area of extreme air and water pollution. The scarcity of water in northeastern México has resulted in this river being considered the only remaining water source for Monterrey. However, Nuevo León is required by law to let some water flow to Tamaulipas for agriculture. The Río Pesquería (Fig. 3) is scheduled to carry all residual water from Monterrey, treated and upgraded to meet agricultural standards. Industry will be obliged to treat water carrying pollutants not eliminated by common sewage treatment to meet standards of the World Health Organization, the U.S. Environmental Protection Agency, or the Secretaría de Desarrollo Social, whichever is more stringent. This water will be given to Tamaulipas.

The Río San Juan proper is the other large river in the basin. This stream will be captured in the new Reservoir El Cuchillo, just upstream of China and before the junction with Río Pesquería, then transferred to a treatment plant and provided to metropolitan Monterrey. There are no known plans to treat pluvial waters, however, which simply flow through towns, receiving garbage, oils and greases, pesticides, detergents, heavy metals, and other matter washed from the streets by rains, which are acidic because of the heavy air pollution in the basin. Although there is another strong program to control such air pollution, acid rain remains a problem that will never be completely solved. Besides, no standard treatment facility gets rid of all these contaminants. Also, most nearby areas in the basin are farmed for crops or cattle, providing their share of pollutants. Consideration should be given to the residual mud accumulated on river and reservoir bottoms over many years of irrational use of pesticides and other chemicals in the basin. A basin-wide restoration plan is needed.

In summary, both the quantity and quality of water in northern México are declining, at a time when the human population is exploding. The population grew from 5.5 to 8.3 million people (53% increase) in the border region between 1980 and 1990 (SEDUE/EPA 1992). Monterrey is expected to more than double in population from 1.9 million in 1980, to a projected 4.0 million by 2000, and to 5.1 million by 2010 (SDU 1986).

Biodiversity: The Fish Fauna of Northern Mexico and Its Endangered and Threatened Species

Biodiversity has not been part of recent considerations for development along the Méxican-U.S. border area (SEDUE/EPA 1992; Eaton & Hurlburt 1992; Schmandt & Mu 1992). Yet México has a continental fish fauna of around 500 species (Miller 1986; Contreras & Almada 1991), 30 of them undescribed forms. The arid and semiarid areas of México are inhabited by approximately 200 fish species; 170 are taxonomically but not ecologically well known, and 18 of the undescribed forms are new discoveries currently being described. The most recent such discoveries are in the following states: Nuevo León, 4 species in 1984 and 2 in 1988; Durango, 2 species in 1988; Coahuila, 2 species in 1990; Chihuahua, 1 species in 1990. Such biodiversity has been neglected by most recent reviewers (Flores & Jeréz 1988, 1989; McNeely et al. 1990), and only Ramamoorthy et al. (1993) deal with fish as an element of biodiversity. None of them refers to the rapid recent increase in fish discoveries.

At least 22 of 29 species we have specifically studied require water that is clean, fresh, highly oxygenated and running, and sand, gravel, or rubble bottom with little or no muddy layer—conditions that are being rapidly lost (Contreras 1975; 1978). When environmental conditions are not adequate because of anthropogenic impact, species disappear locally, and at extremes live in danger of extinction.

This situation was first reported in the 1960s (Miller 1961, 1963; Contreras 1969), when only four species were reported as recent extinctions and 36 species were clearly in danger. The American Fisheries Society's lists of endangered and threatened fish species in North America indicated 67 species for México in 1979 (Deacon et al. 1979); one decade later, this had risen to 123 species (Williams et al. 1989), representing an increase of 83% in 10 years. Extracting Mexican data from tables in Williams et al. (1989), six species were removed because they were found invalid, none of the listed species became extinct, one was upgraded, and five were degraded, so 63 are new threats or new forms already in danger. The number now exceeds 135 (Contreras et al. 1991).

At the beginning of 1989, 11 fish species had gone extinct in nature in northern México (Miller et al. 1989), and four more have done so since (unpublished data). The account of Miller et al. alone represents 47.8% of the 23 extinct fish species reported worldwide by McNeely et al. (1990). We have no world data with which to compare these increased extinction rates witnessed by us since 1989. Locally, one of the most degraded areas of México is Parras, where drying of the springs and species introductions have resulted in the loss of the entire native fish community, including five endemic species and two local populations (Contreras & Maeda 1985).

Fish communities have been monitored in the aforementioned Mexican states since 1963 at localities known from the literature since 1901–1903 (Meek 1902, 1904). By 1975, 27 localities had lost an average of 43.9% of their fish communities, with most of those remaining also showing losses of individuals (Contreras et al. 1976; Contreras 1991a, 1991b). The changes are complex and have not been quantitatively analyzed, although interpretive work continues (Contreras & Lozano 1991). A partial updating for some of those localities to 1982–1985 showed a local loss of about 68% of species (Pérez et al. 1987; Contreras, unpublished data).

Regional Planning and Basin Management

The critical resource in arid and semiarid regions is by definition water. Every action and pollutant release into the environment in any region may become an impact and may be transported downriver to faraway places (Contreras 1975). This includes surface and underground waters, which form a manageable natural unit. Development should be measured against water availability in both quantity and quality (including recirculation, recycling, recharge) on the one hand, and all uses, releases, and losses on the other. Minimum information necessary to measure sustainable availability includes trends in salinity, phreatic water level and river flows, and lowest average recharge. Strong consideration should be given to the fact that polluting water with toxic materials may damage people and biota directly, and that the indirect effects of releasing common salt and other corrosives (such as acids and alkalies) increases the costs and difficulties of treatment or use, be it domestic, agricultural, or industrial, or for biodiversity or total environmental protection.

México needs an integral and rational ecosystem-wide basin management plan for sustainable development, as has been carried out in some other countries for a long time (Cooper 1969; Contreras 1990a; Sadler 1990). This management must rest upon a highly professional environmental impact evaluation system (Contreras 1975, 1980; Medina & Sánchez 1977). Quality control will require biologically based monitoring. If water quality is adequate to support varied and higher life forms, it will generally support human populations as well.

Critical Limits to Development in the Area

During the past few years, many maquiladoras (industries developed by foreign investors due to lower labor costs in México and elsewhere) have opened in México; in the near future they are expected to increase by as many as 3,000–4,000, according to government sources, because of the North American Free Trade Agreement. This will result in increased human populations in an already heavily populated area with inadequate public services and a low-quality, declining water supply. It seems unlikely that the region can withstand such development, unless it is carefully regulated.

Water in the southwestern United States and northern México simply must be managed on the basis of sustainable yield over the long term, without sacrificing quality and biodiversity. If water management were based on a simple average runoff and recharge relative to use, the area would now be in crisis 50% of the time. If use management is based upon the highest anticipated runoff and recharge rather than the average, a permanent crisis will exist, with unstabilized equilibrium and no reserve for emergencies such as a protracted drought. If sustainable yields are established and adhered to, expansion of development will depend upon rational and proactive conservation and recycling of resources, including water. Proper waste management, both in quantity and quality, are essential, both for surface and groundwater supplies. Untreated sewage or other undesirable discharges cannot be allowed into either surface runoff or groundwater recharge, because, ultimately, people downflow will need the water in a form that is satisfactory in both quantity and quality. All these requirements are part of the quality of life (Contreras 1990b).

It cannot be overemphasized that there is a strong need for monitoring and enforcement of water and air quality and garbage disposal. There is also a need for treatment plans for of municipal pluvial discharges, restoration and management plans for the Rio Grande/Rio Bravo basin, and integral ecological monitoring—including inventories of flora and fauna—and status evaluations. This will assure that water quality and quantity are adequate for perpetuation of humans and other life forms within the basin. Due to its complexity, human importance, and size, this case may become a world example if well planned and properly implemented.

This holistic environmental position is even more important now that México faces large development expectations through the Free Trade Agreement, the Border Integral Environmental Plan, and expansion of the maquiladora industry. These will all modernize and internationalize the northern Mexican border but will conflict with the high priority recently given by the Mexican federal government to species conservation.

Conclusions and Recommendations

Under the described scenario, and applying broad ecological and environmental principles, one may reasonably conclude that the availability of adequate water supplies in terms of quantity and quality is the most critical factor limiting human development as we approach the twenty-first century, and not necessarily only for arid and semiarid areas. Population growth must not be allowed to exceed water supply on a long-term basis. Long-range water management policy must include waste disposal, conservation, treatment, recycling, artificial recharge of aquifers, and environmental integrity measured by biodiversity. Biological inventories must be conducted, and the long-term safety and integrity of all life forms must be assumed in any thoughtful water management plan or policy. The result would be conservation of the human environment in its widest dimensions. Conservation of biodiversity will likewise conserve human quality of life.

Acknowledgments

It would be impossible to list everyone who participated in the collection of data and/or specimens, contributed to the thesis, or maintained the fish collection from 1958 to 1992. Intensive revisions of the manuscript were made by E. P. Pister, F. Abarca, G. M. Meffe, and three anonymous reviewers. The maps were produced by a team of illustrators. Some references and tables were compiled by Dora Ortiz. Discussions were held with a number of people at different meetings in northern Mexican states, so their general ideas were considered but not always followed; any failures are ours, but those discussions enlightened our way. We thank them all.

Literature Cited

Brune, G. 1975. Major and historical springs of Texas. Texas Water Development Board Report, Austin, Texas.

Brune, G. 1981. Springs of Texas. Vol. I. Branch-Smith, Forth Worth, Texas.

Contreras-Balderas, S. 1969. Perspectivas de la Ictiofauna en las Zonas Aridas del Norte de México. Memorias del Primer Simposio sobre el Aumento de la Producción de Alimentos en Zonas Aridas. International Center for Arid and Semiarid Land Studies, Texas Technological Publications 3:293–304.

Contreras-Balderas, S. 1975. Impacto ambiental de Obras Hidráulicas. Informe Plan Nacional Hidráulico. Secretaría de Recursos Hidráulicos, México; also, 1985, 2da. ed., Ediciones Imprenta Escolar, FCB/UANL, Monterrey, Mexico, and 1992, 3a. ed.ición, with additional readings, Instituto Tecnológico y de Estudios Superiores de Monterrey, Nuevo León, México.

Contreras-Balderas, S. 1978. Speciation aspects and man-made community composition changes in Chihuahuan desert fishes. Transactions of the First Symposium on the Biological Resources of the Chihuahuan Desert Region, United States Mexico 3:405–431.

Contreras-Balderas, S. 1980. Interacción ecologia—decisión: Nuevos conceptos. Boletín informativo. Centro de Investigaciones Biológicas, Universidad Autónoma de Nuevo León, Monterrey II(5):4–7.

Contreras-Balderas, S. 1990a. Impacto ambiental y manejo de Cuencas en la planificación. Foro Ciudades Industriales y Medio Ambiente (Mimeographed), Monterrey, México.

Contreras-Balderas, S. 1990b. La Calidad de Vida. Foro Ciudades Industriales y Medio Ambiente (Mimeographed), Monterrey, México.

Contreras-Balderas, S. 1991a. Testimony on public hearings of the integrated environmental border plan. Environmental Protection Agency, Laredo, Texas, and Secretaría de Desarrollo Urbano y Ecología, Nuevo Laredo, Mexico.

Contreras-Balderas, S. 1991b. Los peces de la Región Tarahumara Chihuahua/Durango y su uso como indicadores de impacto ambiental. Informe al Centro de Ecología Universidad Nacional Autónoma de México, Mexico City, Mexico.

Contreras-Balderas, S., and P. Almada-Villela. 1991. Fish biodiversity, water availability and regional planning (abstract). 1991 Annual Meeting of the American Society of Ichthyologists and Herpetologists, New York.

Contreras-Balderas, S., and M. L. Lozano-Vilano. 1991. Estado actual de los peces y el agua en el Noreste de México. First Binational Meeting on the Environment, Ciudad Acuña, Coahuila, Mexico.

Contreras-Balderas, S., and M. L. Lozano-Vilano. 1994. New Mexican cyprinid fish *Cyprinella alvarezdelvillari* sp. nv., from Rio Nazas Basin. Copeia, in press.

Contreras-Balderas, S., and A. Maeda. 1985. Estado actual de la ictiofauna nativa de la Cuenca de Parras, Coah., México, con notas sobre algunos invertebrados. Octavo Congreso Nacional de Zoología, Memorias **I**:59–67.

Contreras-B., S., V. Landa, T. Villegas, and G. Rodriguez. 1976. Peces, piscicultura, presas, polución, planificación pesquera y monitoreo en México. Danza de las P. Memorias del Simposio sobre Pesquerías en Aguas Continentales (México) **I**:315–346.

Contreras-Balderas, S., P. Almada-Villela, and D. A. Hendrickson. 1991. Peces amenazados de agua dulce de México (Threatened freshwater fishes of México). Abstract. XI Congreso Nacional de Zoología, Mérida, México.

Cooper, Ch. F. 1969. Ecosistem models in watershed management. Pages 309–324 in G. M. Van Dyne. The ecosystem concept in natural resource management, Academic Press, New York.

Deacon, J. E., G. Kobetich, J. Williams, S. Contreras, et al. 1979. Fishes of North America: Endangered, threatened or of special concern: 1979. Fisheries (Bulletin of American Fisheries Society) **4(2)**:29–44.

Eaton, D. J., and D. Hurlbut. 1992. Challenges in the binational management of water resources in the Rio Grande/Rio Bravo. Pages 1–138 in Policy report 2. U.S.–Mexican Policy Studies Program, University of Texas at Austin.

Edwards, R. E., and S. Contreras-B. 1991. Historical changes in the ichthyofauna of the Lower Rio Grande (Río Bravo del Norte), Texas and México. Southwestern Naturalist **36(2)**: 201–212.

Flores, R. 1990. Hidrogeología de la Comarca Lagunera. Ciclo de Conferencias, Universidad Juárez del Estado de Durango. Gómez Palacio, Durango, México.

Flores-Villela, O., and P. Jerez. 1988. Conservación en México: Síntesis sobre vertebrados terrestres, vegetación y uso del suelo. Instituto Nacional de Investigaciones en Recursos Bióticos/Conservación Internacional, Mexico City, Mexico.

Flores-Villela, O., and P. Jerez. 1989. Mexico's living endowment: An overview of biological diversity. Conservation International & Instituto Nacional de Investigaciones en Recursos Bióticos, Mexico City, Mexico.

INEGI (Instituto Nacional de Estadística, Geografía e Informática). 1981. Síntesis geográfica del estado de Nuevo León. Editorial INEGI, Memoria. Mexico City, Mexico.

INEGI. 1981. Síntesis geográfica del Estado de Zacatecas. Editorial INEGI, Memoria. Mexico City, Mexico.

INEGI. 1983. Síntesis geográfica del Estado de Coahuila. Editorial INEGI, Memoria. Mexico City, Mexico.

INEGI. 1983. Síntesis geográfica del Estado de Tamaulipas. Editorial INEGI, Memoria. Mexico City, Mexico.

INEGI. 1984. Síntesis geográfica del Estado de Baja California (Norte). Editorial INEGI. Mexico City, Mexico.

Maeda, L. 1990. Impacto ambiental regional y la salud en la comarca lagunera. Foro Ciudades Industriales y Medio Ambiente (Mimeographed), Monterrey, México.

Maeda, L. 1992. El hidroarsenicismo Crónico regional un estudio descriptivo. Consejo Cívico Asesor del Departamento de Ecología Municipal, Torreón, México.

McNeely, J. A., et al. 1990. Conserving the world's biological diversity. International Union for the Conservation of Nature, Gland, Switzerland. World Resources Institute, Conservation International, World Wildlife Fund–U.S., World Bank, Washington, D.C.

Medina, J. A., and R. Sánchez. 1977. Impacto ambiental de obras hidráulicas. Plan Nacional Hidráulico, Informes **17**:1–70.

Meek, S. E. 1902. A contribution to the ichthyology of Mexico. Field Columbian Museum Publications 65, Zoological Series **III(6)**:63–121.

Meek, S. E. 1904. The freshwater fishes of Mexico, north of the Isthmus of Tehuantepec. Field Columbian Museum Publications 93, Zoological Series **V**:i–lxiii, 1–252.

Miller, R. R. 1961. Man and the changing fish fauna of the American Southwest. Papers of the Michigan Academy of Sciences **46(1960)**:365–404.

Miller, R. R. 1963. Extinct, rare and endangered American freshwater fishes, XVI International Congress of Zoology **8**:4–11.

Miller, R. R. 1986. Composition and derivation of the freshwater fish fauna of Mexico. Anales de la Escuela Nacional de Ciencias Biològicas, México **30**:121–153.

Miller, R. R., J. D. Williams, and J. E. Williams. 1989. Extinctions of North American fishes during the past century. Fisheries **14(6)**:22–38.

NAFTA. 1992. North American Free Trade Agreement. U.S. Government, Washington, D.C.

PNH (Plan Nacional Hidráulico). 1975. Map of hydraulic resources in North Central México. Dirección General del Plan Nacional Hidráulico. Secretaria de Recursos Hidráulicos, México.

Pérez-Bernal, R., S. Contreras-Balderas, and M. L. Lozano-Vilano. 1987. Cambios de composición de especies de peces en comunidades del Este de México, desde 1903, o La Danza de las P, Revisada. Abstract. Annual Meeting of the Desert Fishes Council, Monterrey, Nuevo León, Mexico.

Ramamoorthy, T. P., R. Bye, A. Lot. 1993. Biological diversity of México: Origins and Distribution. Oxford University Press, New York.

SEDUE/EPA (Secretaría de Desarrollo Urbano/Environmental Protection Agency). 1992. Integrated Environmental Border Plan, Cooperative Plan SEDUE/EPA, Mexico City, Mexico.

SDU (Secretaría de Desarrollo Urbano). 1986. Plan director del desarrollo urbano del area metropolitana de Monterrey. Gobierno del Estado de Nuevo León, Secretaría de Desarrollo Urbano, Monterrey, Mexico.

Sadler, B. 1990. Sustainable development and water resource management. Alternatives **17(3)**:14–19, 21–24.

Schmandt, J., and Xing Ming Mu. 1992. Water and development in the lower Rio Grande Valley. LBJ School of Public Affairs, University of Texas, Austin, Texas.

Tamayo, J. L., and R. C. West. 1964. The hydrography of Middle America. Chapter 3 in Wauchope and West. Handbook of Middle American Indians. Vol. 2. Natural environment and early cultures. University of Texas Press, Austin, Texas.

Williams, J. E., J. E. Johnson, D. A. Hendrickson, S. Contreras-Balderas, J. D. Williams, M. Navarro-Mendoza, D. E. McAllister, and J. E. Deacon. 1989. Fishes of North America: Endangered, threatened or of special concern: 1989. Fisheries **14(6)**:2–21.

The following poem was written after a conversation with Gordon Orians who told of crocodiles being imported to Brazil to provide leather for the fashion industry.

Brazilian Crocodilian Tears

Two Nile Crocodiles
    in Brazil to make handbags and
    shoes,
Escaped their farm,
    to swim the river miles,
    wild,
    with *very* toothy smiles.

One came by,
    with tears in his eyes,
    and asked,
"Won't you please visit with me awhile?
It's a bountiful niche given to us,
    full of food,
    and very crocodilian,
    amazonian,
    amour.

Tomorrow, I may, if I wish,
    nibble pacu fish,
    endemic manatee,
    and, maybe,
    —yes, maybe,
    even thee;
Who knows how far,
    my family will spread,
    in this tropical-green country?"

And as away he swirled,
    to a slow,
    and sassy,
    samba beat,
A thought remained behind—
Now again, we will see,
    if greed's the best,
    long-term management strategy.

R. Eugene Turner
Department of Oceanography and Coastal Sciences
Louisiana State University
Baton Rouge, LA 70803, U.S.A.

Contributed Papers

Livestock Breeds and Their Conservation: A Global Overview

STEPHEN J. G. HALL
Centre d'Etudes Biologiques de Chizé
F 79360 Villiers en Bois, France*

JOHN RUANE
Animal Production and Health Division
Food and Agricultural Organization of the United Nations
Via delle Terme di Caracalla
00100 Rome, Italy**

Abstract: *Of the 3831 breeds or breed varieties of ass, water buffalo, cattle, goat, horse, pig, and sheep believed to exist or to have existed this century, 618 (16%) are estimated to have become extinct. Of the 3213 existing breeds, 475 (15%) are defined as rare. Worldwide, the horse is the species most differentiated into breeds and the water buffalo the least. Breed biodiversity varies markedly between continents, being greatest in Europe and least in South and Central America and Oceania. In the countries of the Old World, numbers of breeds are correlated with human population and with land area, implying that conditions favoring growth in human population also favor diversification of breeds. Peripheral and remote countries have the highest ratios of breeds per million people, implying that remoteness can also promote diversification. Breed extinctions have been most thoroughly documented in Europe and the former U.S.S.R., where the richest countries have lost the highest proportions of their breeds, implying that agricultural development is hostile to breed diversity. Data are particularly lacking from the developing world, and the Global Data Bank for Domestic Livestock of the Food & Agriculture Organization of the United Nations has been set up to collate census and production information.*

Razas de ganado y su conservación una perspectiva global

Resumen: *De las 3831 razas o variedades de razas de burro, búfalo, ganado, cabra, caballo, puerco y oveja que se cree haber existido este siglo, se estima que 618 (16%) se han extinguido. De las 3213 razas que existen hoy, 475 (15%) son definidas como 'raras'. Mundialmente, el caballo es la especie mas diferenciada en razas, y el búfalo menos. Biodiversidad de razas varía fuertemente entre continentes, siendo más marcada en Europa y menos en América del Sur, América Central y Oceania. En los países del Viejo Mundo, el número de razas se correlaciona con el área y el tamaño de las poblaciones humanas, sugiriendo que aquellas condiciones que favoracen el crecimiento de las poblaciones humanas también favorecen la diversificación de razas de animales domésticos. Países remotos o periféricos tienen la relación más alta entre número de razas por millon de personas, sugiriendo que distancia de centros urbanos también favorece diversificación. La extinción de razas ha sido documentada en Europa y la ex-U.S.S.R., donde los países más ricos han perdido una gran proporción de las razas existentes, sugiriendo que desarrollo agrícola es hostil a la diversidad de razas de ganado. La ausencia de información es particularmente marcada en los países en desarrollo, y se ha montado el Banco Mundial de Datos sobre Animales Ganaderos de la FAO para colectar censos e información sobre la producción actual.*

---

* *Current address: Department of Clinical Veterinary Medicine, Madingley Road, Cambridge CB3 0ES, U.K.*
** *Current address: Station de Génétique Quantitative et Appliquée, Institut National de le Recherche Agronomique, 78352 Jouy-en-Josas Cedex, France.*
Paper submitted July 13, 1992; revised manuscript accepted January 11, 1993.

Introduction

Livestock breeds are recognized as genetic resources in the World Conservation Strategy (International Union for the Conservation of Nature and Natural Resources 1980), with rare breeds considered worthy of conservation because the genes and gene combinations they carry may be commercially applicable in the future. Changes in agriculture have led to many breeds becoming rare in the developed world (Hall & Clutton-Brock 1989), and this is also likely to happen in the developing world (Anderson 1992). A world inventory of livestock breeds is needed that could be used in planning for conservation.

The most authoritative equivalent to such an inventory is by Mason (1988), who lists the breeds of ass, water buffalo, cattle, goat, horse, pig, and sheep mentioned in the scientific and agricultural literature. For all breeds, the country or region of origin is given. While "breed" is succinctly defined as "a group of animals that has been selected by man to possess a uniform appearance that is inheritable and distinguishes it from other groups of animals within the same species" (Clutton-Brock 1987), this definition can be expanded to mean "a separately identified (or identifiable) population or group of interbreeding domestic animals. Identification will usually be based on common physical characters such as color, size, shape and also on shared genetic and historical origins. A breed ... is usually associated with a particular ecological zone, geographical area and farming system. Some breeds may, however, be present in several countries. Established crosses between two or more breeds may be recognized as a separate breed, but shifting or transitional crossbred groups are not" (Cunningham 1992).

The first part of this paper is an analysis of the breeds described by Mason (1988). The second part describes the Global Data Bank for Domestic Livestock of the Food & Agriculture Organization (FAO) of the United Nations, which aims to document and characterize the breeds of all the world except Europe but including the former U.S.S.R. This is compatible with the European databank now at Hannover, Germany (Maijala et al. 1984; Simon 1992).

Materials and Methods

The dictionary of Mason (1988) includes all those breeds, varieties of breeds, wild species and feral populations, and extinct breeds encountered in the literature by the compiler (I. L. Mason, personal communication). Over 20% of entries are described as varieties or strains of breeds. For convenience, the term "breed" is used in this paper to cover both the breeds and varieties of breeds described in Mason (1988).

The extinct breeds included probably form an incomplete list of those breeds that have died out or become absorbed by other breeds.

1. Analysis of Livestock Breeds

Breeds were classified as extinct, rare, or common, and by their country of origin. Breeds described as nearly extinct, declining in numbers, or disappearing were defined as rare.

Entries relating to crossbreeds and to groups of animals of indeterminate breed but sharing a characteristic such as coat color or function were disregarded except when they had become stabilized as a new breed. Wild species were disregarded, as were breeds and crossbreeds arising from them, and breeds known to have become extinct more than 100 years ago.

When possible, breeds were assigned to provinces and states of countries. Relationships were explored between number of breeds and land areas, livestock and human populations, and gross national product per head (U.S. dollars). The number of animals of a given species in a continent was divided by the number of breeds of that species to estimate how far the livestock population was differentiated into breeds. Data were taken from the Times Atlas of the World (Times Books 1990), Hunter (1991), and the Food & Agriculture Organization (1991). Regressions and Pearson correlation coefficients were calculated and data transformed as appropriate (Sokal & Rohlf 1981). North, Central and South America, and Oceania were excluded from most analyses because of their relatively small number of native breeds.

2. FAO Global Data Bank

The information that can be stored in the data bank is quite extensive and is described in detail by Ruane (1992). Mason (1988) was the primary source of information for the data bank. Entries referring to unstable crosses between breeds, to interspecies hybrids, or to groups or collections of breeds were excluded. All varieties or strains of breeds as well as wild species and feral populations were included. Additional data on the breeds were retrieved from the literature. A survey was initiated in 1992 to determine which breeds exist in each country and to identify those that are rare or in danger of extinction.

Results

1. Analysis of Livestock Breeds

BREED DISTRIBUTION

The numbers of breeds in each continent are given in Table 1, with breeds classified as extinct ($n = 618$),

Table 1. Summary of world distribution of extinct, rare, and common breeds classified according to continent of origin.*

		Ass	Water Buffalo	Cattle	Goat	Horse	Pig	Sheep	Total	%
Africa	Rare			10		2		4	16	4
	Extinct			22		2		1	25	6
	Common	15	5	159	60	33	6	129	407	
	Total	15	5	191	60	37	6	134	448	
Asia	Rare		2	8	4	14	2	1	31	4
	Extinct			5	1	3	8	2	19	2
	Common	16	55	184	142	71	140	217	825	
	Total	16	57	197	147	88	150	220	875	
Europe	Rare	10		101	29	49	37	109	335	27
	Extinct	5		154	19	58	79	98	413	25
	Common	8	8	209	90	137	76	356	884	
	Total	23	8	464	138	244	192	563	1632	
North & Central America	Rare			8	4	9	5	7	33	16
	Extinct			1	1	4	17	10	33	14
	Common	6	1	57	8	31	31	39	173	
	Total	6	1	66	13	44	53	56	239	
South America	Rare	1		4				1	6	5
	Extinct			19					19	14
	Common	4	2	41	11	21	17	16	112	
	Total	5	2	64	11	21	17	17	137	
Oceania	Rare			1		1	1	2	5	7
	Extinct			2		1	1	5	9	12
	Common			19	6		5	34	64	
	Total			22	6	2	7	41	78	
ex-U.S.S.R.	Rare			9	4	23	2	11	49	15
	Extinct			21	6	20	21	32	100	24
	Common	15	1	51	15	36	33	122	273	
	Total	15	1	81	25	79	56	165	422	
World Totals		80	74	1085	400	515	481	1196	3831	

* The term "common" indicates breeds that are neither extinct nor rare. Percentages are calculated as follows: number rare/number existing; number extinct/total.

rare ($n = 475$), or common ($n = 2738$). Of existing breeds, 15% are defined as rare (ass 11, water buffalo 2, cattle 141, goat 41, horse 98, pig 47, sheep 135). Of these 475 rare breeds, the vast majority (384, or 81%) are native to Europe or to the former U.S.S.R. Of the existing breeds, 134 (4.2%) have multiple countries of origin (by species, ass 4, water buffalo 4, cattle 44, goat 20, horse 24, pig 5, sheep 33; by continent, Africa 51, Asia 48, Europe 24, North, Central and South America 11).

Table 2 presents the ratios of the continental populations of each species to the numbers of breeds of each species. Per continent the ratios were highest for the following species: Africa, goat and pig; Asia, water buffalo and pig; Europe, pig; North and Central America, cattle; South America, sheep and cattle; Oceania, sheep; former U.S.S.R., cattle. Worldwide, the horse is the species most completely differentiated into breeds, and the water buffalo the least.

The extinct breeds, which comprise 16% of all those known in the last 100 years, are presented in Table 3. Of these, 511 (83%) were native to Europe or to the former U.S.S.R.

NUMBERS OF BREEDS, HUMAN POPULATIONS, AND LAND AREAS

Table 4 summarizes for each species (water buffalo being disregarded, and horses and asses combined) corre-

Table 2. Ratios of livestock population (thousand head) to total number of breeds, for each species.

	Ass	Water Buffalo	Cattle	Goat	Horse	Pig	Sheep	Total
Africa	872	500	983	2899	135	2264	1531	1341
Asia	1343	2390	1999	2190	192	2884	1537	1899
Europe	46	47	267	112	17	947	270	294
North & Central America	610	9	2425	1066	321	1675	343	1254
South America	799	600	4123	2125	682	3273	6625	3467
Oceania			1421	326	249	763	5515	3400
ex-U.S.S.R.	20	420	1462	259	75	1409	83	823
World Totals	545	1902	1179	1393	118	1781	995	1078

Table 3. Number of breeds believed to have become extinct within the last 100 years.

	Cattle	Goat	Horse	Pig	Sheep
Africa					
Algeria	3				
Benin	1				
Cameroon	1				
Gambia	1				
Malawi	1				
Nigeria	2				
Rwanda	1				
South Africa	5		2		1
Tanzania	4				
Zimbabwe	3				
Asia					
China					2
Hong Kong			3		
India	2		2		
Japan			1		
Pakistan		1			
Philippines	1				
Taiwan				5	
Turkey	2				
Americas					
Canada			2		1
United States	1	1	2	17	9
Brazil	15				
Chile	1				
Uruguay	1				
Venezuela			2		
Europe					
Austria	16		1		9
Belgium	2				2
Bulgaria	2		2		4
Czechoslovakia	10				1
Denmark	3				1
France	18	2	15	18	33
Germany	29	6	7	10	5
Greece	7				1
Hungary	3		1		
Ireland			1		3
Italy	22		5	24	15
Netherlands					1
Norway	11	5	1		
Poland	5	2		5	1
Portugal	1				
Romania	3		3		
Spain	12		12	8	3
Sweden	4			1	
Switzerland	2	1	4		11
United Kingdom	5	2	4	7	8
Yugoslavia	2		3	5	
ex-U.S.S.R.	22	6	20	21	32
Oceania					
Australia	2		1		1
New Zealand				1	1
Total	224	27	88	126	148
	(21)	(7)	(17)	(26)	(12)

Breeds found in more than one country are listed once for each country, hence the figures given as totals do not necessarily correspond to the numbers in the columns. The numbers of extinct breeds are expressed as percentages (in brackets) of breeds known during the last 100 years. There are also five extinct breeds of ass (Italy 4, Spain 1).

Table 4. Correlation coefficients (Pearson) of numbers of existing breeds of each species (asses and horses combined, buffalo excluded) on (a) log human population and (b) log land area (km^2) of each country.

	Cattle	Equines	Goats	Pigs	Sheep
(*a*) Log Human Population					
Africa	0.51**	0.05 NS	0.49**		0.48**
	(36)	(20)	(24)	(5)	(36)
Asia	0.66***	0.55**	0.62***	0.53*	0.63***
	(28)	(23)	(30)	(13)	(24)
Europe	0.65***	0.72***	0.48*	0.89***	0.62***
	(26)	(25)	(25)	(20)	(26)
(*b*) Log Land Area					
Africa	0.37*	0.02 NS	0.35 NS		0.42*
Asia	0.54**	0.66***	0.47**	0.61*	0.54**
Europe	0.49**	0.56**	0.25 NS	0.57**	0.45*

Only countries (in brackets in upper table; number in each continent) with indigenous breeds are considered.
* $p < 0.05$, ** $p < 0.01$, *** $p < 0.001$.

lations—which were mostly highly significant—of number of breeds per country with log human population and with log land area. Correlations with human population density were not significant.

For each country of the Old World, the number of existing breeds of all species taken together, and the ratio of this number to human population, are given in Tables 5a and 5b. The data for Europe and Africa are also presented in Figures 1 and 2. In these maps, the shading emphasizes differences within the respective continents rather than comparisons between continents. Analyses of other continents are not presented because the presence in each of a single country with a large number of breeds distorted the analysis (existing breeds: U.S.A. 130, Brazil 66, Australia 43).

In the former U.S.S.R. the correlation with log human population of the number of existing breeds (all species combined) in each identifiable province, region, or state was significant ($r = 0.92$, d.f. 28, $p < 0.001$). When this number was divided by human population, peripheral regions—especially in the Caucasus—ranked highest (Table 5b). Likewise, in India, where 19 states had native breeds, the correlation of number of existing breeds with human population was significant ($r = 0.62$, d.f. 18, $p < 0.01$), but the correlation obtained for the states of China was significant only at the 10% level ($r = 0.32$, d.f. 27). In both countries, the states with the greatest numbers of breeds per million people were those bordering neighboring countries (Table 5c).

EXTINCTIONS

In Europe, but not in Africa or Asia, the regression of the proportion of breeds extinct (arcsine transformation) on GNP per head was significant ($p < 0.05$) (Fig. 3). European countries that had allowed more breeds to be

Table 5a. Countries and provinces or states ranked in descending order of number of existing native breeds per million people.

Country	Number of Native Breeds	Breeds per Million People
Africa		
Seychelles	1	14.9
Djibouti	1	10.0
Botswana	9	7.4
Namibia	9	6.9
Gambia	5	6.3
Guinea-Bissau	5	5.4
Mauritania	10	5.3
Somalia	20	3.2
Chad	16	3.0
Mali	21	2.7
Niger	15	2.1
Sudan	51	2.0
Senegal	14	2.0
Libya	7	1.7
Liberia	4	1.6
South Africa	46	1.6
Tunisia	13	1.5
Morocco	36	1.5
Togo	5	1.5
Swaziland	1	1.4
Lesotho	2	1.2
Cameroon	13	1.2
Benin	5	1.1
Kenya	24	1.1
Sierra Leone	4	1.0
Uganda	16	1.0
Angola	9	1.0
Burkina Faso	7	0.8
Zimbabwe	7	0.8
Algeria	16	0.7
Ethiopia	33	0.7
Malawi	4	0.5
Tanzania	14	0.6
Egypt	30	0.6
Ghana	6	0.4
Guinea	3	0.5
Madagascar	5	0.5
Zaire	13	0.4
Mozambique	6	0.4
Nigeria	32	0.3
Rwanda	2	0.3
Côte d'Ivoire	2	0.2
Zambia	1	0.1
Asia		
Yemen	36	195.3
Mongolia	14	6.7
Oman	7	5.8
Bahrain	2	4.8
Bhutan	4	2.9
Israel	12	2.7
Lebanon	5	1.8
Afghanistan	22	1.4
Syria	16	1.4
Nepal	24	1.3
Turkey	55	1.0
Iran	55	1.0
Jordan	3	1.0
Saudi Arabia	11	1.0
Iraq	16	0.9
Pakistan	96	0.9
Cambodia	6	0.8
Sri Lanka	12	0.7
Laos	2	0.5
Malaysia	8	0.5
Cyprus	7	0.5
Vietnam	18	0.3
Philippines	17	0.3
China	236	0.2
India	171	0.2
Taiwan	4	0.2
Thailand	10	0.2
Japan	21	0.2
Indonesia	28	0.2
Bangladesh	17	0.2
Myanmar	6	0.2
North & South Korea	7	0.1

Breeds found in more than one country were included for each country.

lost than would be expected are above the regression line; those with fewer than expected are below.

2. FAO Global Data Bank for Domestic Livestock

Currently (December 1992) there are 2056 breeds and varieties in the data bank (ass 63, water buffalo 70, cattle 567, goat 258, horse 247, pig 245, sheep 606). For 37% of these breeds there are data on population size. Of these, 207 number fewer than 5000 animals or are known to be rare or endangered. Of the entries, 38% include some information on production (such as adult body size, birth weight, milk traits, fleece weight, litter size). Thirty percent of entries have information on both population size and production characteristics.

In April 1992 the European data bank contained 707 entries (ass 5, cattle 206, goat 54, horse 110, pig 92, sheep 240) (Simon 1992). The numbers of entries in the two data banks are not comparable because in the former there is only one entry per breed, even if the breed has multiple countries of origin, while in the latter, there is one entry per breed per country of occurrence.

Discussion

Of the 3831 livestock breeds or varieties known this century, 618 are now extinct, leaving 3213 existing breeds of which 475 are defined as rare. All these are very large numbers, compared with the number ($n = 647$) of threatened and endangered wild mammalian species (World Conservation Monitoring Centre 1992). This livestock biodiversity must be catalogued and its evolution understood if it is to be properly conserved.

There is no general hypothesis explaining why breed biodiversity is greater in some continents than in others.

Table 5b. For legend, see Table 5a.

Europe		
Faeroe Islands	3	64
Iceland	6	24
Corsica	4	20
Crete	4	12
Wales	24	11
Balearic Islands	6	11
Norway	20	8.8
Malta	3	8.7
Bulgaria	65	8.1
Switzerland	29	7.2
Scotland	33	6.9
Albania	21	6.7
Slovenia	8	6.2
Austria	16	5.5
Portugal	56	5.5
Sardinia	8	4.8
Ireland	13	4.8
Greece	39	4.7
Canary Islands	6	4.3
Spain	109	3.7
Italy	119	3.6
France	126	3.5
Denmark	12	3.1
Sweden	19	3
Hungary	25	2.7
Sicily	13	2.7
Netherlands	37	2.6
Croatia	10	2.6
Czechoslovakia	28	2.5
England	102	2.5
Macedonia	4	2.5
Belgium	17	2.1
Finland	10	2
Romania	40	2
Poland	57	1.9
Germany	78	1.7
Montenegro	1	1.6
Serbia	11	1.3
Northern Ireland	1	1.3
Bosnia Hercegovnia	5	1.1
Former Soviet Union		
Gorna-Altai	2	10.4
Dagestan	16	8.9
Estonia	10	6.4
Armenia	11	3.4
Georgia	18	3.3
Tuva	1	3.2
Turkmenistan	11	3.4
Buryat	3	2.9
Yakut	3	2.8
Azerbaijan	19	2.7
Kirghizia	11	2.6
Tadzhikstan	14	2.7
Karachaevo-Cherkassy	1	2.4
Komi	3	2.4
Lithuania	7	1.9
Kazakhstan	31	1.9
Latvia	5	1.9
Khakass	1	1.8
North Ossetian	1	1.6
Chuvash	2	1.5
Mari	1	1.3
Kabardino-Balkar	1	1.3
Uzbekistan	19	0.9
Byelorussia	7	0.7
Russian Federation (excepting areas listed here)	99	0.7
Ukraine	31	0.6
Bashkir	1	0.3
Moldavia	1	0.2

Coverage in the literature may be uneven, yet many comparisons are probably realistic. For example, the U.S.A. has 130 native breeds and the U.K. has 160, though the former has 38 times the land area and four times the human population. Considering the number of breeds in relation to the livestock populations of the continents (Table 2), it appears that breed diversification is least obvious in South America and Oceania. In both continents this is largely due to the small number of sheep breeds.

Worldwide, the horse is the species most thoroughly differentiated into breeds and the water buffalo the least. An early priority would be to check how differentiated horse breeds really are, and whether water buffalo can be divided further into breeds. The large proportion (51 out of 448) of African breeds with multiple countries of origin suggests, too, that the possibility of local varieties of these breeds should be examined. Other priorities can be identified from Table 2, such as investigating whether hitherto undescribed breeds of sheep exist in South America.

Breed Diversification

Within each continent, breed diversification is most marked in populous countries, as shown by the correlations between human population size and number of breeds. This suggests the hypothesis that the conditions that have favored growth in numbers of people have also favored diversification of breeds. When the number of breeds is expressed per million people, it becomes clear that peripheral and remote countries and provinces have disproportionately large numbers of breeds. Sometimes (as in Seychelles, Faeroe Islands, Gorno-Altai) this is because there are very few people and there happen to be one or a few native breeds, but there are several peripheral countries—particularly in Europe (Fig. 1)—where breed numbers are high in absolute terms. In the former U.S.S.R. it is also the peripheral areas such as the states of the Caucasus that have the greatest breed diversity when assessed in this way. In Asia and in Africa (Fig. 2) semi-arid or arid countries such as Mongolia, Yemen, Oman, and those of the Sahel, Botswana and Namibia have the greatest proportional diversity of breeds. In both China and India, border states and provinces—which have harsh terrain—have the greatest breeds to people ratios.

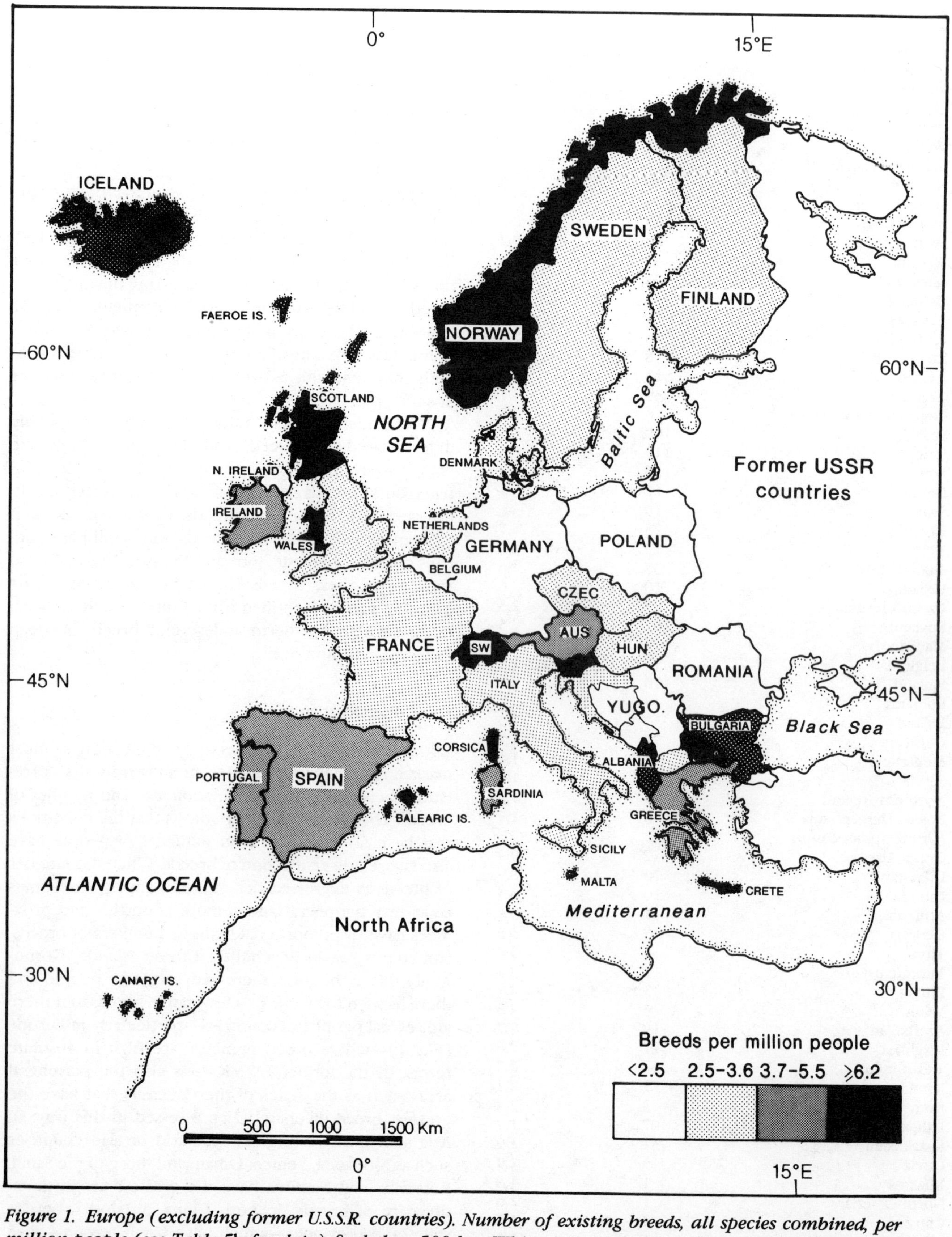

Figure 1. Europe (excluding former U.S.S.R. countries). Number of existing breeds, all species combined, per million people (see Table 5b for data). Scale bar: 500 km. White dots on black = 6.2 or more breeds per million people; black dots on white = 3.7–5.5 breeds per million people; hatched areas = 2.5–4.8 breeds per million people; white areas = 2.5 or fewer breeds per million people.

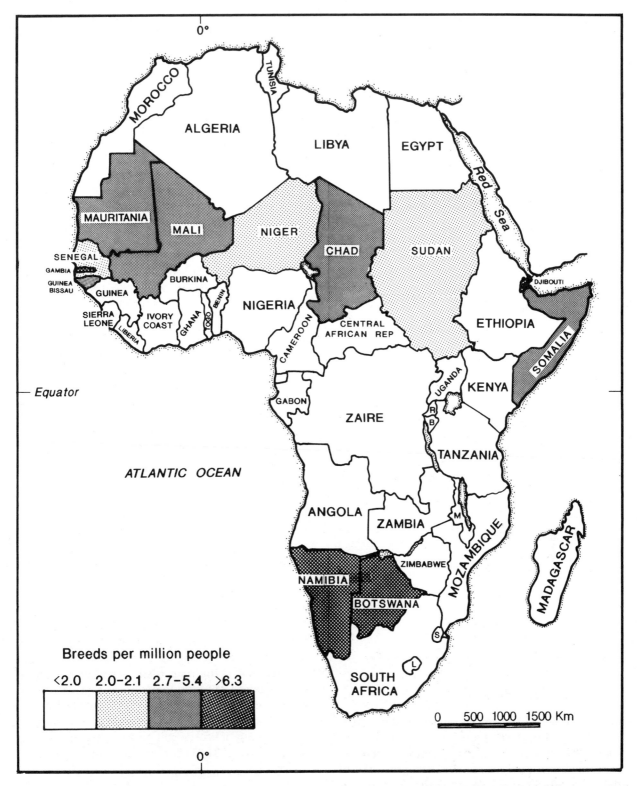

Figure 2. Africa (excluding Seychelles). Number of existing breeds, all species combined, per million people (see Table 5a for data). Scale bar: 1000 km. White dots on black = 6.3 or more breeds per million people; black dots on white = 2.7–6.2 breeds per million people; hatched areas = 2.0–2.2 breeds per million people; white areas = 2.0 or fewer breeds per million people.

Table 5c. For legend, see Table 5a.

China		
Qinghai	7	1.7
Tibet	2	0.9
Xinjiang Ugyur	13	0.9
Ningxia	3	0.7
Inner Mongolia	13	0.6
Shanxi	14	0.5
Shaanxi	15	0.5
Jiangxi	15	0.4
Zhejiang	16	0.4
Fujian	9	0.4
Gansu	6	0.3
Shanghai	4	0.3
Yunnan	10	0.3
Hubei	13	0.3
Guanxi-Zhuang	10	0.3
Heilongjiang	8	0.2
Guangdong	15	0.2
Guizhou	7	0.2
Jilin	5	0.2
Beijing	2	0.2
Hebei	11	0.2
Anhui	10	0.2
Jiangsu	10	0.2
Liaoning	5	0.1
Hunan	7	0.1
Sichuan	12	0.1
Shandong	8	0.1
Henan	5	0.1
India		
Sikkim	2	5.3
Himachal Pradesh	3	0.6
Manipur	1	0.6
Jammu Kashmir	4	0.6
Rajasthan	22	0.5
Gujarat	18	0.4
Punjab	7	0.4
Maharashtra	25	0.3
Orissa	11	0.3
Tamil Nadu	19	0.3
Haryana	5	0.3
Karnataka	10	0.2
Madhya Pradesh	12	0.2
Assam	4	0.2
Andhra Pradesh	9	0.1
Uttar Pradesh	16	0.1
Delhi	1	0.1
Kerala	2	0.1
Bihar	7	0.1

These observations suggest another hypothesis, that breed diversification is also promoted in areas that are geographically isolated. It is quite possible that even within a small area such as Britain both processes of breed diversification have operated. Wales and Scotland are relatively isolated, with large numbers of breeds per million people (Table 5b), while England, with a large human population, has a large absolute number of breeds.

Breed Rarity and Extinction

Individual breeds become threatened when the husbandry systems to which they are adapted are modified or replaced (Hall 1990). Only 19% of the world's recognized rare breeds are native to countries outside Europe or the former U.S.S.R.; clearly, high priority must be given to identifying rare breeds in continents where they have not been recorded. The FAO Global Data Bank is already accumulating this information. Of non-European breeds, 207 have been identified as rare (the number inferred from Mason [1988] is 140). It is predicted that when the data bank is complete the non-European rare breeds of the world will total 350.

The proportions of breeds that have become extinct this century are highest in Europe and the former U.S.S.R., probably due to rapid and centralized agricultural development. This hypothesis is strengthened by the finding that in Europe relatively rich countries have lost proportionately more of their breeds. Extinctions may have been under-recorded elsewhere; it is now being suggested (Anderson 1992) that many breeds, particularly of cattle, will become extinct in the developing world through the importation of semen of exotic breeds as part of aid programs. Historically, pigs and cattle seem to be the species most prone to breed extinction. Many of the species that have become extinct in Europe and the former U.S.S.R. may have done so as a consequence of the development of intensive husbandry (pigs) and of artificial insemination (cattle).

Breed Conservation

The FAO Global Data Bank for Domestic Livestock will be of fundamental importance to conservation. It provides an inventory of breeds for each country, and the means to identify and monitor endangered populations. Documentation of breed characteristics will also enable the planning strategies for development and utilization of indigenous breeds.

The Global Data Bank will also help to form and test hypotheses of the evolution of breeds. At present, such hypotheses are lacking, and they will undoubtedly be necessary if taxonomic distinctiveness becomes accepted as the criterion for conservation of a breed, as recommended by Hall (1990). This question of a theoretical framework for the assessment of priorities for the conservation of breeds will be addressed in a forthcoming review (S. J. G. Hall, in preparation).

Acknowledgments

J. Clutton-Brock, J. G. Hall, I. L. Mason, M. G. Murray, and P. Murphy kindly commented on the manuscript. D.

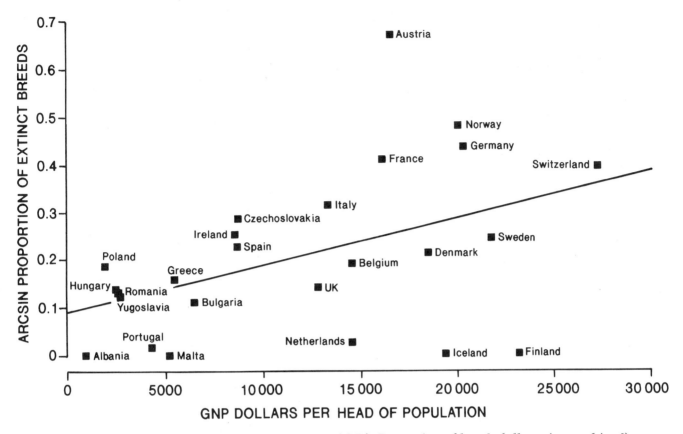

Figure 3. Europe (excluding countries of the former U.S.S.R.). Proportion of breeds (all species combined) recorded that have become extinct, in relation to GNP/GDP per head. The regression was significant at the 5% level.

Hughes drew the figures and M. Lahr provided the Spanish summary.

Literature Cited

Anderson, C. 1992. A gene library that goes 'moo.' Nature, London **355:**382.

Clutton-Brock, J. 1987. A natural history of domesticated mammals. Cambridge University Press and British Museum (Natural History), Cambridge and London, England.

Cunningham, E. P. 1992. Group A: Priority setting. Pages 91–93 in J. E. O. Rege and M. E. Lipner, editors. African animal genetic resources: Their characterisation, conservation and utilisation. Proceedings of the research planning workshop held at the International Livestock Centre for African (ILCA), Addis Ababa, Ethiopia, February 19–21, 1992. ILCA, P.O. Box 5689, Addis Ababa, Ethiopia.

Food & Agriculture Organization. 1991. Production yearbook, vol. 44 (1990). FAO statistics series no. 99. Food and Agricultural Organization of the United Nations, Rome, Italy.

Hall, S. J. G. 1990. Genetic conservation of domestic livestock. Pages 289–318 in S. R. Milligan, editor. Oxford reviews of reproductive biology 12. Oxford University Press, Oxford, England.

Hall, S. J. G., and J. Clutton-Brock. 1989. Two hundred years of British farm livestock. British Museum (Natural History), London, England.

Hunter, B., editor. 1991. The statesman's year-book. 128th edition, 1991–1992. Macmillan, London, England.

International Union for the Conservation of Nature and Natural Resources. 1980. World conservation strategy. Living resource conservation and sustainable development. Gland, Switzerland.

Maijala, K., A. V. Cherekaev, J.-M. Devillard, Z. Reklewski, G. Rognoni, D. L. Simon, and D. E. Steane. 1984. Conservation of animal genetic resources in Europe. Final report of an European Association for Animal Production (E.A.A.P.) working party. Livestock Production Science **11:**3–22.

Mason, I. L. 1988. A world dictionary of livestock breeds, types and varieties. 3rd edition. CAB International, Wallingford, England.

Ruane, J. 1992. The global data bank for domestic livestock. Pages 49–54 in J. E. O. Rege and M. E. Lipner, editors. African animal genetic resources: Their characterisation, conservation and utilisation. Proceedings of the research planning workshop held at International Livestock Centre for Africa (ILCA), Addis Ababa, Ethiopia, February 19–21, 1992. ILCA, P.O. Box 5689, Addis Ababa, Ethiopia.

Simon, D. L. 1992. Summary of breeds in the E.A.A.P. animal genetic data bank, Hannover. Livestock Production Science **32**:92, 93.

Sokal, R. R., and F. J. Rohlf. 1981. Biometry. 2nd edition. Freeman, San Francisco, California.

Times Books. 1990. The Times atlas of the world—comprehensive edition. 8th edition. Bartholomew and Times Books, Edinburgh, Scotland and London, England.

World Conservation Monitoring Centre. 1992. Global biodiversity: Status of the earth's living resources. B. Groombridge, editor. Chapman and Hall, London, England.

Contributed Papers

Can Extractive Reserves Save the Rain Forest? An Ecological and Socioeconomic Comparison of Nontimber Forest Product Extraction Systems in Petén, Guatemala, and West Kalimantan, Indonesia

NICK SALAFSKY
BARBARA L. DUGELBY
JOHN W. TERBORGH

School of the Environment and Center for Tropical Conservation
Duke University
3705-C Erwin Road
Durham, NC 27705, U.S.A.

Abstract: *We compare existing nontimber forest product extraction systems in Petén, Guatemala, and West Kalimantan, Indonesia, to identify key ecological, socioeconomic, and political factors in the design and implementation of extractive reserves. Ecological parameters include the spatial and temporal availability of harvested products and the sustainability of harvesting practices from both a population and an ecosystem perspective. Socioeconomic and political factors include the presence or absence of well-defined resource tenure rights, physical and social infrastructure, markets, and alternative land uses. We conclude that although extractive reserves can play a significant role in preserving tropical forests as a part of a broader land-use spectrum, their effectiveness is highly dependent on prevailing local ecological, socioeconomic, and political conditions. Ultimately, extractive reserves should be regarded as one component of an overall approach to the problem of tropical deforestation.*

Pueden las reservas extractivas salvar la selva tropical lluviosa?

Resumen: *Comparamos los sistemas extractivos de productos no madereros existentes en Petén, Guatemala, y Kalimantan del oeste, Indonesia, con la finalidad de identifiear los factores ecológico, socioeconómico, y politicos, chaves pora el diseño y la implementacion de reservas extractivas. Los parámetros ecológicos incluyen la disponibilidad espacial y temporal de los productos cosechados, asi como la sostenibilidad de las prácticas de cosecha tanto desde una perspectiva poblacional, como del ecosistema. Los factores socioeconómicos y políticos incluyen la presencia o ausencia de derechos de propiedad de recursos bien definidos, infraestructura física y social, mercados, y usos alternativos de la tierra. Concluimos que si bien las reservas extractivas pueden desempeñar un rol significativo en la preservación de las selvas tropicales como parte de un espectro más amplio de uso de la tierra, su efectividad es altamente dependiente de las condiciones ecológicas, socioeconómicas y políticas existentes. En ultima instancia, las reservas extractivas deben ser consideradas como un componente de una propuesta completa para solucionar el problema de la deforestación tropical.*

Address correspondence to J. Terborgh.
Paper submitted August 1, 1991; revised manuscript accepted April 7, 1992.

Introduction

Beginning with the efforts of the National Council of Rubber Tappers in Brazil, considerable excitement has been generated over the past few years among the conservation and development community about the prospects of establishing extractive reserves that can maintain biodiversity while simultaneously providing a sustainable economic return to local peoples and governments (Fearnside 1989a; Schwartzman 1989; Allegretti 1990). This excitement was fueled by an article in *Nature* (Peters et al. 1989) proposing that the long-term financial return from the harvest of nontimber forest products found in a hectare of Amazonian rain forest far outweighed the net benefits of timber production or agricultural conversion of the same area of land. The findings from Peters et al. and similar studies (Alcorn 1989; Anderson & Jardim 1989; Prance 1989; Anderson 1990; Gómez-Pompa & Kaus 1990; Peters 1990) fueled the hopes of many conservationists, who found in these results an alluring mix of ecological, economic, and social justifications for preserving rain-forest lands in a relatively pristine condition.

Several authors, including Peters et al. (1989), have attempted to temper this enthusiasm for extractive reserves by pointing out that hypothetical calculations of the income to be derived from an average hectare of extractive reserves have significant limitations (Fearnside 1989a; Vasquez & Gentry 1989; Browder 1990a, 1990b, 1992; Pinedo-Vasquez et al. 1990). The unique mix of ecological, socioeconomic, and political conditions existing at each potential reserve site makes generalization extremely risky. Only an analysis of a cross-section of tropical forest regions will support broad conclusions about the wide-scale applicability of the extractive reserve concept.

Unfortunately, the complex nature of nontimber forest product extraction systems currently precludes detailed quantitative comparisons of these systems. Nevertheless, the impending destruction of the world's remaining tropical forests, and the pressing needs of policy and decision makers for sound management guidelines, make it imperative that both the promise and limitations of the extractive reserve concept be fully understood.

In this paper, we compare nontimber forest product extraction systems in Guatemala and Indonesia to identify ecological, socioeconomic, and political parameters that need to be considered in the design and implementation of extractive reserves. For rhetorical purposes, we present evidence that these parameters facilitate sustained nontimber forest product extraction in the Petén and inhibit it in Kalimantan. It is not our intention, however, to imply that nontimber forest product extraction is perfectly suited to Petén or that it has no role to play in Kalimantan. Instead, our analysis seeks to underscore the importance of local conditions to the ultimate success or failure of an extractive reserve, while simultaneously recognizing the influence of national and international economic and political factors. Although this paper focuses primarily on market-oriented extractive reserves, many of the issues we discuss in the context of commercial nontimber forest products may apply as well to other land uses, such as collection of nontimber forest products for household consumption or small-scale timber extraction.

Study Sites, Methods, and Background Descriptions

Study Sites and Methods

Descriptions of the two extractive systems in this paper are based on research visits to Guatemala and Indonesia in 1990, supplemented by previous research experience in both these regions. In Guatemala, field work was conducted in the Department of Petén, in the villages of Uaxactun and Carmelita and in harvesting camps near El Mirador, all of which are located in the newly created Maya Biosphere Reserve (Fig. 1). In Indonesia, field work was conducted in the province of West Kalimantan, focusing on the villages of Keranji and Benawai Agung, both of which border Gunung Palung National Park (Fig. 2).

We spent several weeks in each country, living in the study villages, accompanying workers into the forest on harvesting trips, and conducting standardized interviews with the members of households engaged in nontimber forest product extraction, processing, and trade. We also spoke extensively in each country with storekeepers and traders, with local, regional, and national governmental authorities, and with scientists conducting research on these topics.

Background Descriptions of Extractive Systems

PETÉN

Nontimber forest products have been exploited in the Petén since at least the time of the Mayan civilization. According to recent evidence, the Mayas relied heavily on numerous nontimber forest products for subsistence and trade during much of the first millennium A.D. (Nations & Nigh 1980; Voorhies 1982; Gómez-Pompa & Kaus 1990). Today, many *Peteneros* still rely on a highly diversified portfolio of forest products to meet food, fuel, fodder, construction, and medicinal needs (Heinzman & Reining 1989).

Three primary products—chicle (latex from *Manilkara zapota* trees), xate (pronounced shá°te—fronds from *Chamaedorea spp.* palms), and allspice (fruits from *Pimenta dioica* trees)—form the basis of an extensive export-oriented extraction system (Table 1). Chicle extraction began in the late nineteenth century

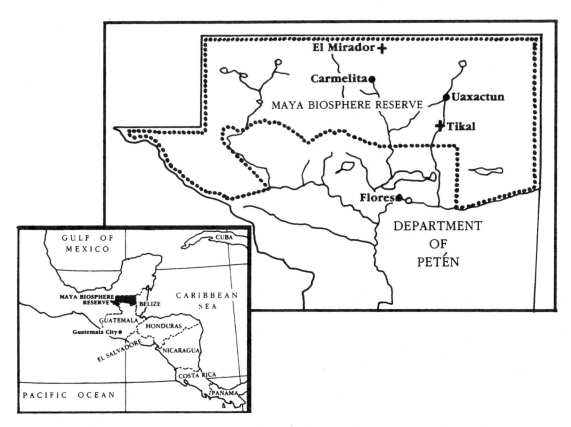

Figure 1. Location of study villages in the Maya Biosphere Reserve, Department of Petén, Guatemala.

and peaked in the 1930s and 1940s, when P. K. Wrigley built airstrips in Petén to harvest chicle, used in the production of chewing gum. In recent years, with the advent of synthetic latex, chicle extraction has declined to some degree. Allspice and xate extraction, however, have grown substantially since 1960 (Schwartz 1990). Today, extraction of these three products has developed into a major industry, employing harvesters who collect the products from the forest, contractors who supply the harvesters and bring the products to regional transportation hubs, and processors who prepare the products for export. It has been estimated that this extractive industry provides part- or full-time employment for over 7000 people in Petén and an annual export revenue of US $4–7 million to the Guatemalan economy (Nations 1989).

WEST KALIMANTAN

Nontimber forest products have undoubtedly been exploited for centuries in Kalimantan by indigenous Dayak tribes and various immigrant settlers. As in Guatemala, local villagers harvest many different products from the forest (Burkhill 1935; deBeer & McDermott 1989). The extraction systems employed in West Kalimantan are less formally organized than those in the Petén. Nevertheless, a number of products are collected on a part-time basis by villagers who go into the forest for periods lasting from a half-day to a month. Harvested products are generally sold to local shopkeepers and middlemen. No figures are currently available concerning the value of extractive industries in the region. Government statistics for Indonesia as a whole, however, estimate that the annual value of nontimber forest products exported in the years 1983 to 1987 ranged between US $18 and $48 million excluding rattan, and US $127 and $238 million including rattan (deBeer & McDermott 1989). The value of products consumed domestically, although undocumented, is probably even higher.

Ecological Factors

Density of Exploited Species

A primary factor in determining the success of an extractive system is the density at which the desired species occur in the forest (Peters 1990). The density of a given plant (or animal) species tends to be inversely related to the overall diversity of the ecosystem. Species density affects the *search time* necessary to locate target individuals, the *travel time* needed to move between these individuals, and the *carrying time* needed to bring the gathered product back to a central collecting point. A basic tenet of optimal foraging models of behavior holds that as the density of a given species decreases (and thus search, travel, and carrying times increase),

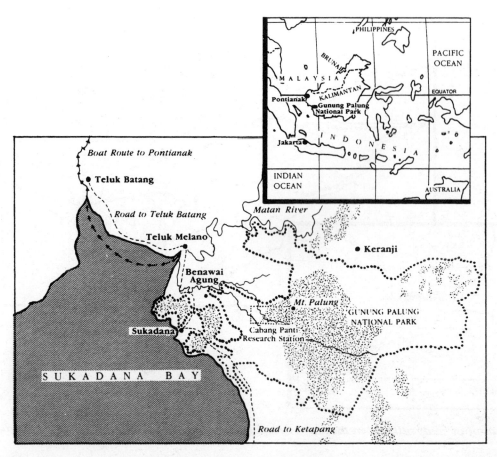

Figure 2. Location of study villages near Gunung Palung National Park, West Kalimantan, Indonesia.

the overall return from the product decreases and hence the product should be less preferred (Charnov 1976; Krebs & McCleery 1984).

PETÉN

Subtropical forests such as those of the Petén contain a relatively low tree species diversity (on the order of 50–100 species per ha) and a correspondingly higher mean density of individuals per species (Lindell 1937; Foster & Brokaw 1982). A recent survey of three sites in Petén found densities of 47, 33, and 23 chicle trees (>10 cm dbh) per ha; 9, 31, and 12 allspice trees (>10 cm dbh) per ha; and 2279 and 2479 xate plants per ha in active harvesting zones (only the latter two sites have numbers reported for *xate*) (Cabrera Madrid et al. 1990). The high densities of these plants enable harvesters to minimize search, travel, and carrying times and thus to collect large amounts of the product over short periods of time. We followed one xate harvester who was on average able to cut fronds (1–5 per plant) from 211 plants per hour over the course of a morning's work.

KALIMANTAN

The tropical forests of Kalimantan, by contrast, have high species diversity (on the order of 150–225 species per ha) and a correspondingly low density of individual taxa (Wallace 1878; Whitmore 1984; Gentry 1988). Low densities of producing plants have a profoundly negative impact on the ability of villagers to harvest many potential forest products efficiently. In general, to overcome the high costs of search, travel, and carrying time, harvesters are restricted to the most valuable products, or those that have a predictable and/or patchy distribution. For example, gaharu, a resin that is derived from the diseased heartwood of several *Aquilaria* species and used in perfume production, can bring as much as several thousand dollars per tree in local markets (deBeer & McDermott 1989; Dixon et al. 1991). This product has a sufficient value to induce harvesters to travel throughout the forest in search of infected trees (Jessup & Peluso 1986). For the majority of potential products, however, the economic return may not justify the time required to find them in the forest.

Temporal Availability of Products

Another important factor in determining the success of an extractive system is the temporal availability of products. Ecological constraints such as fruiting phenology and economic constraints such as seasonal demand can dictate that a given product be only periodically har-

Table 1. Selected forest products with a major role in local extraction systems.

Common Name	Scientific Name	Form	Part	Uses	Harvest Season	Final Market	Harvesting Method
Petén							
Chicle	*Manilkara zapota* (Sapotaceae)	canopy tree	latex	chewing gum & glue	Sep–Feb	export to Japan, US, Europe	tapped from tree in rainy season
Xate	*Chamaedorea* spp. (Arecaceae)	understory palm	leaf	floral arrangements for weddings	Mar–Jun	export to US, Europe	cut from plant
Allspice	*Pimenta dioica* (Myrtaceae)	mid-canopy tree	ripe fruit	culinary & pickling spice	Jul–Aug	export to US, Europe, Russia	picked from felled branches
West Kalimantan							
Gaharu	*Aquilaria* spp. (Thymeliaceae)	mid-canopy tree	resin from infected heartwood	perfume for weddings & funeral rituals	continuous?	export to Saudi Arabia, Asia	cut from felled trees and then boiled out
Medang	*Litsea* spp. (Lauraceae)	mid-canopy tree	bark	aromatic mosquito repellent coils	continuous?	domestic urban areas, export to S.E. Asia	stripped from felled trees, cut, dried and pressed
Illipe nuts	*Shorea* spp. (Dipterocarpaceae)	emergent tree	seed	cooking oil & chocolate	masting periods (3–7 yrs)	domestic & export to S.E. Asia	gathered from ground dried and pressed
Ironwood	*Eusideroxylon zwageri* (Lauraceae)	canopy tree	timber	termite & rot resistant wood	continuous?	local & regional	hand felled, floated out on rafts
Damar	*Dipterocarpus* spp. (Dipterocarpaceae)	emergent tree	exuded resin	caulk for ship construction	continuous?	local & regional	collected from cuts in bark
Rattan	*Calamus* spp. (Arecaceae)	climbing liana	stem	furniture & construction	continuous?	regional & export	cut and pulled down from canopy
Wild fruits	various	understory & canopy trees	fruits & seeds	food	masting & seasonal	local & regional	picked or collected off the ground

vestable. An ideal extractive system should be based on a mix of products whose availability and demand periods are staggered so as to sustain harvest activities throughout the year.

PETÉN

In Petén, the phenology of and demand for the three main nontimber forest products allows them to be exploited at different times of the year in a complementary fashion. Chicle can be harvested only during the rainy season (August to January) when relative moisture, temperature, and rate of transpiration enhances the flow of latex through the trees (Karling 1934). Although xate palm fronds are harvested year-round, the peak demand period (in part created by the spring wedding season in the U.S. and Europe) occurs between March and June. Finally, allspice fruits can be harvested only in July and August when the crop is fully ripened, a period that falls between the end of peak xate demand and the onset of chicle collection. The sequential nature of these harvest seasons enables harvesters to earn steady cash income throughout the year while moderating the demands placed on continually available resources such as *xate*.

KALIMANTAN

In contrast, the forests of Kalimantan and Southeast Asia as a whole are characterized by extreme phenological cycles. Many species of trees produce fruit only once every 3–5 years during a *masting season*, at which time many taxa fruit simultaneously. Masting is hypothesized to be proximally cued by brief periods of warm, dry days (or associated cool, clear nights) and ultimately driven by the need to satiate seed predators (Janzen 1974; Ashton et al. 1988). According to this hypothesis, masting limits the numbers of fruit- and seed-eating animals in the forest by restricting food availability to brief periods, when the animals are swamped by more fruit than they can possibly consume.

This phenological pattern severely affects human fruit and seed "predators" in much the same way it does animals. Extraction systems and markets for many perishable fruit products are extremely difficult to maintain, given that the fruit may be available for only one month every several years. To make matters worse, during these periods of availability, local markets soften because everyone can easily obtain their own supply from the forest. In addition, the glut causes prices to bottom out in urban markets (Pinedo-Vasquez et al.

1990). The harvest of wild fruits also has to compete with that of related domestic fruits that follow the same phenological cycle. For example, wild durian (*Durio* spp.) are occasionally harvested by villagers but are rarely transported to urban markets, since more valuable domestic durians are available at the same time.

This phenological dilemma can be avoided to some degree by gathering nonperishable fruits. One example can be seen in the harvest of illipe nuts, seeds of *Shorea* spp., which are harvested during intense masting periods and pressed for their oils. Because the seeds can be converted into an unperishable commodity, a large industry has developed that provides an important though sporadic source of cash income to Dayak villagers. The large temporally (and spatially) clumped patches of seeds produced by the masting trees also offset some of the problems arising from low population densities of target species.

Product and Ecosystem Sustainability

Finally, it is important to consider the long-term sustainability both of populations of harvested species and of the overall ecosystem. Organisms in a complex ecosystem such as a tropical forest are interlinked in ways that may frustrate naive efforts at management. The removal of a given product from a forest may upset delicate natural balances, which could then reverberate throughout the ecosystem (Gilbert 1980; Terborgh 1988).

Extractive activities are most likely to be unsustainable if they result in killing or damaging individuals of the target species, especially if it is a long-lived organism such as a canopy tree. Extraction may also be unsustainable if it affects reproduction, thus depressing regeneration rates. For example, harvesting of dispersed seeds may inhibit seedling survival and population regeneration (Janzen 1970).

Harvesting in areas of high seed density (such as beneath the crown of the mother tree), however, may not have as negative an impact on regeneration, especially if seedling survival is largely density dependent (seedling survival is inhibited under conditions of high seed density). Fortunately, human and animal seed predators generally prefer these high density patches. Recent empirical work in Madre de Dios, Peru, reveals that Brazil nut (*Bertholletia excelsa*) harvesters collect nuts only under the most productive individual trees, ignoring many others (E. Ortiz, personal communication). Moreover, they search only under the crown and ignore more peripheral zones where the seeds most important to recruitment might be expected to lie. If these widely dispersed seeds escape harvesters, then extractive activities may have little effect on regeneration (Peters 1990).

On the level of the overall ecosystem, extractive activities will ultimately be unsustainable if critical nutrients, such as nitrogen and phosphorus, are removed at rates greater than that at which they are being replenished (Likens et al. 1977). The prevailing paradigm is that tropical forests on very poor soils are delicately balanced ecosystems sustained by highly efficient nutrient recycling mechanisms (Jordan 1985; Stallard 1988). Many nontimber forest products contain substantial concentrations of limiting nutrients, and over the long term it is possible that the export of these nutrients may be unsustainable.

PETÉN

The extractive industries of the Petén have prospered for decades because the target resources have so far not been overexploited. Harvesting practices do not appear to damage plants or jeopardize population recruitment if done correctly. Individual chicle trees have been tapped for decades with seemingly minimal harmful effects, although inexperienced harvesters may fatally injure trees by cutting too deep (Karling 1934). Likewise, xate harvest, which does not take reproductive parts of plants and leaves enough fronds to support continued growth, theoretically seems sustainable. Cabrera-Madrid et al. (1990), however, compared xate palm densities in harvested areas with the protected forests of Tikal National Park and found that there was an average of 2279 and 2478 stems per ha in two harvested areas, whereas there was an average of 4506 stems per ha in the protected forest. It is more difficult to determine the effect of harvesting practices on the ecosystem as a whole, but harvested products probably contain minimal concentrations of critical nutrients.

KALIMANTAN

In Kalimantan, there is little or no available evidence concerning the long-term sustainability of nontimber forest product harvest practices. One might suspect, however, that some of the current harvest practices may be unsustainable, as they either destroy the harvested individual or remove reproductive parts. For example, harvests of both gaharu and medang (bark from *Litsea* spp. used in the production of mosquito repellant coils), as currently practiced, involve cutting down the tree to obtain the desired product. Likewise, the illipe nut harvest could be adversely impacting *Shorea* reproduction. Although reproductively active *Shorea* trees produce vast crops of seeds, only a few survive insect and vertebrate seed predators, even fewer are successfully dispersed to suitable germination sites, and fewer yet survive to become saplings and then adults (Howe & Smallwood 1982). Removal of illipe nuts may reduce the number of seeds going though these natural evolutionary bottlenecks to a level that may diminish effective replacement. Harvested products in Kalimantan, such as illipe nuts (which like other seeds are nutrient rich), could also conceivably constitute an unsustainable drain on the overall ecosystem over the long term.

Socioeconomic and Political Factors

Resource Tenure and Conservation Incentives

An important problem in many extractive systems is the lack of incentives for individuals to conserve available resources for long-term use. Instead, individuals respond to perverse incentives to overharvest resources that are often rooted in existing land and resource tenure regimes (McNeely 1988). Many forest products are open-access resources that are publicly owned, yet their use is not governed by formal or even informal rules (Ciriacy-Wantrup & Bishop 1985). In such situations, individual harvesters have little or no incentive to conserve or manage the resource. In contrast, traditional societies with established rules governing resource use operate under a managed system of common property that can potentially provide incentives for conservation (Berkes et al. 1989).

PETÉN

Although most of the land and resources in the Maya Biosphere Reserve are publicly owned, there is some precedent for resource tenure within the nontimber forest product industries. This tenure appears to be expressed in informal rules or understandings among contractors and harvesters that guide their interactions and harvesting practices. The limited number of contractors, and their informal monopolies over traditional harvesting areas, seem to give them a long-term interest in preserving the resource base. Likewise, many of the harvesters work with one another and with one contractor on a long-term basis, and therefore have incentives to cooperate in maintaining productivity (Schwartz 1990). As an example, several *xateros* mentioned to us that they would not generally harvest leaves within 15 minutes' walk of the camp; instead, these were to be reserved for harvesters who needed a few leaves to make up a full bundle at the end of the day.

KALIMANTAN

Parts of Kalimantan have well-documented common property regimes (see Jessup & Peluso 1986). In the natural forests surrounding Gunung Palung National Park, however, there appears to be little or no evidence of resource tenure or common property rules governing nontimber forest product extraction. Instead, resource extraction undergoes pronounced boom and bust cycles in which a resource is discovered, exploited, and then driven to near extinction (Gentry & Vasquez 1988).

The gaharu wood industry provides one of the best examples of this cycle. Gaharu was first harvested in the more densely populated regions of Southeast Asia (de-Beer & McDermott 1989). As the stock of trees became depleted, the market shifted to more remote areas, such as Kalimantan. When a market developed in the Gunung Palung region in the mid 1980s, villagers suddenly began searching for gaharu. Most of the best quality wood was harvested by teams of professionals who swept the forest on month-long harvesting trips. Many others undertook harvesting trips as well, which generally resulted in the harvest of lower quality wood and even the felling of undiseased trees. As a result, the regional populations of gaharu were depleted within a few years (even in Gunung Palung National Park, where harvesting was illegal), and the industry moved on to other even more remote areas. The over-harvesting of gaharu, driven by its high value, provides an extreme example of the boom and bust cycles to which the extractive industry is prone. Similar yet less precipitous cases of resource depletion have occurred, however, with many other extractive products such as ironwood (a termite-resistant silca-impregnated wood from the species *Eusideroxylon zwageri*) and medang.

Physical and Social Infrastructure

Extraction of products for both household consumption and market sale requires that harvesters be able to transport products from the forest to the point of consumption or sale. This need for transport is especially important with regard to commercially-oriented products, since markets for these goods normally exist only in large towns or regional trading centers. Accordingly, the success of extractive systems often depends on the availability of developed physical (roads, trading boats) and social (middlemen, export companies) infrastructures that enable harvesters to transport and sell their products.

Physical infrastructure is of greatest importance when harvested products are delicate or perishable. Many potentially marketable forest fruits, for example, are not found in regional or international markets, in part because spoilage during transport cannot be avoided. Latexes and fibers, on the other hand, can survive rough handling and storage for long periods if transport is not readily available.

Social infrastructure, consisting of middlemen and traders, is also crucial (Jessup & Peluso 1986). It is difficult for many harvesters who live in remote villages to travel to regional markets. Accordingly, the presence of traders and storekeepers enables harvesters to sell their products at or near the village or harvesting site. In addition, the middlemen may provide credit and/or supplies, which enable villages living a hand-to-mouth existence to go on extended harvesting trips. Although the typical combination of low product prices and high interest rates on supplies ensures that economic rents accrue mostly to middlemen, in their absence the harvesters would earn nothing. This is not to say that such debt peonage is necessarily a desirable situation (Fearnside 1989a). Harvesters may nonetheless prefer its conditions to those offered by other more risky subsistence alternatives.

PETÉN

The extractive industries of Petén are based on commodities that are relatively durable and easily transported. In their processed form, chicle, allspice, and xate all last over three months in a marketable condition and are not very susceptible to transportation damage. In addition, the processing and transportation infrastructures established during the heyday of chicle extraction, consisting of forest camps, warehouses, airstrips, roads, and air and ship transportation routes, are still used today for the movement of all nontimber forest products.

Moreover, a well-established patron-client system exists that facilitates the organization and financing of thousands of harvesters and the transportation of their products. Middlemen contractors serve as patrons to teams of harvesters, whom they transport to harvest camps within the forest. The patrons loan harvesters money to pay for food and other supplies needed by their families during their absence. Harvesters also buy food on credit from the patron while in the forest, often at a 100% to 500% mark-up (Heinzman & Reining 1988). At the end of each week, harvesters are paid for their production, less loaned money. Over the past few years, competition among contractors for laborers has enabled harvesters to gain more leverage and, accordingly, better wages and working conditions.

KALIMANTAN

In Kalimantan, by contrast, there are few roads, and most transport is by water. Products harvested from the forest generally have to be small and light enough to be transported from the forest in small canoes that may have to be carried around obstructions in the streams. For some highly valuable products such as ironwood, it becomes worthwhile to expend enormous efforts to transport the wood (which is so heavy that it sinks and has to be lashed to rafts of floating wood). For many other products, however, the expected return may not compensate for the transport costs.

Although arrangements with middlemen exist in the Gunung Palung region, they tend to be somewhat informal. For example, medang harvesters generally go to the forest on their own for up to ten days at a time to harvest the bark, having first obtained supplies from a local shopkeeper at high interest rates. Although it is difficult to document, the lack of competition among middlemen for harvesting territories may in part be responsible for restricting extractive activities.

Product Demand

For many extractive industries, perhaps the most important impediment is the lack of demand for harvested products either in commercial markets or for household consumption (Pinedo-Vasquez et al. 1990). This problem is compounded by the presence of synthetic substitutes for many products and the fact that many products that seem ideal in theory may conflict with established cultural taboos and preferences (deBeer & McDermott 1989). In general, products with a low demand elasticity (a unit-free measurement of change in demand in relation to change in price) will have steady markets, whereas products with a high elasticity will tend to be in high demand only when they can be obtained at low cost. It is also important to distinguish between demand originating in ephemeral fads from that based on long-term needs (Browder 1992).

PETÉN

The extractive reserve system in Petén has been successful because it was developed to meet an existing market demand. Chicle extraction, for example, was directly driven by the demand for chewing gum. When petroleum-based latex substitutes became available, the bottom fell out of the market and has only recently revived as some gum manufacturers have resumed using natural latex, and as other industrial uses for chicle (such as glue) have been developed. Similarly, annual extraction of allspice reportedly depends on (of all things) the success of the Russian herring catch, since the spice is a major component of the pickling brine.

KALIMANTAN

In Kalimantan as well, products such as gaharu that have a significant demand have been extensively exploited. Other readily available products, however, have been less exploited due to insufficient demand. For example, damar resin, an exudate collected from *Dipterocarpus* spp., is sold in many urban and village markets for use as a shipping caulk, but the overall demand is not great. Accordingly, the price paid to collectors makes extensive harvesting uneconomical. Furthermore, the price of damar is ultimately constrained by the price of synthetic caulking materials, which may be more abundant in large urban areas. Similarly, game animals could potentially be sustainably (or unsustainably) harvested in the Gunung Palung region (Caldecott 1988), but they are not because Islamic law prohibits consumption of most wild animals. Animals such as the rusa deer (*Cervus unicolor*) that are not taboo, however, are readily eaten when available.

Pressures for Alternative Land Uses and Political Power

The final socioeconomic factor to consider in the design of extractive systems is the extent to which other land uses, such as timber production or conversion to agriculture, may economically and politically outcompete the extractive industries. Economic benefit-cost models can be used to compare the long-term net present value of maintaining extractive reserves versus using the land for timber production or agriculture (Peters et al. 1989). These models have limitations, however, in that they may make unrealistic assumptions about the various ecological and economic factors outlined above; they do not always account for consumption of products by local peoples and they may not be appropriate to forestry situations (Fearnside 1989b). Furthermore, there may be many cases in which the models demonstrate the exact opposite of what conservationists want to hear, namely that timber production may be the most profitable use of forest lands, especially in the face of high discount rates and logging incentives created by distorted tax systems and capital markets (Repetto 1988; Fearnside 1989b). Finally, extractive industries may be unable to compete with plantations or agroforestry systems that can produce similar products through a more intensive use of the land.

The limitation of the economic benefit-cost approach is especially significant in light of current political realities. It may be possible to show that a given extractive system would have a higher net present value than a logging operation on the same area of land. Nonetheless, if the benefits of the extractive reserve accrue to local villagers and small-scale middlemen, whereas the economic rents from logging would be captured by urban elite with strong political connections, it is not hard to guess which system would win out. Similarly, land-hungry agricultural transmigrants from densely populated areas generally have greater political influence than harvesters living in remote forest villages. Although recent events in Eastern Europe graphically demonstrate the potential for rapid political reform, extractive reserve design should take into account existing political realities.

Extractive reserves can generally support only low human population densities. Fearnside (1989a) found that experienced brazil-nut harvesters in the Amazon used 300–500 ha of forest for each family (a population density of 1.0–1.7 persons per km^2). Needless to say, it may be extremely difficult for land planners to justify extractive reserves that require tens or hundreds of hectares of forest to support one household in countries that have extremely high population densities.

Ultimately, even if extractive reserves are more economically sensible than competing land-use alternatives, there is no guarantee that the society as a whole will use economic efficiency as a decision-making criterion.

PETÉN

In the Petén, the extractive system has persisted in part because some of the most powerful people in the region (including members of the military) capture the majority of rents from extracted products. Logging has had some impact on deforestation rates over the past decades. There is also increasing pressure to convert forest to agricultural lands. Nonetheless, the government has temporarily halted timber concessions and banned immigration into the newly created Maya Biosphere Reserve. In part, this decision indicates that maintaining the forest as an extractive reserve is in the interests of politically powerful individuals. The extent of this commitment may be tested in the near future, given recent reports that the loggers, who have exhausted the mahogany and cedar trees that were their primary timber sources, are now beginning to cut chicle trees illegally.

KALIMANTAN

In Kalimantan, by contrast, extractive reserve rents accrue to Dayak and Melayu villagers and to Chinese traders and storekeepers, who can be very wealthy but have little or no political power. Instead, political power is held by the owners of timber companies and by the Javanese and Balinese transmigrants. A stark example of the effects of these power dynamics can be seen in one of our study villages, which is located in the midst of a recently exploited timber concession. Over the past decade, the forest surrounding the Dayak village has been stripped away, effectively eliminating many extractive activities that were occurring there. In the initial years, villagers were minimally employed by the timber company, but today the few jobs that are available are held by Javanese immigrants.

Logging in Kalimantan is big business, generating enormous revenues (Gillis 1988; Repetto 1988). Wood is the major export in the region; the value of Indonesian timber exports in the early 1980s was roughly an order of magnitude greater than that of nontimber forest products (Gillis 1988). Although no explicit study of the potential value of extractive reserves in the region has been performed, it seems unlikely that current levels of nontimber forest product extraction will be comparable, even over the long term. Over time, it may be possible for indigenous groups to organize and obtain power as the Brazilian rubber tappers union has done, but this is a difficult process that is compounded by the Southeast Asian cultural reluctance to defy existing authorities. Overall, it does not seem realistic to imagine that Dayak villagers will be able to compete effectively with urban elite and multi-national timber corporations for access to the forest.

Perhaps the ultimate constraint on the extraction of nontimber forest products from natural forest in the Gunung Palung region is that village residents have an alternative source for many forest products. Over the past centuries, residents of this area have developed extensive forest gardens, a multi-species agroforestry system that produces a wide variety of products similar to those harvested from natural forest. Harvest of products from these managed agroforestry systems has a number of advantages over extraction from natural forests, in that the forest gardens (1) offer enhanced densities of desired species, (2) have phenological cycles released from the masting pattern (and thus produce fruit more often), (3) are to some degree protected against animal competitors, (4) enjoy clearly defined property rights, (5) are located near villages, and (6) provide products that are culturally preferred and of a higher quality. For example, although many fruits such as the langsat (*Lansium* spp.) grow in both forest and gardens, there is little or no incentive to harvest the wild fruits for either market sale or household consumption since they generally ripen at the same time as those in the forest gardens. Similarly, the secure tenure rights of the gardens make them the only place in the region near settlements in which relatively large ironwood trees can be found.

Discussion

Table 2 presents a summary of the ecological and socioeconomic factors that determine the potential and limitations of a nontimber forest product extraction system. As demonstrated by the above examples, the ecological and socioeconomic conditions of Petén generally favor the creation of extractive reserves. Extractive systems in Petén have the potential to be economically sustainable while preserving relatively intact forest. In Kalimantan, on the other hand, prevailing ecological and socioeconomic conditions make it unlikely that extractive reserves will play a major part in saving the rain forest. This is not to say that nontimber forest product extraction has no role to play in this region, but rather that it will have to be a part of a diversified multiple-use management plan.

As outlined in Table 2, many of the ecological limitations on extractive reserves could ultimately be overcome by the design and implementation of careful management regimes. For example, low species densities in a forest could be artificially enhanced by enrichment plantings of desired species (Gómez-Pompa & Kaus 1990; Viana 1990; Leighton et al. 1991). Phenological limitations could be offset by developing new products that can be harvested in off-periods (Heinzman & Reining 1988; Dixon et al. 1991) or by finding genetic strains of existing products that produce fruit at desirable times (Peters 1991). Maintenance of population levels could be enhanced through management techniques and educational efforts ensuring that sufficient reproduction occurs (Leighton et al. 1991; Robinson & Redford 1991). And, finally, ecosystem sustainability could be enhanced through reduction of nutrient removals by techniques such as husking fruits while still in the forest (Jordan 1985).

Socioeconomic factors limiting extractive reserves can also be mitigated through various management regimes. For example, resource tenure could be granted

Table 2. Summary of extraction systems in Petén and W. Kalimantan (see text for detailed explanations of rankings).

Parameter	Petén	Kalimantan	Significance	Potential management options
Ecological Factors				
density of exploited species	high	low	determines search and travel times	enrichment plantings
temporal availability of products	regular	irregular (masting)	determines steadiness of harvesting income	artificial selection
sustainability (species level)	moderate (?)	low (?)	determines long-term viability of exploited species	seedling plantings, bottleneck removal
sustainability (overall system)	moderate (??)	??	determines long-term viability of overall system	reduction of nutrient loss
Socioeconomic and Political Factors				
resource tenure	partially specified	open-access	determines incentives for resource conservation	legalize and enforce tenure rights
physical infrastructure	moderately developed	minimally developed	determines ability to transport products to market	build infrastructure
social infrastructure	highly developed	moderately developed	determines opportunities for harvesters to find work	co-ops and incentives for fair middlemen
market demand	moderate to high	low to high	determines degree to which products will be harvested	marketing techniques
political power of industry participants	moderate	minimal	determines degree to which extractive system can be maintained	political action and education
pressure for alternative land use	minimal	strong	determines competition to convert forest lands	holistic land management strategies
existence of alternative agroforestry system	no	yes	determines labor and resources invested into forest system	holistic land management strategies

to open-access resources, or existing traditional rights could be recognized (Bromley & Cernea 1989). Infrastructure could be enhanced to improve transportation (Browder 1990b). Cooperatives could be organized to allow harvesters to transport their products to market without losing most of the rents to middlemen (Jessup & Peluso 1986). Marketing techniques could be employed to increase both domestic and export demand for nontimber forest products (deBeer & McDermott 1989; Dixon et al. 1991). Political action and education could be used to obtain political power for disenfranchised forest villagers and to formally recognize the benefits they may be receiving from the forest (Allegretti 1990). And, finally, holistic land management strategies could be used to obtain the maximum possible sustained economic return from a region (McNeely & Miller 1984; McNeely et al. 1990).

In considering any one of these potential management strategies, it is very important to distinguish between what is technically feasible and what is politically and economically feasible. For example, an enrichment planting of a given species may be biologically feasible, but if the trees require decades to mature and produce a salable product, the incentive to plant the trees will be minimal. It is also important to consider that the relevant factors may be interlinked in complex ways. For instance, building roads into a forest to enhance access to products may lead to overharvesting and also enhance rates of deforestation by loggers and agriculturalists. Accordingly, management strategies will have to be carefully designed and implemented to suit the unique ecological and cultural context of each local situation.

Ultimately, any proposed land use, such as an extractive reserve, needs to be considered as part of a comprehensive regional strategy. Extractive reserves are not stand-alone solutions, but they can effectively complement other land uses. Existing land-use patterns in the study villages in Kalimantan provide one model of such a land-use spectrum that encompasses farms, home gardens, forest gardens, timber and nontimber extraction

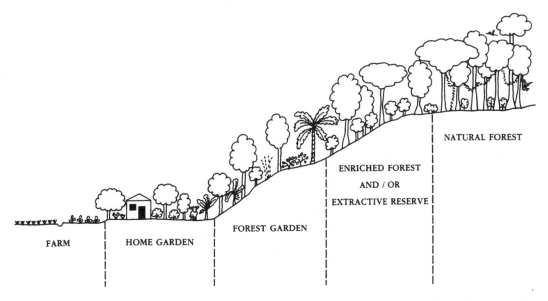

Figure 3. The land-use spectrum in the Gunung Palung region. The following is a list of some products and services generated by the different components of the land-use spectrum. Products are listed by English name or, where no common English name exists, by regional and/or local name with scientific name, following Sastrapradja et al. (1977a, 1977b, 1977c, 1978).

Farms: rice, soybeans, cassava, various vegetables, grass (for tethered cattle).
Home gardens: bananas, papaya, pineapple, hot peppers, rambutan (*Nephilium lappaceum*), ornamental flowers and shrubs, chickens, ducks, goats.
Forest gardens: durian (*Durio zibenthinus*), rubber, coffee, bamboo, sugar palm (*Arenga sacharifera*), pakawai (*Durio* spp.), mangosteen (*Garcinia mangostana*), langsat and duku (*Lansium* spp.), jambu (*Szygium* spp.), terap (*Artocarpus elasticus*), nangka (*Artocarpus heterophyllus*), cempadak (*Artocarpus integer*), jeluti (*Artocarpus* spp.), machang (*Mangifera foetida*), kuweni (*Mangifera* spp.), belimbing (*Averhoa* spp.), lengkeng (*Euphoria longen*), rambai (*Baccaurea motleyana*), gandaria (*Bouea macrophylla*), tengkawang (*Shorea* spp.), and more.
Enriched forests and/or extractive reserves (potentially): various fruits, timber, rattan, gaharu, damar, medang, illipe nuts, medicinal plants, game animals.
Natural forests: new fruit species, new strains of existing fruits, pollinators, ground water recharge, carbon sink, eco-tourism, and option, existence and spiritual values.

areas, and natural forest (Fig. 3). This diverse arrangement of natural and managed habitats provides a wide range of products and environmental services to local human inhabitants, while buffering core conservation areas. Likewise, in Guatemala, the Maya Biosphere Reserve (Nations 1989) provides another model of holistic land-use planning that encompasses agriculture, timber and nontimber forest product extraction, ecotourism, and protected forest.

Conclusions

Based on our observations in Petén and Kalimantan, we offer the following conclusions concerning the extractive reserve concept and appropriate uses for tropical forest lands.

- Extractive reserves are not the panacea that some people would have them be and by themselves will not save the world's rain forests.
- Extractive systems can, however, be an important component of a broader land-use spectrum that includes agricultural plots, forest gardens, timber and nontimber extraction areas, and conservation forest and provides a variety of products and services.
- Local ecological, economic, political, and cultural conditions need to be understood and incorporated into decisions concerning the feasibility, role, location, and extent of extractive reserves, as well as the design, implementation, and management of such systems.
- Various management strategies and policy decisions can facilitate the development of successful extractive systems, but they need to be carefully selected so as not to cause unintended negative effects.

Acknowledgments

N. Salafsky and B. L. Dugelby would like to thank the National Council of Protected Areas (CONAP), the National Commission on the Environment (CONAMA), the Center for Conservation Studies, University of San Carlos (CECON), and the Tropical Agronomic Research Center (CATIE) for sponsoring and supporting our work in the Petén, and the Indonesian National MAB Committee-LIPI, the Center for Research and Development in Biology-LIPI (Puslitbang Biology-LIPI), and the Forest Conservation Service (PHPA) for sponsoring our visit to Kalimantan. We would like especially to thank the residents of the villages of Uaxactun, Carmelita, Keranji, and Benawai Agung, who took us into their homes and put up with our questions.

In addition, we are grateful for the assistance and advice that we received from all of those who helped us along the way, including A. Lehnhoff, H. Rivera, I. Ponciano, B. Heinzman, M. Cabrera Madrid, S. López, C. Reining, A. Solórzano, and M. McLemore in Guatemala, and T. Heinald, D. Sastrapradja, Soetikno W., J. Sugardjito, E. Sumaradja, H. Prayitno, Amir, M. Leighton, Ah-Yen, C. Knab, M. Kusneti, H. Makinuddin, Tadyn, C. Webb, N. Yonkow, S. Zens, I. Koswara, E. Hammond, C. Peters, and E. Saffran in Indonesia.

Funding for our travels was provided by the Sally Hughes-Schrader Travel Grant of the Duke Sigma-Xi Chapter and the Duke Graduate School, the Lazar Fellowship Fund of the Duke School of Forestry and Environmental Studies International Group, the Duke–University of North Carolina Latin American Studies Program, the Josiah Charles Trent Memorial Foundation, and the Explorer's Club. Support for writing this paper was provided by the Duke University Center for Tropical Conservation under a cooperative agreement with the U.S. Agency for International Development.

We thank J. Browder, L. Curren, C. Danks, C. Knab, M. McKean, R. Oren, G. Paoli, D. Peart, F. Putz, N. Schwartz, and three anonymous referees for reading and providing helpful comments on various drafts of this manuscript, and V. Salafsky for drawing and typesetting the figures and maps.

Literature Cited

Alcorn, J. B. 1989. An economic analysis of Huastec Mayan forest management. Pages 182–206 in J. O. Browder, editor. Fragile lands of Latin America: strategies for sustainable development. Westview Press, Boulder, Colorado.

Allegretti, M. H. 1990. Extractive reserves: an alternative for reconciling development and environmental conservation in Amazonia. Pages 252–264 in A. B. Anderson, editor. Alternatives to deforestation: steps toward sustainable use of the Amazonian rain forest. Columbia University Press, New York, New York.

Anderson, A. B. 1990. Extraction and forest management by rural inhabitants in the Amazon estuary. Pages 65–85 in A. B. Anderson, editor. Alternatives to deforestation: steps toward sustainable use of the Amazonian rain forest. Columbia University Press, New York, New York.

Anderson, A. B., and M. A. G. Jardim. 1989. Costs and benefits of floodplain forest management by rural inhabitants in the Amazon estuary: a case study of Açai palm production. Pages 114–129 in J. O. Browder, editor. Fragile lands of Latin America: strategies for sustainable development. Westview Press, Boulder, Colorado.

Ashton, P. S., T. J. Givinish, and S. Appanah. 1988. Staggered flowering in the Dipterocarpaceae: new insights into floral induction and the evolution of mast fruiting in the aseasonal tropics. American Naturalist 132:44–66.

Berkes, F., D. Feeny, B. J. McCay, and J. M. Acheson. 1989. The benefits of the commons. Nature 340:91–93.

Bromley, D. W., and M. M. Cernea. 1989. Management of common property natural resources: some conceptual and operational fallacies. World Bank, Washington, D.C.

Browder, J. O. 1990a. Extractive reserves will not save tropics. BioScience 40:626.

Browder, J. O. 1990b. Beyond the limits of extraction: tropical forest alternatives to extractive reserves. The Rainforest Harvest Conference, London, England.

Browder, J. O. 1992. The limits of extractivism. BioScience 42:174–182.

Burkhill, I. H. 1935. A dictionary of the economic products of the Malay Peninsula. Crown Agents for the Colonies, London, England.

Cabrera Madrid, M., R. Heinzman, S. López, C. Reining, and A. Solórzano. 1990. Non-timber forest products in the Maya Biosphere Reserve: results of ecological and socioeconomic surveys and recommendations for management and investigations. Unpublished draft report.

Caldecott, J. 1988. Hunting and wildlife management in Sarawak. International Union for the Conservation of Nature and Natural Resources, Gland, Switzerland.

Charnov, E. L. 1976. Optimal foraging: the marginal value theorem. Theoretical Population Biology 9:129–136.

Ciriacy-Wantrup, S., and R. Bishop. 1985. Common property as a concept in natural resource policy. Natural Resources Journal 4:173–727.

deBeer, J. H., and M. J. McDermott. 1989. The economic value of non-timber forest products in Southeast Asia with an emphasis on Indonesia, Malaysia, and Thailand. Netherlands Committee for the International Union for the Conservation of Nature and Natural Resources, Amsterdam, The Netherlands.

Dixon, A., H. Roditi, and L. Silverman. 1991. From forest to market: a feasibility study of the development of selected non-timber forest products from Borneo for the U.S. market. Project Borneo/Harvard Business School, Cambridge, Massachusetts.

Fearnside, P. M. 1989a. Extractive reserves in Brazilian Amazonia. BioScience 39:387–393.

Fearnside, P. M. 1989b. Forest management in Amazonia: the need for new criteria in evaluating development options. Forest Ecology and Management 27:61–79.

Foster, R. B., and N. Brokaw. 1982. Structure and history of the vegetation of Barro Colorado Island. Pages 67–82 in E. G. Leigh, A. S. Rand, and D. M. Windsor, editors. The ecology of a tropical forest: seasonal rhythms and long-term changes. Smithsonian Institution Press, Washington, D.C.

Gentry, A. H. 1988. Changes in plant community diversity and floristic composition on evolutionary and geographic gradients. Annals of the Missouri Botanical Garden 75:1–34.

Gentry, A. H., and R. Vasquez. 1988. Where have all the *Ceibas* gone? A case history of mismanagement of a tropical forest resource. Forest Ecology and Management 23:73–76.

Gilbert, L. 1980. Food web organization and the conservation of neotropical diversity. Pages 11–34 in M. E. Soulé and B. A. Wilcox, editors. Conservation biology: an evolutionary-ecological perspective. Sinauer Associates, Sunderland, Massachusetts.

Gillis, M. 1988. Indonesia: public policies, resource management, and the tropical forest. Pages 43–113 in R. Repetto and M. Gillis, editors. Public policies and the misuse of forest resources. Cambridge University Press, Cambridge, England.

Gómez-Pompa, A., and A. Kaus. 1990. Traditional management of tropical forests in Mexico. Pages 45–64 in A. B. Anderson, editor. Alternatives to deforestation: steps toward sustainable use of the Amazonian rain forest. Columbia University Press, New York, New York.

Heinzman, R. M., and C. S. Reining. 1988. Sustained rural development: extractive forest reserves in the Northern Petén of Guatemala. United States Agency for International Development, Guatemala City, Guatemala.

Heinzman, R. M., and C. S. Reining. 1989. Non-timber forest products in Belize and their role in a biosphere reserve model. Unpublished report. Institute of Economic Botany, New York Botanical Garden, Bronx, New York.

Howe, H. F., and J. Smallwood. 1982. Ecology of seed dispersal. Annual Review of Ecology and Systematics 13:201–28.

Janzen, D. H. 1970. Herbivores and the number of species in tropical forests. American Naturalist 104:501–528.

Janzen, D. H. 1974. Tropical blackwater rivers, animals, and mast fruiting by the *Dipterocarpaceae*. Biotropica 6:69–103.

Jessup, T. C., and N. L. Peluso. 1986. Minor forest products as common property resources in East Kalimantan, Indonesia. Common Property Resource Management, National Academy Press, Washington, D.C.

Jordan, C. F. 1985. Nutrient cycling in tropical forest ecosystems. John Wiley and Sons, Chichester, England.

Karling, J. S. 1934. Dendrograph studies on *Achras zapota* in relation to the optimum conditions for tapping. American Journal of Botany 21:161–193.

Krebs, J. R., and R. H. McCleery. 1984. Optimization in behavioral ecology. Pages 91–121 in J. R. Krebs and N. B. Davies, editors. Behavioral ecology: an ecological approach. Sinauer Associates, Sunderland, Massachusetts.

Leighton, M., P. C. Schulze, and D. R. Peart. 1991. Appraisals of enrichment planting in selectively logged forest in Kalimantan: preliminary analysis of economic and ecological variables. Conference on Interactions of People and Forests in Kalimantan. New York Botanical Garden, Bronx, New York.

Likens, G. E., F. H. Bormann, R. S. Pierce, J. S. Eaton, and N. M. Johnson. 1977. Biogeochemistry of a forested ecosystem. Springer-Verlag, New York, New York.

Lindell, C. L. 1937. The vegetation of Petén. Carnegie Institute of Washington, Washington, D.C.

McNeely, J. A. 1988. Economics and biological diversity: using economic incentives to conserve biological resources. International Union for the Conservation of Nature and Natural Resources, Gland, Switzerland.

McNeely, J. A., and K. A. Miller (eds.) 1984. National parks, conservation, and development: the role of protected areas in sustaining society. Smithsonian Institution Press, Washington D.C.

McNeely, J. A., K. R. Miller, W. V. Reid, R. A. Mittermeier, and T. B. Werner. 1990. Conserving the world's biodiversity. International Union for the Conservation of Nature and Natural Resources, Gland, Switzerland.

Nations, J. D. 1989. La Reserva del al Biosfera Maya, Petén: estudio tecnico. Report to the Guatemalan National Council of Protected Areas (CONAP), Guatemala City, Guatemala.

Nations, J. D., and R. B. Nigh. 1980. The evolutionary potential of Lacondon Maya sustained-yield tropical forest agriculture. Journal of Anthropological Research 36:1–30.

Peters, C. M. 1990. Population ecology and management of forest fruit trees in Peruvian Amazonia. Pages 86–98 in A. B. Anderson, editor. Alternatives to deforestation: steps toward sustainable use of the Amazonian rain forest. Columbia University Press, New York, New York.

Peters, C. M. 1991. Population ecology and management of illipe nut in a mixed dipterocarp hill forest. Conference on Interactions of People and Forests in Kalimantan. New York Botanical Garden, Bronx, New York.

Peters, C. M., A. H. Gentry, and R. O. Mendelsohn. 1989. Valuation of an Amazonian rain forest. Nature 339:655–656.

Pinedo-Vasquez, M., D. Zarin, P. Jipp, and J. Chota-Inuma. 1990. Use-values of tree species in a communal forest reserve in Northeast Peru. Conservation Biology 4:405–416.

Prance, G. T. 1989. Economic prospects from tropical rainforest ethnobotany. Pages 61–74 in J. O. Browder, editor. Fragile lands of Latin America: strategies for sustainable development. Westview Press, Boulder, Colorado.

Repetto, R. 1988. The forest for the trees: government policy and the misuse of forest resources. World Resources Institute, Washington, D.C.

Robinson, J. G., and K. H. Redford. 1991. Sustainable harvest of neotropical forest animals. Pages 415–429 in J. G. Robinson and K. H. Redford, editors. Neotropical wildlife use and conservation. University of Chicago Press, Chicago, Illinois.

Sastrapradja, S., S. H. Lubis, E. Djajasukma, H. Soetarno, and I. Lubis. 1977a Sayur-sayuran (Vegetables). Lembaga Biologi Nasional (LBN-LIPI), Bogor, Indonesia.

Sastrapradja, S., N. Wulijarni-Soetijptu, S. Danimihardja, and R. Soejono. 1977b. Ubi-ubian (Root and Tuber Crops). Lembaga Biologi Nasional (LBN-LIPI), Bogor, Indonesia.

Sastrapradja, S., U. Sutisna, G. Panggabea, J. P. Mogea, S. Sukardjo, and A. T. Sumarto. 1977c. Buah-buahan (Fruits). Lembaga Biologi Nasional (LBN-LIPI), Bogor, Indonesia.

Sastrapradja, S., S. Danimihardja, R. Soejono, N. W. Soetjipto, and M. S. Prana. 1978. Tanaman Industri (Industrial Plants). Lembaga Biologi Nasional (LBN-LIPI), Bogor, Indonesia.

Schwartz, N. B. 1990. Forest society: a social history of Petén, Guatemala. University of Pennsylvania Press, Philadelphia, Pennsylvania.

Schwartzman, S. 1989. Extractive reserves: the rubber tappers' strategy for sustainable use of the Amazon rainforest. Pages 150–165 in J. O. Browder, editor. Fragile lands of Latin America: strategies for sustainable development. Westview Press, Boulder, Colorado.

Stallard, R. F. 1988. Weathering and erosion in the humid tropics. Pages 225–246 in A. Lerman and M. Meybeck, editors. Physical and chemical weathering in geochemical cycles. Kluwer Academic Publishers, Dordrecht, The Netherlands.

Terborgh, J. W. 1988. The big things that run the world—a sequel to E. O. Wilson. Conservation Biology 2:402–403.

Vasquez, R., and A. H. Gentry. 1989. Use and misuse of forest harvested fruits in the Iquitos area. Conservation Biology 3:350–361.

Viana, V. M. 1990. Seed and seedling availability as a basis for management of natural forest regeneration. Pages 99–115 in A. B. Anderson, editor. Alternatives to deforestation: steps toward sustainable use of the Amazonian rain forest. Columbia University Press, New York, New York.

Voorhies, B. 1982. An ecological model of the Early Maya of the Central Lowlands. Pages 000–000 in K. V. Flannery, editor. Maya subsistence. Academic Press, New York, New York.

Wallace, A. R. 1878. Tropical nature and other essays. Macmillan, London, England.

Whitmore, T. C. 1984. Tropical rain forests of the Far East. Clarendon Press, Oxford, England.

Notes

Assessing the Economic Value of Traditional Medicines from Tropical Rain Forests

MICHAEL J. BALICK

Institute of Economic Botany
The New York Botanical Garden
Bronx, NY 10458, U.S.A.

ROBERT MENDELSOHN

Yale University
School of Forestry and Environmental Studies
New Haven, CT 06511, U.S.A.

Introduction

In recent years, increasing attention has been given to the value of the tropical rain forest as a source of non-timber market products. Although estimates exist for the value of select forest products (Peters et al. 1989; Tobias and Mendelsohn 1990), many have yet to be quantified. One important class of products that has not yet been valued is tropical pharmaceuticals. Several recent essays have noted that tropical forests are a rich source of unknown chemicals that may eventually prove useful to medicine (Abelson 1990; Oldfield 1989). In addition, traditional medicines are currently the basis for much of the primary health care delivered in tropical nations (Farnsworth et al. 1985). For example, traditional practitioners provide up to 75% of the primary health care needs of rural people in Belize (R. Arvigo, personal communication). Local forests are the source of the plants processed into therapies used in traditional medical systems (Balick 1990). In this paper, we quantify the value of the forests for their therapeutic products, using data from Belize, Central America.

Current methods for the harvest of medicinal plants from forests and fallows involve both destructive and nondestructive practices. For example, destructive methods of harvest include stripping a tree completely of its bark, or cutting it to facilitate harvest. Removing the roots or tubers from a woody or herbaceous plant can also result in its demise. Nondestructive methods include removal of some percentage of the leaves or peeling small strips of bark, to avoid girdling the tree. Based on our observations in Belize, the process of gathering medicinal plants often resembles the harvesting of trees for timber, a more destructive approach. Although this process can be highly destructive for a specific site, provided the harvested area is sufficiently small and that harvests occur over long enough rotations, we suspect that the overall process could be sustainable. It is this approach of long rotations and clearing that we evaluate in this paper. Experiments are underway, however, in Belize to extract medicines more continuously from a plot by removing small amounts of plant material from each tree. As we learn more about the possibilities of this alternative extraction method, it too can be evaluated from both an ecological and an economic viewpoint.

Methods

In order to quantify the value of managing forests as a source of traditional medicines, we began with an inventory of plant material in specific plots. We utilized two sample plots from secondary hardwood forests in the Cayo district of Belize that are representative of the surrounding region. Plot 1 is 0.28 ha in size and plot 2 is 0.25 ha in size. Plot 1 is approximately 30 years old and is located in a valley at about 200 m in elevation. Plot 2 is approximately 50 years old and is located on a

Paper accepted December 4, 1991.

ridge in the foothills of the Maya mountains at about 350 m in elevation. Clearing plot 1, we were able to collect 86.4 kg dry weight of marketable medicinal plant material. The species collected are listed in Table 1. In plot 2, we collected 358.4 kg dry weight of medicine; the composition of which is described in Table 2. Thus, extrapolating on a per hectare basis, we found 308.6 and 1433.6 kg dry weight of medicines on the two plots, respectively.

Results

Local herbal pharmacists and healers purchase unprocessed medicine from small farmers at a rate of $2.80/kg (all values are expressed in U.S. dollars). Multiplying the quantities of medicine found per hectare above by this price suggests that clearing a hectare would yield the farmer between $864 and $4,014 of gross revenue. The farmer, however, has costs he must bear to harvest this material. The collection of plant material required seven man days on plot 1 and 20 man days on plot 2. On a per hectare basis, harvesting required 25 man days on plot 1 and 80 man days on plot 2. Given the local wage rate of $12/day, total harvest costs of the two plots are $300 and $960, respectively. Subtracting these costs from gross revenue, the net revenue from clearing a hectare is consequently $564 and $3,054 on each of the plots.

Not enough information is available to know what rotation age is optimal for collecting medicines. It is not

Table 1. Medicinal plants harvested from valley forest plot (no. 1) in Cayo, Belize.

Common Name	Scientific Name	Use[a]
Bejuco Verde	*Agonandra racemosa* (DC.) Standl.	Sedative, laxative, "gastritis," analgesic
Callawalla	*Phlebodium decumanum* (Willd.) J. Smith	Ulcers, pain, "gastritis," chronic indigestion, high blood pressure, "cancer"
China Root	*Smilax lanceolata* L.	Blood tonic, fatigue, "anemia," acid stomach, rheumatism, skin conditions
Cocomecca	*Dioscorea* sp.	Urinary tract ailments, bladder infection, stoppage of urine, kidney sluggishness and malfunction, to loosen mucus in coughs and colds, febrifuge, blood tonic
Contribo	*Aristolochia trilobata* L.	Flu, colds, constipation, fevers, stomach ache, indigestion, "gastritis," parasites

[a] *Uses listed are based on disease concepts recognized in Belize, primarily of Maya origin, that may or may not have equivalent states in Western medicine. For example, kidney sluggishness is not a condition commonly recognized by Western-trained physicians, but is a common complaint among people in this region.*

Table 2. Medicinal plants harvested from ridge forest plot (No. 2) in Cayo, Belize

Common Name	Scientific Name	Use[a]
Negrito	*Simaruba glauca* DC.	Dysentery & diarrhea, dysmenorrhea, skin conditions, stomach and bowel tonic
Gumbolimbo	*Bursera simaruba* (L.) Sarg.	Antipruritic, stomach cramps, kidney infections, diuretic
China root	*Smilax lanceolata* L.	Blood tonic, fatigue, "anemia," acid stomach, rheumatism, skin conditions
Cocomecca	*Dioscorea* sp.	Urinary tract ailments, bladder infection, stoppage of urine, kidney sluggishness and malfunction, to loosen mucus in coughs and colds, febrifuge, blood tonic

[a] *Uses listed are based on disease concepts recognized in Belize, primarily of Maya origin, that may or may not have equivalent states in Western medicine. For example, kidney sluggishness is not a condition commonly recognized by Western-trained physicians, but is a common complaint among people in this region.*

clear how much time should elapse between each harvest on a specific plot. However, assuming that we use the current age of the forest in each plot as a rotation length, we can calculate at least an estimate of the present value of harvesting medicine sustainably into the future. The present value of an infinite stream of harvests every t years (beginning with a standing forest) can be calculated from a standard Faustman formula:

$$V = R/(1 - e^{-rt}),$$

where R is the net revenue from a single harvest and r is the real interest rate. We assume the real interest rate is 5% for this calculation. Given a 30 year rotation in plot 1 and substituting the appropriate values of r and R into the above equation suggests that the present value of medicine in plot 1 is $726/ha. Making a similar calculation for plot 2 but extending the rotation to 50 years yields a present value of $3,327/ha.

Discussion

These estimates of the value of using tropical forests for the harvest of medicinal plants compare favorably with alternative land uses in the region. For example, estimates of the value of intensive agriculture in the Brazilian rainforest are $339/ha (Florschutz 1983) and milpa (corn, beans, and squash) in Guatemalan rainforest are $288/ha (Heinzman and Reining 1988). Even the most successful pine plantations proposed for the tropics expect to yield only $3,184/ha (Sedjo 1983). We also identified commercial products such as allspice (*Pimenta dioica*), copal (*Protium copal*), chicle (*Manilkara zapota*), and construction materials (beams for houses) in the sample plots that could be harvested and added to the total value of the plots. Thus, our data suggest that protection of at least some areas of rainforest as extractive reserves for medicinal plants appears to be economically justified. We feel that a periodic harvest strategy is a realistic and sustainable method of utilizing the forest, based on our evaluation of the flow of medicinal plant materials. For example, within a 50 ha parcel of forest similar to the second plot we analyzed, it would appear that one could harvest and clear one hectare per year indefinitely.

People skeptical of the efficacy of traditional medicine in primary health care delivery systems may argue that these figures overestimate the value of tropical forest medicines because local people should rely on commercial factory-produced pharmaceuticals instead. To substitute Western medicine for traditional healers would require a substantial increase in health expenditures for a country like Belize. Further, there is increasing acceptance that, in primary health care delivery systems, traditional medicines provide effective modalities for many conditions (Farnsworth et al. 1985; Akerele 1988).

The analysis used in this study is based on current market data. The estimates that these forests are worth $726 and $3,327/ha for their medicinal plants could change based on local market forces. For example, if knowledge about tropical herbal medicines becomes even more widespread and their collection increases, prices for specific medicines would fall. Similarly, if more consumers become aware of the potential of some of these medicines or if the cost of commercially produced pharmaceuticals becomes too great, demand for herbal medicines could increase, substantially driving up prices. Finally, destruction of the tropical forest habitats of many of these important plants would increase their scarcity, driving up local prices. We have already observed this scenario in Belize with some species, especially those in primary and secondary forest habitats. We predict that the value of tropical forests for the harvest of nontimber forest products will increase relative to other land uses over time as these forests become more scarce.

Conclusion

We expect that the results of this study will stimulate follow-up studies to quantify the stock and growth of plant medicines in primary and secondary forests. Systems for the sustainable collection of plant medicines and other nontimber products from the tropical forest need to be documented and developed for use on a much broader scale. Tropical forests are a source of medicine for hundreds of millions of people in the developing world. Combining the present value of medicine with that of other sustainable nontimber forest products provides a compelling and quantifiable argument for the conservation and careful management of tropical and subtropical forests.

Acknowledgments

We thank H. Bormann, J. Brown, D. Daly, E. Elisabetsky, J. Gordon, R. Herdt, C. Gyllenhaal, S. Mori, C. Peters, and M. Pinedo for their comments on this manuscript. We are grateful to R. Arvigo and G. Shropshire of the Ix Chel Tropical Research Center, Ltd., Belize for assistance with field logistics, as well as several anonymous informants for sharing with us their knowledge about medicinal plants. This project was supported by grants from the U.S. Agency for International Development, The Metropolitan Life Insurance Foundation, the U.S. National Institutes of Health/National Cancer Institute, The Nathan Cummings Foundation, The Overbrook Foundation, and The Rex Foundation. We also thank the Belize Forestry Department and Belize Environmental Center for their collaboration with this project.

Literature Cited

Abelson, P. H. 1990. Medicine from plants. Science **247**, 513.

Akerle, O. 1988. Medicinal plants and primary health care: an agenda for action. Fitoterapia **59**, 355–363.

Balick, M. J. 1990. Ethnobotany and the identification of therapeutic agents from the rainforest. Pages 21–23 in D. J. Chadwick and J. Marsh, editors. Bioactive compounds from plants, J. Wiley and Sons, Chichester, England.

Farnsworth, N. L., O. Akerele, A. S. Bingel, D. D. Soejarto, and Z. G. Guo. 1985. Medicinal plants in therapy. Bull. WHO **63**:965–981.

Florschutz, G. 1983. Análise economica de estabelecimentos rurais de Municipio de Tome-Açu, Pará: Un estado de caso. Documentos **19** EMBRAPA/CPATU Belem, Brazil.

Heinzman, R., and C. Reining. 1988. Sustained rural development: extractive forest reserves in the northern Peten of Guatemala. A report to U.S. AID/Guatemala, contract number 520-0000-0-00-8532-00.

Oldfield, M. L. 1989. The value of conserving genetic resources. Sinauer Associates, Sunderland, Massachusetts.

Peters, C. P., A. H. Gentry, and R. O. Mendelsohn. 1989. Valuation of an Amazonian rainforest. Nature **339**:655–656.

Sedjo, R. 1983. The comparative economics of plantation forestry: a global assessment. Resources for the Future, John Hopkins Press, Baltimore, Maryland.

Tobias, D., and R. Mendelsohn. 1990. The value of recreation in a tropical rainforest reserve. Ambio **20**:91–93.

Editorial

The Business of Conservation

A large part of the natural world has been damaged or destroyed by unregulated commerce. Now, various groups of conservationists are trying to save some of the most spectacular remnants of nature—both species and ecosystems—with more commerce. The idea is an interesting one, and it is always nice to turn the tables and use the methods of exploiters to prevent more serious exploitation, but the risks are high and not everyone in the business of saving by selling seems to have given much thought to them.

I first came across commercial conservation in one of its early and most dubious forms, the farming of sea turtles, a phoenix-like enterprise that always springs up from the ashes of its last bankruptcy. Because of its biological and economic complexity, the farming of sea turtles nicely illustrates many of the problems inherent in the commercialization of conservation and is worth examining in some detail.

The theory of sea turtle farming goes something like this: green turtles (*Chelonia mydas*) are in trouble because they are taken from the nesting beaches and adjacent waters in huge numbers as adults and used for leather, oil, soup cartilage, and meat. The eggs are taken for their alleged aphrodisiac properties and for baking. This catch, coupled with the take at the distant feeding grounds, if unchecked will lead to extinction. How much better to take the surplus production of eggs, a minority of the total laid each season, hatch them, raise them to market size, and sell them. This will displace wild-caught animals from the market because of the easy, guaranteed availability of the farm-raised turtles and the higher, more uniform quality of the domestic product. As a bonus, we can set aside a few breeders to supply eggs in the future, thus creating a closed system with no net drain on wild populations.

From the beginning, sea turtle farming was strongly opposed by the late Archie Carr, the preeminent turtle biologist of this century, and by a sizeable group of his students and colleagues, including me. The reasons we gave for this opposition, and which we still give as the battle continues a quarter of a century later, are rooted in a mix of biology and economics. Here are a few of the more important ones.

The first problem is the unavoidable expense of coping with the biology of a nondomesticated species, a frequent worry for conservation entrepreneurs. Green turtles mature slowly, especially after the hatchling stage when they switch to a herbivorous diet. Animals of some populations may take nearly a half-century to reach sexual maturity—they are not exactly broiler chickens. To speed this up, captive turtles must be fed expensive food enriched with animal protein, something akin to trout chow. Feeding efficiency is low as pellets dissolve or are washed away before the turtles can eat them.

Holding tanks are expensive to build and circulating sea water is expensive to filter and pump. Fencing off bays or inlets instead of using tanks doesn't work because the fences wash away in storms; if they don't, the impoundments become heavily polluted. In both tanks and impoundments, crowded turtles get sick and have to be treated, which is another expense.

The second problem concerns reproduction, always the measure of the success of any conservation plan. Keeping a breeding stock of green turtles for a large operation is economically (and biologically) impractical. Even assuming that all biological obstacles to producing fertile eggs in captivity are overcome—a very large assumption—I once calculated that it would take 32,000 kg (70,300 lbs) of adults, mostly females, to supply just 50,000 eggs per year. The space required for 293 mature sea turtles and the cost of feeding and maintaining them makes this notion of breeding stock a manifest absurdity.

So turtle farms are really turtle ranches, according to the CITES definition. They are not self-maintaining. To survive, green turtle ranches must take eggs from the dwindling wild populations every year. As a justification of this destructive raiding, some yearling turtles have been released, with much publicity, to compensate for the eggs taken; a small number of yearlings presumably equalling the natural survivorship of many eggs. But because of the complex life cycle and migrations of green turtles, there is not a shred of evidence that released, ranch-raised yearlings ever find and join the right reproductive pool, or any reproductive pool. Moreover, because sex is determined nonchromosomally by incubation temperature of the eggs, it is quite possible that groups of released turtles have inappropriate sex ratios.

The third problem is one of market stimulation, a problem that is common to many kinds of profit-making conservation. To sell their products and keep up the necessary cash flow, turtle ranches have to produce a stable demand for green turtle meat, for green turtle oil for cosmetics, for green turtle leather, for dried, varnished yearlings as tourist souvenirs, for little hatchlings preserved in lucite blocks, and for any other preparation of the animal that can be sold. Because of the high costs of ranching sea turtles, all these products must be priced as luxury items. But the high prices and especially the developed, stable markets make it extremely attractive to poach wild turtles and to find ways of slipping them into commerce. Poaching is always cheaper than ranching. There may not be any way to protect an endangered species once the world market is stirred up, even in the name of conservation. The power of global demand erodes all safeguards. The problem is one of scale: the world market is just too big for most natural systems. Thus, the commercial ranching of green turtles inevitably brings us around again on the downward spiral—a little closer to the extinction of the remaining populations. By no stretch of the imagination is this conservation.

Each of these three sets of problems in the ranching of sea turtles is generated by an interaction between biology and economics. Do similar situations exist in other examples of species conservation as a business venture, with or without the introduction of global marketing? After all, few commercial species that can be raised in captivity have the intractable biology of sea turtles. In the pages of *Conservation Biology*, Geist

(2(1):15, 1988) has warned against the sale of wildlife products, but Grigg (3(2):194, 1989) has advocated the commercial harvesting of kangaroos as a way of conserving them and their habitat. Iguanas are being ranched; whether this interesting cottage industry succeeds in making a profit while preserving iguana habitat and wild iguanas remains to be seen.

The commercial exploitation of entire ecosystems—as opposed to single species—is again very problematic as a conservation technique, for many of the same reasons. For years, the preservation of the New Jersey Pine Barrens has been justified by some conservationists as a way of saving the enormous aquifer of pure, fresh water that lies beneath. But if that water is ever exploited on a large scale, the water table will fall, and even a small drop will cause extensive changes in the ecosystem, including major loss of wetlands.

In the past few years, "ecotourism" has been widely advocated by conservationists, with little knowledge of its effects in most cases. Similarly, the campaign to sell nuts and fruits from tropical rainforests has preceded any serious attempt to find out whether the exploited forests can sustain these worldwide sales, however well-intentioned. An article in this journal by Vasquez and Gentry (3(4):350, 1989) suggests that the collection of forest-harvested fruits for market may not be sustainable in the highly diverse forests around Iquitos, Peru. On the other hand, Peters et al. (3(4):341, 1989), in the same issue, find that "oligarchic," Amazonian forests of low tree diversity may be able to sustain careful, "market-oriented extraction of fruits" indefinitely.

In this issue of *Conservation Biology* (p. 128), the article by Balick and Mendelsohn is a pioneering effort to quantify the value of traditional medicines extracted from rainforests in Belize. The authors are well aware of the problem of sustainability and base their calculations on long rotations, small harvested areas, and local use of the forest products. I hope that others with less ecological knowledge do not ignore the authors' caveats and extrapolate from their data to promote commercial extraction on an inappropriate scale. As in the case of green turtle ranching, local forests may not be able to survive the demands of the world market.

Specific biological knowledge is needed to determine the feasibility of each venture in commercial conservation. On this biological basis, some projects, such as sea turtle ranching, ought to be ruled out from the start. Others, such as the use of forests to provide traditional medicines, may prove feasible. Nevertheless, even in these latter cases the economic scale of the operation is likely to be limiting: there will be few schemes of commercial conservation that can supply the world market without destroying the species and ecosystems they purport to save.

David Ehrenfeld

The Role of Foreign Debt in Deforestation in Latin America

RAYMOND E. GULLISON
ELIZABETH C. LOSOS
Department of Ecology and Evolutionary Biology
Princeton University
Princeton, NJ 08544, U.S.A.

Abstract: *Much current controversy exists about the role that foreign debt plays in deforestation in Latin America. In an attempt to familiarize concerned biologists with this issue, we present an overview of the proposed positive and negative effects that foreign debt has had on neotropical forest loss. From the literature, we identify three main hypotheses as to how the large external debts of developing Latin American countries may contribute to deforestation. We find that (1) countries have not increased exports of tropical timber and beef in response to rising debt; (2) external debts have contributed to economic stagnation and an associated increase in poverty in Latin America, which in turn has caused the degradation of marginal lands, but the role of debt in this process cannot be isolated from other important contributing factors; and (3) while debt payments have probably led to governmental budget cutbacks in environmental spending, historically spending in these areas has not been high. On the positive side, we found that (1) innovative debt-for-nature swaps have traded devalued debt for a commitment and funds to create and protect nature reserves; and (2) debt could be exchanged for forestry and agricultural sectoral reform, which would have very large positive effects on the conservation and management of forests.*

El rol de la deuda externa en la deforestación de Latino América

Resumen: *En el presente existe mucha controversia sobre el rol que la deuda externa juega en la deforestación de América Latina. En un intento de familiarizar a biólogos interesados en este tema, nosotros presentamos una reseña sobre los efectos positivos y negativos que la deuda externa ha tenido en la pérdida de selvas Neotropicales. A partir de la literatura sobre el tema, nosotros identificamos tres hipótesis principales sobre como las grandes deudas externas de los países en vias de dasarrollo de América Latina contribuirían a la deforestación. Encontramos que (1) los países no han incrementado las exportaciones de madera tropical y carne como respuesta a la creciente deuda, (2) las deudas externas han contribuido al estancamiento económico y al asociado incremento de la pobreza en América Latina, lo que a su vez ha causado la degradación de las tierras marginales; sin embargo, el rol de la deuda en este proceso no puede ser aislado de otros importantes factores cotribuyentes, y (3) si bien los pagos de la deuda probablemente han producido recortes en el presupuesto de gastos relativos a medio ambiente, históricamente los gastos en estas áreas no han sido altos. Desde un punto de vista positivo, (1) el innovativo troque de deuda por naturaleza ha intercambiado deuda devaluada por un compromiso y fondos para crear y proteger reservas naturales y (2) la deuda puede ser intercambiada por una reforma en los sectores de bosques y agricultura, la cual puede tener un gran efecto positivo en la conservación y manejo de los bosques.*

Address correspondence to R. E. Gullison.
Paper submitted June 19, 1991; revised manuscript accepted April 2, 1992.

The *U.K. Friends of the Earth* is lobbying British banks because it believes that alleviation of international debt will reduce causes of rain forest destruction. Their slogan is "Friends of the Earth say **Stamp out the debt, not the rain forest.**" (Tropical Timbers 1990)

The high rate of deforestation in the tropics has emerged as one of the most important environmental crises of present time and has led concerned people to search for causes and solutions to this problem. Overpopulation and poverty have been cited as major causes of deforestation, but these are problems neither quickly nor easily rectified by the actions of developed countries. Environmentalists in developed countries have turned their attention to other ways in which developed countries influence deforestation in developing nations. In particular, the high level of foreign debt incurred by developing nations starting in the late 1970s (see Fig. 1) has attracted attention because developed nations have played a powerful role as international lenders. If debt is playing a hand in causing tropical deforestation, then developed countries could potentially do much to alleviate this pressure by reducing the debt burden of developing countries.

Initial evidence that deforestation and indebtedness are linked came from correlations between indebtedness and deforestation among countries (for example, see Fig. 2 and Ayres 1989). Large countries, such as Brazil, have very high foreign debts and very high levels of forest loss. However, these simple analyses ignore the fact that both deforestation and indebtedness will co-

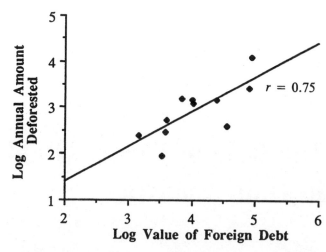

Figure 2. Average log annual amount of forest cut (in 1,000s ha) is highly correlated with average log value of foreign debt ($1,000,000) for 11 Latin American countries in the period 1980–1985. (Data from FAO Production Yearbooks; World Bank 1988–1989.)

vary strongly with the size of a country's population. When we standardize these measurements for population size, the correlation disappears (see Fig. 3). A more careful analysis is required to identify the role indebtedness in deforestation.

Many hypotheses have been proposed that draw a causal link between debt and deforestation. We have found these hypotheses incomplete and confusing, and usually presented with little evidence. Since future environmental policy may depend heavily on how debt is perceived to influence deforestation, we feel it is important that biologists be acquainted with the predominant

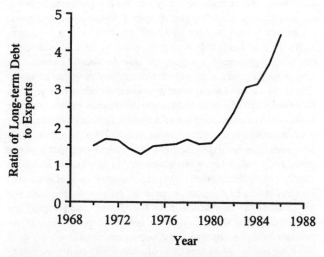

Figure 1. The average debt burden (ratio of long-term foreign debt to exports) from 1970 to 1987 for 11 Latin American countries. Beginning in 1980, indebtedness has increased rapidly (data: from World Bank 1988–1989). Countries included are Argentina, Bolivia, Brazil, Colombia, Costa Rica, Ecuador, Mexico, Paraguay, Peru, Uruguay, and Venezuela. Unless otherwise noted, these countries form the data base for the remaining figures.

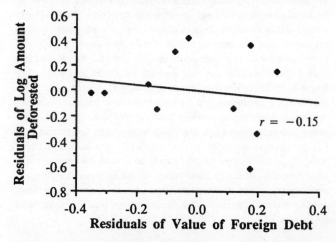

Figure 3. The same data as in Fig. 2, but both variables regressed first against population size and the residuals plotted against one another. Once the effect of population size is removed, the correlation between debt and deforestation disappears. (Data from FAO Production Yearbooks; World Bank 1988–1989.)

arguments. In the first part of this essay, we distill from the literature the main hypotheses as to how debt may cause deforestation in Latin America. Only one of these hypotheses invokes a direct link between the debt and deforestation—that indebted countries are exploiting their tropical forests to earn foreign exchange for debt payments (Potter 1988; Corson 1990). Two other hypotheses propose indirect links. Under one scenario, debt increases poverty, which in turn forces poor people to use marginal lands such as tropical forests (World Resources Institute 1989). The final hypothesis is that high debt payments cause cutbacks in the funding of environmental programs (World Wildlife Fund 1988). We describe each of these hypotheses in some detail and discuss available evidence for them. In the second part of this essay, we look at some positive effects that debt reduction has had on tropical forest conservation, and we discuss a potential approach that links debt reduction to natural resource conservation.

The Role of Debt in Deforestation

Export Promotion

The simplest hypothesis relating foreign debt to deforestation is that highly indebted nations increase exploitation of forests to earn foreign exchange for debt payments (the "export promotion" hypothesis). This may be caused by an internal decision by the government of the indebted nation, or by external conditions imposed upon a country by a multilateral lending agency such as the International Monetary Fund as part of a larger group of austerity measures (Lombardi 1985). If this hypothesis is true, we expect increasing indebtedness to lead to greater exports of natural resources. We consider two ways that export promotion could lead to increased deforestation: first, debt causes the increased export of tropical timbers; second, debt causes the increased export of beef. Increased beef production is largely realized by clearing forests for new pastures.

PROMOTION OF TIMBER EXPORTS

We wished to determine the extent to which exports of tropical timbers are related to levels of indebtedness within Latin American countries. The approach we took was to use stepwise regression to identify variables that are significant predictors of timber production in nine Latin American countries. As the measure of indebtedness in our regression equations, we use the ratio of long-term foreign debt over the value of exports (Devlin 1989). This is a better measure of indebtedness than the absolute value of the foreign debt because it standardizes the amount of debt by a country's ability to earn foreign exchange and make payments.

We had difficulty in choosing a meaningful measure of timber exports to use in our analysis. Timber exports take many forms, from sawlogs to veneer to furniture, so that it is extremely difficult to calculate the volume of tropical woods that are exported each year. Fluctuations in timber prices make it impossible to use the value of timber product exports to determine the volume of wood exported each year. These problems limited us to testing for effects of indebtedness on total roundwood production (the total volume of wood cut for timber and fuelwood). If indebtedness is causing governments to promote timber exports to any significant degree, this should be reflected in an increase in roundwood production.

We used stepwise regression to determine the effect of indebtedness on roundwood production for each of nine Latin American countries from 1976 to 1985. We included as independent variables in the stepwise regression equation (1) change in ratio of long-term debt to exports, (2) change in gross national product (GNP, a measure of economic growth), (3) change in world timber prices, (4) change in value of forest product exports (exports of wood in all forms), and (5) a trend variable, time. Stepwise regression allows only those variables that have a significant effect on predicting variation in the dependent variable to enter the regression equation. The stepwise regressions were repeated with a one-year time lag in roundwood production to allow for delayed effects of the independent variables on roundwood production.

The results of the regression analyses for each country are shown in Tables 1 and 2. Of the variables considered, changes in world timber prices most frequently entered the regression equations. Most important, indebtedness entered only one regression equation (for Paraguay), and in this case the value is negative, showing that as indebtedness increased roundwood production actually decreased. Thus our data provide no support for the export promotion hypothesis and suggest that other factors, particularly world timber prices, more strongly influence roundwood production.

Finally, to place this argument in proper context, we consider to what degree forest product export earnings could contribute to debt reduction. Figure 4 shows the proportion of total value of long-term debt that forest exports constitute. At their peak in 1980, forest product export earnings approached 1% of total long-term debt for all of Latin America, and the proportion has steadily decreased since then, to 0.43% in 1985. Forest product export earnings are so small in relation to debt that it is difficult to believe that they could play an important role in debt servicing.

PROMOTION OF BEEF EXPORTS

Although we find no evidence that foreign debt causes increased deforestation by causing increased timber exports, it could do so indirectly by causing increased beef

Table 1. Stepwise regression of factors likely to influence roundwood production (no time lag) in nine Latin American countries (1976–1985).

Country	Forest exports	GNP	Loans/ exports	Timber price	Time	R^2	p-value
Bolivia	—	—	—	—	0.151	0.313	0.027
Brazil	0.027	—	—	0.022	−0.003	0.763	0.004
Chile	—	—	—	0.123	—	0.278	0.037
Costa Rica	—	—	—	—	−94.58	0.246	0.041
Paraguay	—	—	−0.119	—	—	0.348	0.016
Peru	—	—	—	0.084	—	0.199	0.071
Colombia						no variables significant	
Ecuador						no variables significant	
Mexico						no variables significant	

Data presented are beta coefficients of variables that contributed significantly to the regression equation, and R^2 and p-values (F-value > 3.0 to enter the equation). Indebtedness did not enter any of the equations, except for Paraguay. (Data from FAO Production Yearbooks [1977–1986], FAO Trade Yearbooks [1977–1986], International Financial Yearbook [1977–1986].)

exports, as much new pasture land comes from cleared forests. Beef production has been steadily rising since the early 1970s, and conversion of forested land to pasture is a major cause of deforestation in some countries. Despite rising beef prices, however, the proportion of total beef produced that is exported (see Fig. 5) has not followed the same increasing trend as indebtedness has. It may be true that conversion to pasture land is a major cause of deforestation in the New World, but beef exports have not increased in response to the rapidly increasing debt burdens that Latin American countries have suffered. Increased beef production has instead gone to domestic markets.

As with timber exports, the magnitude of earnings from beef exports are very small compared with the amount of foreign debt owed (see Fig. 4), limiting the importance these funds can play in reducing foreign debt.

We conclude from our analyses that foreign debt owed by Latin American countries is not leading to increased deforestation through promotion of timber and beef exports. It is unlikely that debt reduction would lead to a decrease in exports and a corresponding decrease in deforestation. Of the factors considered in our analysis, roundwood production seems influenced most by world timber prices rather than by indebtedness. In the case of beef, an ever increasing proportion is going to domestic markets. Furthermore, earnings from both timber and beef exports are very small compared with the amount of debt owed and seem unlikely to play any significant role in debt reduction.

Debt-Induced Marginalization

This hypothesis proposes that debt-induced economic problems cause increased poverty, which in turn forces people to use and degrade marginal lands, such as tropical forests. Because of the difficulty in identifying both the role that debt plays in economic deterioration and the role of poverty in influencing land-use patterns, we limit the discussion in this section to a clear elucidation of the hypothesis and an examination of the difficulties involved in collecting data to test it.

Step 1. Debt-induced economic deterioration. It is commonly agreed that economic deterioration in the 1980s has led to increased poverty (Iglesias, as quoted in Durning 1989) and to widespread reversals in child health levels, nutrition, and education (Cornia et al.

Table 2. Stepwise regression, as in Table 1, but with a one-year time lag in roundwood production.

Country	Forest exports	GNP	Loans/ exports	Timber price	Time	R^2	p-value
Bolivia	—	3.83	—	—	—	0.201	0.061
Brazil	0.030	—	—	—	−0.002	0.547	0.017
Chile	0.139	—	—	—	—	0.207	0.089
Colombia	0.009	0.173	—	—	40.51	0.663	0.023
Costa Rica	—	—	—	—	−94.58	0.246	0.041
Ecuador	—	—	—	0.162	—	0.525	0.003
Mexico	—	—	—	−0.01	—	0.461	0.005
Paraguay						no variables significant	
Peru						no variables significant	

Indebtedness did not enter any of the regression equations. (Data from FAO Production Yearbooks [1977–1986], FAO Trade Yearbooks [1977–1986], International Financial Yearbooks [1977–1986].)

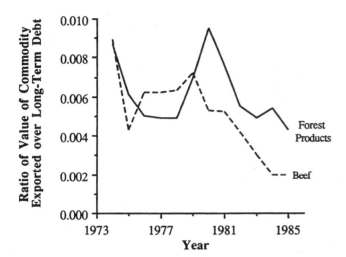

Figure 4. Total annual forest product earnings and beef export earnings as a proportion of total long-term foreign debt in Latin America. Beef and forest product earnings comprise only a very small proportion of long-term debt, limiting the role that they can play in debt reduction. (Data from FAO Trade Yearbooks; World Bank 1988–1989.)

1987). It is unclear, however, to what extent debt payments and austerity measures imposed on debtor countries have contributed to this economic downturn. Increasing debt burdens do correlate well with increasing economic stagnation in Latin American countries, but separating the effect of debt from other factors leading to an economic recession can be very difficult (Ribe et al. 1990). Moreover, the debt crisis is partially a product of poor economic performance. Countries with poor economies do not have the foreign exchange to make debt payments. The extent to which the debt crisis is a cause of economic deterioration, rather than simply a result of it, is not clear.

Step 2. Poverty harms the environment. The second half of this argument concerns how increasing poverty influences land use by the poor. Unemployment and economic hardship encourage people to move to marginal lands. Hungry families attempt to grow food on any land available, regardless of the fragility of the soil or the suitability of crops. Often the poor do not have the opportunity to let land lie fallow or to undertake reforestation. Instead, short-term needs force landless families to destroy rain forest plots, cultivate already degraded lands, plow steep mountain slopes, and overgraze fragile pastures (Durning 1989; Corson 1990). This process becomes a vicious cycle, as degraded ecosystems offer diminishing yields. The poor are pushed even further into more marginal lands. Poverty also may lead to increased family size and population growth, which results in even greater pressure on the environment.

Two complicating factors arise when we try to generalize about the interaction of poverty and land use. First, in certain cases, an increase in economic well-being can also accelerate the use of marginal lands. In northern Madagascar, for example, when poor farmers get access to small amounts of capital and a cash crop such as vanilla, the shift from subsistence farming to cash crop cultivation tends to increase deforestation (Dr. A. Jolly, personal communication). Second, poverty and population pressure can cause migration from rural areas to urban shanty-towns and slums (Corson 1990). It is unknown whether increasing poverty results in a net movement of people to marginal agricultural land or to urban centers.

Environmentalists have put steps one and two together, invoking the law of transitivity: The debt burden intensifies economic depression and poverty. Poverty leads to environmental destruction. Thus, elimination of debt should decrease environmental degradation. Although we accept the general logic supporting both relationships, there is insufficient data to evaluate the importance of debt in causing increased poverty and the role that poverty plays in causing an increase in the use of marginal lands. Furthermore, factors other than poverty and debt may be more important in causing the use of marginal lands. For example, the lack of productive, available land in Latin America results largely from the land tenure system and inequitable land ownership. Only 7% of the population owns 93% of the land (Corson 1990). Reducing or eliminating Latin American debt probably would not have a profound impact on the degradation of marginal lands because there is simply no other land available for landless peasants to colonize under current land tenure arrangements.

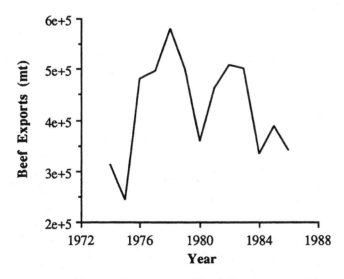

Figure 5. The total amount of beef that is exported (in metric tonnes) is cyclical and shows a decreasing trend in the 1980s, suggesting that Latin American countries are not increasing beef exports to pay for debt service. (Data from FAO Trade Yearbooks; FAO Production Yearbooks).

Debt-Induced Funding Shortages for Environmental Programs

The third hypothesis linking debt to deforestation is that debt-induced governmental budget cutbacks have reduced the money spent on environmental programs. Severe cuts in spending on basic health and education have occurred (UNICEF figures cited in Cornia et al. 1987), so we assume that environmental expenditures have been similarly affected. With or without financial cut-backs in the 1980s, the current level of funding for environmental programs is only a small fraction of what is necessary. The World Resources Institute (1989) estimated financial needs for environmental programs in developing countries at 20–50 billion per year (1–2% of GNP). Thus, if government expenditures were already insignificant compared to what is needed for environmental programs, then reductions in spending caused by indebtedness are unlikely to have increased deforestation substantially.

A further complicating factor is that debt-induced budget cutbacks in other sectors may have actually decreased deforestation by reducing funds for projects such as highway and dam construction (Ledec 1989). Economic models have demonstrated that high debt payments decrease investment in capital, which may lead to a decrease in timber extraction and road construction (Hansen 1989). Thus the net effect of debt on government expenditures may actually have been to decrease deforestation.

Although debt may not have led to deforestation through decreased environmental expenditures, it is clear that indebtedness now stands as an obstacle to initiating new environmental programs. However, simple debt alleviation is no guaranty that governments will spend new capital on environmental improvement programs rather than on environmentally damaging ones.

In summary, there is little evidence supporting the three main hypotheses concerning how the debt burden may be causing deforestation in Latin America. Nevertheless, various methods of debt reduction currently play or potentially could play significant roles in new environmental initiatives in Latin America.

Using Debt for Environmental Initiatives

High levels of foreign debt give creditor organizations a tool for encouraging changes in the environmental programs of debtor nations. To this date, debt forgiveness and debt reduction have been exchanged for land protection and local currency for conservation programs. Here we describe these debt-for-nature swaps and suggest how additional action could be taken to influence natural resource use policy.

Debt-for-Nature Swaps

Debt-for-nature swaps involve the exchange of foreign debt for commitments to and/or funds for conservation in the indebted country (von Moltke 1987; Cody 1988; Fuller & Williamson 1988; Hansen 1989; Ledec 1989). The prerequisite for a swap is that international financial institutions currently trade a country's debt among themselves at a discounted "secondary market value"; this value represents the expected probability that the country will repay the debt. A nongovernmental organization (NGO) from a developed country, working in conjunction with an NGO in the debtor country, obtains the debt at the secondary market value. The NGOs then swap the debt for an environmental program valued at a higher price than its secondary market value. Thus, the impact of every dollar invested by the NGOs is increased by a multiplier effect.

Many developing nations wish to increase their commitment to environmental protection, but they give other concerns—including debt servicing—higher priority for their limited funds. Debt-for-nature swaps offer an opportunity for governments to allocate more funds for conservation while also reducing their external debt to some extent (Cody 1988). Additional benefits accrue to environmental NGOs. Not only do they receive increased funding, but they also develop connections with national and international financial, legal, and political communities (Fuller & Williamson 1988).

There are problems associated with debt-for-nature swaps. These exchanges increase domestic spending, which can be inflationary. To reduce inflationary effects, the size of swaps can be restricted, and long-term, interest-bearing bonds can be issued that spread out payments over time (Cody 1988; Fuller and Williamson 1988; Harwood 1988). More seriously, some countries view these swaps as an encroachment on their sovereignty; the swaps set host government's priorities and take needed domestic resources away from other social programs (Cody 1988; Intrados 1988a). However, debt-for-nature swaps usually fund programs already directly or indirectly approved by the government, but that are lacking funds. Furthermore, swaps cannot be arranged without the participation and consent of debt country governments (Cody 1988; Ayres 1989). In fact, local NGOs usually initiate the programs (World Wildlife Fund 1988).

Latin American debt-for-nature swaps have taken place in Bolivia, Costa Rica, Ecuador, and Mexico. Others are being considered in Brazil, Chile, Guatemala, Mexico, Argentina, the Dominican Republic, and Peru (Intrados 1988a; Ayres 1989; Hansen 1989, Anonymous 1991). These swaps have significantly benefited conservation programs in the countries where they have been tried. However, debt-for-nature swaps are far less successful as a tool for debt reduction. As of early 1990, six private debt-for-nature swaps had retired $100 million in debt, which is only about 0.00008% of total external debt in developing countries.

Sectoral Reforms and Debt Buybacks

Government policies concerning land-use have recently been identified as one of the most significant factors contributing to deforestation in the tropics (Repetto & Gillis 1988). In Brazil, for example, tax laws strongly encourage agricultural settlement in Amazonian rain forests (Mahar 1989). Improving tax policies in certain sectors such as agriculture, forestry, energy, and water resources not only may reduce ecological damage but also may decrease heavy fiscal and economic losses within developing countries (Mahar 1989; World Resources Institute 1989). The World Resources Institute (1989) proposes that sectoral reforms be linked to foreign debt reduction.

Forestry sectoral adjustment programs typically promote improved natural forest management, increased investment in forest plantations, and stronger forest conservation plans. They also specify policy changes critical to improving the design and collection of forest resource fees, correcting incentives to forest-based processing industries, and reducing subsidies to competent land uses such as cattle ranching and agricultural settlements. Such changes can raise government revenues, despite reduced forest exploitation. Furthermore, they can potentially increase net foreign exchange earnings by placing higher taxes on enterprises generating foreign exchange (World Resources Institute 1989).

A study of forestry sector reform in Cote d'Ivoire documented that such reforms could raise $150 million per year. Though no similar studies have been done for Latin American countries, the World Resource Institute estimates that taxes, credits, and other economic policies have created fiscal losses ranging from $500 million to over $1 billion per year in Brazil (World Resources Institute 1989).

If reforms are so beneficial, why have they not already been carried out? To a large extent, powerful private interest groups close to governments have been effectively able to block the implementation of sectoral reform in forestry and agriculture (Mahar 1989). Implementing these reforms could be politically more feasible, however, if new earnings from forestry sector reforms could be used by heavily indebted countries to buy back their own debt at a favorable discount from creditors (only rarely are governments allowed to buy their own debt back on the secondary market). Local elites with special interests would find it more difficult to oppose pressure from multilateral financial institutions for tax reforms that would benefit the government and the environment, and decrease their government's debt burden. The amount of debt that could be retired through sectoral reforms is potentially much greater than could be retired by debt-for-nature swaps. An additional benefit would be that sectoral reforms are not inflationary.

Conclusion

> It doesn't make any sense . . . to forgive any or all of [the Latin American] debt if they don't start again on a fresh footing, with sound economic and environmental strategies. (Randy Curtis of the Nature Conservancy, quoted in Intrados 1989b)

From examining the proposed hypotheses that relate foreign debt to deforestation in Latin America, we come to the following conclusions:

(1) Heavily indebted countries do not appear to be promoting timber and beef exports in response to increasing indebtedness. Debt is unlikely to be contributing to deforestation in this manner.

(2) Economic deterioration linked to large foreign debts is likely causing deforestation by increasing poverty, which increases the use of marginal lands. However, it is difficult to quantify how much debt has actually increased poverty. In addition, it is unclear whether increased poverty results in a net movement of people to urban centers or to marginal lands.

(3) Debt-induced funding shortages are preventing the initiation of environmental programs, but spending in these areas has traditionally been very low.

We conclude that the role of debt in causing deforestation has been overstated. There is little rigorous evidence for any of the hypotheses linking debt to deforestation. The debt now impedes new environmental initiatives, however, as there is simply very little money available for conservation programs and natural resource management. It is important to make the distinction between debt causing deforestation and debt acting as one of many barriers to implementing solutions to the deforestation problem. In the first case, unilateral debt reduction by lending agencies will result in less deforestation. In the second case, however, debt reduction is not sufficient to reduce deforestation but must be tied with policy reforms that address the roots of the problem. It is clearly important that we gain a better understanding of the relationship between debt and deforestation to aid in policy formation.

On the positive side, the following debt-related measures have been taken or can potentially be taken to alleviate environmental problems in Latin America:

(1) Debt purchased on the secondary market has been exchanged for land for national parks and funding for environmental programs in innovative debt-for-nature swaps.
(2) Debt reduction could be linked to reform of sectoral policies concerning natural resource use, addressing directly what some consider to be one of the major causes of deforestation in the tropics.

These initiatives could go far in correcting the environmental framework in which development takes place in Latin America.

Acknowledgments

We thank the many people who have read and made suggestions on earlier versions of this manuscript. In particular, we thank Kristin Ardlie for her excellent suggestions for ways to improve both the content and style of this essay. We thank Carlos Urzua for providing the impetus for this essay, and for his continuing advice and support. During the period this essay was written, R. Gullison was supported by an NSERC 1967 pre-doctoral fellowship, and E. Losos was supported by an NSF pre-doctoral fellowship and a Pew Charitable Trust fellowship in the Program in Integrated Approaches to Sustainable Development and Conservation.

Literature Cited

Anonymous. 1991. Unique debt swap to protect forest. Canopy 1–2.

Ayres, J. M. 1989. Debt-for-equity swaps and the conservation of tropical rain forests. Trends in Ecology and Evolution **4:**331–332.

Cody, B. 1988. Debt-for-nature swaps in developing countries: an overview of recent conservation efforts. Congressional Research Service Report for Congress. The Library of Congress, Washington, D.C.

Cornia, G. A., R. Jolly, and F. Stewart. 1987. Adjustment with a human face, vol. 1. Clarendon Press, Oxford, England.

Corson, W. H. 1990. The global ecology handbook: what you can do about the environmental crisis. Beacon Press, Boston, Massachusetts.

Devlin, R. 1989. Debt and crisis in Latin America: the supply side of the story. Princeton University Press, Princeton, New Jersey.

Durning, A. B. 1989. Poverty and the environment: reversing the downward spiral. Worldwatch Paper 92. Worldwatch Institute, Washington, D.C.

Food and Agricultural Organization of the United Nations (FAO). 1977–1986. FAO Production Yearbooks, vols. 30–39. Rome, Italy.

Food and Agricultural Organization of the United Nations (FAO). 1977–1986. FAO Trade Yearbooks, vols. 30–39. Rome, Italy.

Fuller, K. S., and D. F. Williamson. 1988. Debt-for-nature swaps: a new means of funding conservation in developing nations. International Environment Reporter **0149:**301–303.

Hansen, S. 1989. Debt for nature swaps—overview and discussion of key issues. Ecological Economics **1:**77–93.

Harwood, J. 1988. Nature swaps. Taxes International **93:**3–12.

International Monetary Fund (IMF). 1977–1986. International Financial Statistics, vols. 30–39. Washington, D.C.

The Intrados Group. 1988a. Debt-for-nature swaps. Swaps **2:**1–8.

The Intrados Group. 1988b. An Interview with the Nature Conservancy's Randy Curtis. Swaps **3:**7–10.

Ledec, G. 1989. International debt and environmental degradation in developing countries: summary of key linkages. Draft Memorandum. Washington, D.C.

Lombardi, R. W. 1985. Debt trap: rethinking the logic of development. Praeger Publishers, New York, New York.

Mahar, D. J. 1989. Deforestation in Brazil's Amazon region: magnitude, rate, and causes. Pages 87–116 in G. Schramm and J. J. Warford, editors. Environmental management and economic development. Johns Hopkins University Press (for The World Bank), Baltimore, Maryland.

Potter, G. A. 1988. Dialogue on debt: alternative analyses and solutions. Center of Concern.

Reppetto, R., and M. Gillis (editors). 1988. Public policies and the misuse of forest resources. Cambridge University Press, Cambridge.

Ribe, H. S., Carvalho, R., Liebenthal, P. Nicholas, and E. Zuckerman. 1990. How adjustment programs can help the poor: the World Bank's experience. World Bank Discussion Paper 71. The World Bank, Washington, D.C.

Tropical Timbers. 1990. Bank Role **5:**2.

von Moltke, K. 1987. Debt for nature: an overview. World Wildlife Fund, Washington, D.C.

World Bank. 1988–1989. External debt of developing countries. Washington, D.C.

World Resources Institute. 1989. Natural endowments: financing resource conservation for development. Washington, D.C.

World Wildlife Fund. 1988. Debt-for-nature swaps: a new conservation tool. World Wildlife Fund Letter **1:**1–9.

Financial Considerations of Reserve Design in Countries with High Primate Diversity

JOSÉ MARCIO AYRES
Wildlife Conservation International
New York Zoological Society
Bronx, New York 10460, U.S.A.

RICHARD E. BODMER*
Departamento de Zoologia
Museu Paraense Emilio Goeldi
Caixa Postal 399, 66.040
Belém, Pará, Brazil

RUSSELL A. MITTERMEIER
Conservation International
1015 18th Street, N.W.
Suite 1000
Washington, D.C. 20036, U.S.A.

Abstract: *Many developing countries with high primate diversity are still forming wildlife reserves. However, financial resources for conservation in these countries are often limited. In this paper we show that the biology of reserve design in terms of island biogeography and minimum viable populations can be compatible with measures that minimize maintenance costs, namely perimeter surveillance. Large protected areas have substantially less perimeter to patrol than many small reserves of the same total area. Likewise, circular reserves have less perimeter than square or rectangular ones. Generally, countries with high primate diversity are using desirable strategies in design of nature reserves in terms of both finances and conservation; they have relatively large reserves and large amounts of area protected in relation to country size. However, some key primate-containing countries, such as Madagascar, Indonesia, and southeastern Brazil, require many small reserves to insure preservation of endemic taxa.*

---

* Correspondence should be addressed to this author.
Paper submitted September 23, 1989; revised manuscript accepted July 12, 1990.

Resumen: *Muchos de los países en desarrollo con una alta diversidad de primates estan aún instituyendo reservas de vida silvestre. Sin embargo, los recursos financieros para la conservación en éstos países son, con ftecuencia, limitados. En este artículo mostramos que la biología del diseño de la reserva en los términos de las islas biogeográficas y el mínimo de poblaciones viables puede ser compatible con medidas que minimizen el costo del mantenimiento, por ejemplo, la vigilancia en los límites del área. La áreas protegidas exténsas tienen substancialmente ménos perímetro que vigilar que muchas reservas pequeñas con la misma área total. De la misma manera, las reservas circulares tienen ménos perímetro que las cuadradas o las rectangulares. Generalmente, los países con una alta diversidad de primates están utilizando estrategias deseables en el diseño de reservas naturales en términos tanto de financiamiento como de conservación; ellos tienen reservas relativamente grandes y grandes extensiones de áreas protegidas en relación al tamaño del país. Sin embargo, algunos países claves en cuanto a los primates, como Madagascar, Indonesia y el sur-este de Brasil, requieren de muchas reservas pequeñas para asegurar la preservación de los taxa endémicos.*

Introduction

Protected areas should conserve nature by protecting species of interest (i.e., endangered species), preserving entire ecosystems, and maintaining maximum biological diversity (Soulé & Simberloff 1986; Western 1986; Wilson 1988; Leader-Williams et al. 1990). Conservation biologists have paid much attention to design of protected areas in terms of size, number, and shape (Diamond 1975; Lovejoy et al. 1986). Theories of island biogeography and minimum viable populations are used to predict consequences of large versus small protected areas and usually conclude that large reserves protect species, ecosystems, and biodiversity with greater certainty than smaller ones (Soulé & Simberloff 1986).

Financial resources for protected areas are often limited, especially in developing countries (Leader-Williams & Albon 1988). Thus, when establishing wildlife reserves, conservation biologists must consider costs of maintaining protected areas together with biological theory. In this paper, we consider design of protected areas that reconcile biology with financial considerations of reserve protection. Numbers of nonhuman primate species are well documented and countries with a high primate diversity are used to demonstrate desirable reserve designs in terms of both finances and conservation.

Methods

Financial costs of protecting reserves are assumed to be directly proportional to distance patrolled. This assumption does not take into consideration costs of management and educational programs in buffer zones.

Data for 121 countries were analyzed for country size, number of reserves, and total area protected from WRI (1988). Countries with no protected area were excluded (N = 23).

Number of reserves and total area protected, corrected for country size, were determined by calculating the residuals from the slope of the best fitting least squares line on logarithmic transformed data as

$$\log_{10} Y = b \cdot \log_{10} X + \log_{10} k$$

where X = country size (in total area), Y = total area protected (for the first analysis) and number of reserves (for the second analysis), b = the allometric exponent, and k = the allometric coefficient (Wilkinson 1987). Negative residuals indicate relatively less area protected or fewer number of reserves, whereas positive residuals represent relatively greater values.

Primate conservation in relation to reserve design was examined using the 15 countries with the greatest number of primate species (ranging from 22 to 55 species) taken from an updated list of Mittermeier and Oates (1985). Number of primate species, corrected for forest habitat, was calculated using residuals from the slope of the best-fitting least squares line on logarithmic transformed data of total area of closed forest (from WRI 1988) and number of primate species. Positive residuals represent relatively greater primate diversity in relation to amount of forest and negative residuals relatively smaller.

Countries with high primate diversity were plotted onto the regression lines of total area protected and number of reserves, calculated from all 121 countries. Thus, key primate countries were compared with the world average in terms of total area protected and number of reserves (corrected for country size).

Results

Size and Shape of Reserves

Patrolling of protected areas is divided into perimeter and internal surveillance. Patrolling perimeters of several large reserves requires vigilance over much smaller distances than patrolling perimeters of many small reserves with the same total area. However, internal surveillance between several large reserves and many small ones requires vigilance over similar distances.

For example, a square reserve of 10,000 km^2 will only require 400 km of perimeter surveillance, whereas 100 smaller reserves of equal size and having the same total area will require 4,000 km of perimeter patrol, an order of magnitude greater. However, internal patrolling of the large reserve with a straight transect every one km will require 9,900 km of internal surveillance, whereas the small reserves using the same straight transects every one km will require 9,000 km of surveillance. This only amounts to a 10% difference of internal distance patrolled between the large versus small reserves. Thus, variation in the distance patrolled of reserves of different sizes, but having the same total area, depends almost entirely on perimeter surveillance.

The relationship between area and perimeter changes with geometrical shapes. For example, circular reserves have a smaller perimeter than square reserves, and square reserves in turn have a smaller perimeter than rectangular reserves. In addition, perimeter does not increase linearly with area, as demonstrated above. If both shape and area are constant, total perimeter increases with increasing number of areas as a power function of

$$Y = a \cdot X^b (a > 0; X \geq 1)$$

where Y = total perimeter, a = a constant indicating the minimum perimeter, b = ½, and X = number of reserves.

Actual Reserves

Larger countries tend to have less area protected relative to country size than smaller countries, because the slope of the logarithmic plot between country size and total area protected (b = 0.89) is less than 1.0 (r = 0.73; p < 0.001) (Fig. 1a). Likewise, larger countries have fewer reserves relative to country size than smaller ones, again because the slope between country size and number of reserves (b = 0.35) is considerably smaller than 1.0 (r = 0.45; p < 0.001) (Fig. 1b).

Countries with positive residuals for total area protected and negative residuals for number of reserves (i.e., a few large reserves) are minimizing the perimeter of reserves and have relatively smaller distances to patrol, yet they have large amounts of area protected in relation to country size (type A). Conversely, countries with positive residuals for both total area protected and number of reserves (i.e., many small reserves) have relatively greater distances to patrol and still have large amounts of area protected, corrected for country size (type B). Countries with negative residuals for both total area and number of reserves have the least distance to patrol, because they have relatively few reserves and relatively small amounts of area under protection (type C). Countries with negative residuals for total area protected and positive residuals for number of reserves have relatively large distances to patrol, yet have relatively little area protected (type D).

Strategies of type A and C are most cost-effective, and types A and B have relatively large areas protected. Thus, in terms of both finances and conservation, strategy type A is most desirable, because of proportionally smaller perimeters to patrol and relatively large areas under protection.

Primate Conservation

Wild populations of nonhuman primates occur in more than 90 countries. However, the majority of the 230+ species of primates occur in few tropical-belt countries (Mittermeier & Oates 1985) (Table 1). For example, Brazil, Zaire, Cameroon, Madagascar, and Indonesia have 67% of the world's primates species. The relationship between area of closed forest and number of primate species (b = 0.11; r = 0.71; p < 0.001) indicates that Brazil has the greatest primate diversity relative to forest habitat followed by Madagascar, Equatorial Guinea, and Cameroon (Fig. 2).

In general, countries of high primate diversity are using desirable financial and conservation strategies when plotted onto the global regression lines of total area protected and number of reserves, calculated from all 121 countries (Fig. 3). Of the key primate countries, 47% are of type A; they are using the most desirable strategy by having relatively large amounts of protected area divided into proportionally large reserves (Fig. 4). Four countries have relatively large areas protected, but divided into small reserves (type B), three countries have relatively small amounts of protected area divided into proportionally large reserves (type C), and only one country has both relatively small amounts of protected area divided into small reserves (type D). Of the four countries with the greatest primate diversity relative to closed forest only Cameroon is using both the most desirable financial and conservation strategies.

Most countries with high primate diversity (73%) have more area protected in relation to country size than the world average. However, Madagascar and Equatorial Guinea are among the top three countries in primate diversity relative to closed forest, yet they have less area protected relative to country size than other key primate countries and the world average.

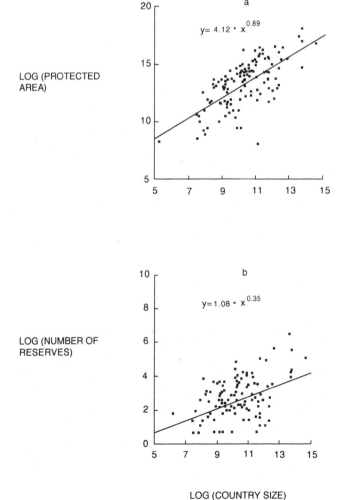

Figure 1. Logarithmic plots between country size (N = 121) and (a) total area protected and (b) number of reserves.

Table 1. Number of primate species, area of closed forest, percentage of endemic primates, and rank of relative primate diversity. Data are from an updated list of Mittermeier and Oates (1985) and WRI (1988).

Country	Number of primate species	Area of closed forest (km^2)	Percentage of endemic primates	Diversity rank
Brazil	52	3,574,800	35	1
Madagascar	28	103,000	93	2
Eq. Guinea	21–22	12,950	0	3
Cameroon	28–29	179,200	0	3
Uganda	19	7,650	0	5
Zaire	29–32	1,057,500	6–7	6
Nigeria	23	59,500	4	7
Indonesia	33–35	1,138,950	44–50	8
Colombia	27	464,000	11	9
Cen. Afr. Rep.	19–20	35,900	0	10
Angola	18–19	29,000	0	11
Peru	27	696,800	7	12
Congo	22	213,400	0	13
Gabon	19	205,000	0	14
Bolivia	17–18	440,100	0	15

Discussion

Financial costs of maintaining protected areas must be considered together with biology. Overall, financial considerations examined in this paper are in accordance with theories of island biography and minimum viable populations in that large reserves are more desirable than smaller ones (Soulé & Wilcox 1980; Soulé & Simberloff 1986).

Patrolling reserves using both perimeter and internal surveillance is a major cost for maintaining protected areas. The distance patrolled during internal surveillance, unlike perimeter surveillance, changes little with increasing numbers of protected areas; thus, financial considerations are more dependent on perimeter distances. Perimeter surveillance can be minimized by reducing the number of reserves but maintaining the same total area protected, or by making reserves as circular as possible. However, many reserves use natural boundaries, such as rivers, that expedite surveillance more than preconceived circular shapes.

Surveillance of reserves is not the only cost for successfully protecting areas, and patrolling should often be performed by local communities adjacent to reserves, not outsiders. When communities patrol reserves this increases awareness and responsibility for protected areas by these communities, brings in cash income that would otherwise be acquired by exploiting natural habitats, and avoids conflict with external authorities. Rational management of renewable resources in buffer zones, and educational and health programs are additional costs that must be considered for protected areas (Bodmer et al. 1990). Financially, it would appear to be more cost-effective to have one large well-trained team working in a large reserve than many small teams scattered throughout small reserves.

Recommendations made in this paper should only be considered together with zoogeographic models that insure preservation of endemic taxa (Wetterberg et al., 1976). Designing a few large reserves has the drawback of not protecting all areas of species endemism, which is caused either by natural geographic distributions of species or extensive human development and habitat destruction. For example, Madagascar has the greatest number of endemic species of primates (28 of the 30 primate species occurring on Madagascar are endemic),

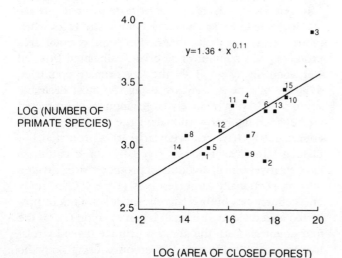

Figure 2. Logarithmic plot between total area of closed forest and the 15 countries with the greatest number of primate species. 1 is Angola, 2 Bolivia, 3 Brazil, 4 Cameroon, 5 Central African Republic, 6 Colombia, 7 Congo, 8 Equatorial Guinea, 9 Gabon, 10 Indonesia, 11 Madagascar, 12 Nigeria, 13 Peru, 14 Uganda, and 15 Zaire.

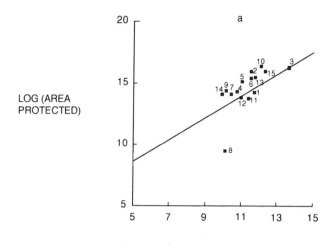

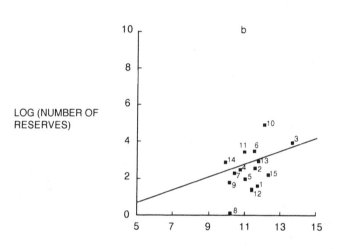

Figure 3. Key primate countries plotted onto the regression lines for (a) total area protected and (b) number of reserves from all 121 countries. 1 is Angola, 2 Bolivia, 3 Brazil, 4 Cameroon, 5 Central African Republic, 6 Colombia, 7 Congo, 8 Equatorial Guinea, 9 Gabon, 10 Indonesia, 11 Madagascar, 12 Nigeria, 13 Peru, 14 Uganda, and 15 Zaire.

and its strategy of having relatively more smaller reserves than other key primate countries may insure the protection of this greater biodiversity (Pollock 1986; Richard & Sussman 1987). Similarly, Indonesia has hundreds of islands with endemic fauna and their strategy of having many small reserves helps protect these endemic taxa (Marsh 1987). Also, in the Atlantic forests of southeastern Brazil where habitat destruction has seriously degraded forests, many small reserves have been set aside in the remaining patches of natural habitat (Ayres 1989). However, theories of biogeography and minimum viable populations predict higher extinction rates

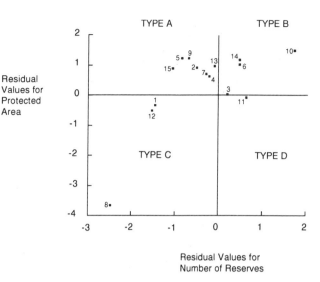

Figure 4. Residual values of total area protected and number of reserves from the 15 key primate countries. Countries (corrected for size) of type A have large amounts of area protected and large reserves, type B large amounts of area protected and small reserves, type C small amounts of area protected and large reserves, and type D small amounts of area protected and small reserves. 1 is Angola, 2 Bolivia, 3 Brazil, 4 Cameroon, 5 Central African Republic, 6 Colombia, 7 Congo, 8 Equatorial Guinea, 9 Gabon, 10 Indonesia, 11 Madagascar, 12 Nigeria, 13 Peru, 14 Uganda, and 15 Zaire.

for these small reserves and our results predict relatively greater costs.

Many developing countries with high primate diversity in Africa, Southeast Asia, and South America are still planning protected areas. The major problem is how to effectively protect these reserves. Economic factors are important, because the amount of money invested in conservation is correlated with success (Leader-Williams & Albon 1988). In this paper we attempted to combine ideas on financing protected areas with biological theories, in light of the limited resources available for developing countries with high primate diversity.

Acknowledgments

We thank E. Bennett, D. J. Chivers, A. Harcourt, and N. Leader-Williams for their useful comments on earlier versions of the manuscript. JMA was supported by a postdoctoral scholarship from the CNPq (Brazilian National Research Council) and IBM-Brazil.

Literature Cited

Ayres, J. M. 1989. A questão ecológica na Amazônia. Pages 129–136 in IDESP, editor. Estudos e problemas Amazônicos:

história social e economica e temas especiais. Secetaria do Estado de Educação, Belém, Brazil.

Bodmer, R. E., J. W. Penn, T. G. Fang, and L. Moya I. 1990. Management programmes and protected areas: the case of the Reserva Comunal Tamshiyacu-Tahuayo, Peru. Parks 1:21–25.

Diamond, J. M. 1975. The island dilemma: lessons from modern biogeographic studies for the design of natural reserves. Biological Conservation 7:129–146.

Leader-Williams, N., and S. Albon. 1988. Allocation of resources for conservation. Nature 336:533–535.

Leader-Williams, N., J. Harrison, and M. J. B. Green. 1990. Designing protected areas to conserve natural resources. Science Progress. 74:189–204.

Lovejoy, T. E., R. O. Bierregaard, Jr., A. B. Rylands, et al. 1986. Edge and other effects of isolation on Amazon forest fragments. Pages 257–285 in M. Soulé, editor. Conservation biology: the science of scarcity and diversity. Sinauer Associates, Sunderland, Massachusetts.

Marsh, C. W. 1987. A framework for primate conservation priorities in Asian moist tropical forests. Pages 343–354 in C. W. Marsh and R. A. Mittermeier, editors. Primate conservation in the tropical rain forest. Alan R. Liss, New York.

Mittermeier, R. A., and J. F. Oates. 1985. Primate diversity: the world's top countries. Primate Conservation 5:41–48.

Pollock, J. 1986. Towards a conservation policy for Madagascar's eastern rain forests. Primate Conservation 7:82–86.

Richard, A. F., and R. W. Sussman. 1987. Framework for primate conservation in Madagascar. Pages 329–341 in C. W. Marsh and R. A. Mittermeier, editors. Primate conservation in the tropical rain forest. Alan R. Liss, New York.

Soulé, M. E., and D. Simberloff. 1986. What do genetics and ecology tell us about the design of nature reserves? Biological Conservation 35:19–40.

Soulé, M. E., and B. A. Wilcox. 1980. Conservation biology: an evolutionary-ecological approach. Sinauer Associates, Sunderland, Massachusetts.

Western, D. 1986. Introduction: primate conservation in the broader realm. Pages 343–353 in J. G. Else and P. C. Lee, editors. Primate ecology and conservation. Cambridge University Press, Cambridge, England.

Wetterberg, G. B., M. T. J. Padua, C. S. de Castro, and J. M. C. de Vasconcellos. 1976. An analysis of nature conservation priorities in the Amazon. Instituto Brasileiro de Desenvolvimento Florestal, Brasilia.

Wilkinson, L. 1987. Systat: the system for statistics. SYSTAT, Evanston, Illinois.

Wilson, E. O. 1988. Biodiversity. National Academy Press, Washington, D.C.

World Resources Institute. 1988. World resources 1988–89. Basic Books, New York.

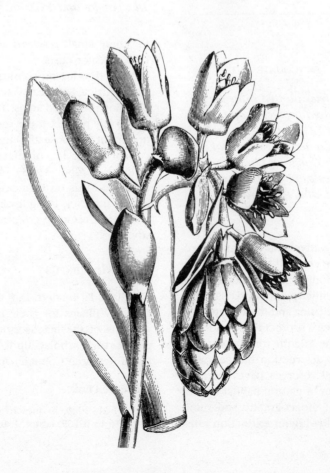

Cheese, Tourists, and Red Pandas in the Nepal Himalayas

PRALAD B. YONZON*
Department of Wildlife
University of Maine
Orono, Maine 04469, U.S.A.

MALCOLM L. HUNTER, JR.
Department of Wildlife
University of Maine
Orono, Maine 04469, U.S.A.

Abstract: *Langtang National Park in the Nepal Himalayas provides resources for about 30,000 people living in and around the park. The park is the site of two cheese factories that produce 14,000 kg of cheese per year for marketing in Kathmandu. The factories collect about 140,000 liters of milk annually and use well over 100,000 kg of fuel wood to process the milk into cheese. Loans and one-year advance payments encourage farmers to maintain large herds of* chauri *(a yak-cattle hybrid), and in many areas overgrazing has resulted. The presence in the area of large herds of* chauri, *their herders, and dogs has led to the death of many red pandas, a species that is probably on the verge of extinction in Langtang. It is estimated that there are fewer than 40 red pandas and these are isolated in four or more populations. Their fecundity is low and mortality is high. This problem might be solved by reducing cheese production and restricting the number of* chauri *while commensurately increasing the price of cheese so that farmers' income from milk could remain the same. Current cheese prices are already too high for Nepali consumers (US $4.30/kg), but cheap by the standards of the Western tourists who are the sole market. Thus, increasing prices would have little impact on the cheese market, but could be translated into a substantial benefit for the farmers, whose use of grazing lands must ultimately be sustainable, and for the red pandas and other wildlife that must share the mountain landscapes with the farmers.*

Resumen: *El Parque Nacional Langtang en las Himalayas-Nepal provee de recursos a 30,000 personas que viven dentro o alrededor del parque y es la ubicación de dos fábricas de queso con una producción de 14,000 kg. anuales para vender en Kathmandu. Las fábricas colectan casi 140,000 litros de leche al año y utilizan más de 100,000 kg de madera para transformar la leche en queso. Préstamos y pagos con un año de adelanto estimulan a los campesinos a mantener grandes manadas de* chauri *(un híbrido de ganado vacuno y yak) y en muchas áreas esto ha ocasionado el sobrepastoreo. La presencia de grandes manadas de* chauri, *los pastores y los perros han ocasionado la muerte de muchos pandas rojos, una especie que probablemente esté a punto de extinguirse en el área de Langtang. Se estima que hay ménos de 40 pandas rojos y que éstos estan aislados en cuatro o más poblaciones. Su fecundidad es baja y su mortalida es alta. Este problema podría ser resuelto reduciendo la producción de queso, restringiendo el número de* chauri, *e incrementando el precio del queso de tal manera que el ingreso que reciban los campesinos por la leche sea el mismo. El precio actual del queso es ya muy alto para los consumidores de Nepal ($4.30/kg en dólares americanos), pero bajo para el común de los turistas quienes son el único mercado. Así, el incremento en el precio del queso tendría poco impacto en el mercado, pero podría traducirse en un beneficio substancial para los campesinos, cuyo uso de las tierras de pastoreo debe al final ser sustentable, y para los pandas rojos y la vida silvestre que deberán compartir las montañas con los campesinos.*

---

* *Correspondence should be addressed to this author. Present affiliation: Nepal Conservation Training and Research Institute, King Mahendra Trust for Nature Conservation. Present address G.P.O. Box 2448 Kathmandu, Nepal.*
Paper submitted Nov. 14, 1989; revised manuscript accepted July 11, 1990.

Introduction

In recent years, it has become evident that nature protection strategies developed in North America and Europe cannot be imposed rigidly in developing countries and that innovative approaches are needed. Many protected areas in poorer countries have begun to allow limited exploitation of natural resources by local people, thus reducing the conflicts between parks and their neighbors and ultimately bolstering protection of the parks (Miller 1984; Sherpa 1988; Upreti 1985). Nepal has been in the forefront of this movement to achieve the twin goals of conservation and development (Lehmkuhl et al. 1988; Mishra 1982). However, it has been easier to avoid conflicts in Nepal's Terai (lowland) parks than in the mountain parks. The Terai parks were historically uninhabited because of malaria, but the mountain parks are characterized by many generations of human settlement and a complex, fragile physical environment. When Langtang National Park (1710 km^2) was established in 1977 between Kathmandu and the border with Tibet (People's Republic of China), it encompassed several villages. Currently about 30,000 people living in or adjacent to the park rely upon its resources, which include fuel wood and fodder.

We present case studies of two villages in the park, Syabru (population: 475 people) and Langtang (553 people) to describe nonsustainable agricultural development activities that have had a significant impact on wildlife (Fig. 1). Virtually all families are engaged in

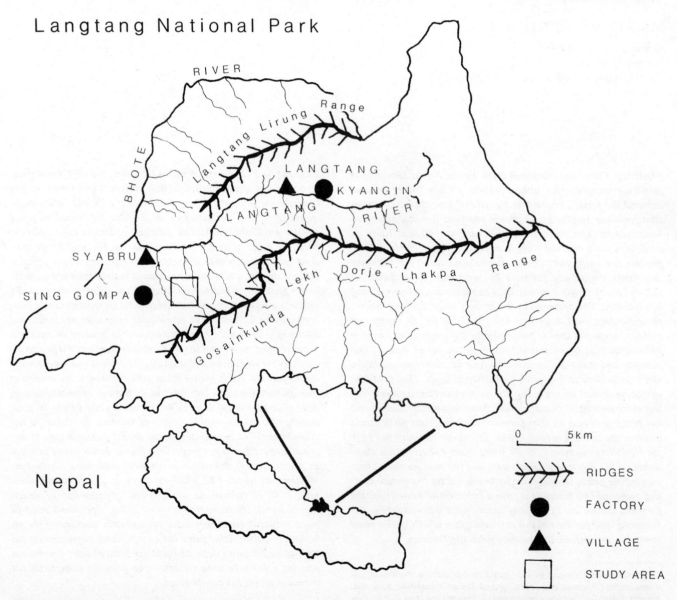

Figure 1. Langtang National Park, Nepal. Cheese factories located at Sing Gompa and Kyangin are associated with the villages of Syabru and Langtang, respectively. The red panda study area, outlined in a square, is east of Sing Gompa; locally it is called Cholang-Dhokache.

farming, and 60% of the active population (excluding those too young or too old) participate in farming directly (Gurung 1988). However, no family owns more than 1.5 ha of tillable land and a food deficit prevails (Gurung 1988). This fact, plus low soil fertility, a cold climate, and short growing seasons, has made livestock rearing an important, low-risk economic supplement to grain farming (Guillet 1981). The preferred livestock has traditionally been *chauri,* a crossbreed of yak (*Bos grunniens*) and hill cow (*Bos taurus*), which provide food, milk products that can be bartered, and manure. Since 1953 Langtang farmers have been pressured to become involved in a government-sponsored agricultural development scheme. Specifically, they have been encouraged to produce large quantities of milk to be processed into cheese for marketing in Kathmandu. This has led to excessive fuel wood cutting and grazing, and detrimental impacts on wildlife. Apparently the introduction of a cash economy has made the use of park resources unsustainable.

We studied these impacts as part of a two-year research project on the conservation and ecology of the red panda (*Ailurus fulgens*), an uncommon species that survives in Langtang National Park in small, isolated populations. This paper is based on (1) biological studies of the red panda (Yonzon 1989; Yonzon et al., in preparation; Yonzon & Hunter, in press); (2) ancillary studies of village economics and their impact on nearby vegetation (Gurung 1988; Shrestha 1988; Yonzon 1988); (3) examination of the records of the Dairy Development Corporation and the Agricultural Development Bank of Nepal; and (4) data extracted from previous studies of cheese factories at Kyangin (Durham University Himalayan Expedition 1977) and Sing Gompa (Timmerman & Platje 1987).

Cheese Production

Historically, milk produced in excess of family needs was processed by traditional methods to produce butter, *chhurpi* (protein extracted from buttermilk), and other buttermilk products such as *sherghum.* These products were bartered in Tibet in exchange for commodities such as salt, sheep and goats, and ornaments. Introduced technology now permits most milk produced in the valley to be processed into cheese and butter. Occasionally, a few families make traditional use of milk.

The Dairy Development Corporation (DDC) first produced cheese in Langtang on an experimental basis in 1953 with technical assistance from the United Nations Food and Agriculture Organization (FAO) (Nembang 1987). Later, the Swiss Association for Technical Assistance (SATA) helped to establish permanent cheese factories. Functionally independent since 1964, the DDC now has two cheese factories in the Langtang Valley, one at Kyangin at 3840 m for Langtang village cattle and the other at Sing Gompa at 3254 m for Syabru. These two cheese factories produce about 20% of the ca. 71,000 kg of cheese processed yearly in the mountains of Nepal.

Although cheese production in the Langtang Valley has been heralded as a "very successful" project because of good economic conceptualization at the beginning and a smooth transition from Swiss to Nepali management (Hagen 1987), there remains the all-important question of whether the benefits are sustainable. The Kyangin factory is permitted by the park authorities to use 46,000 kg of fuel wood annually, but a recent study on its environmental impacts indicated that it is using over 100,000 kg of fuel wood (Timmerman & Platje 1987), much of it green wood cut from live trees near the treeline. In addition, villagers use 169,000 kg and tourists and their porters 44,000 kg, yielding a total of 313,000 kg of fuel wood consumed annually. The annual growth increment of the upper Langtang Valley forest was estimated to be 213,000 kg; thus it is estimated that 100,000 kg of fuel wood are overharvested every year (Timmerman & Platje 1987).

Traditional methods of processing milk also require fuel wood, but the introduced cheese-making technology is theoretically 25% more efficient in its energy requirement than the traditional method of making butter and buttermilk. In practice, the modern methods ultimately require more fuel wood than the traditional methods. This is best illustrated by a study of the Sing Gompa cheese factory, where processing 72,000 l of milk per annum requires an estimated 4650 kg of fuel wood by the traditional method and 3720 kg for processing by the introduced method (Durham University Himalayan Expedition 1977). However, an additional 5400 kg of fuel wood are required for the cooking and heating requirements of the cheese factory staff (five people). Thus, the introduced method requires 95% more fuel wood by bringing technical staff to the Valley. Notably, per capita consumption of fuel wood by factory staff, 1080 kg/person/year, was 3.5 times greater than the average consumption, 306 kg/person/year, of Kyangin villagers (Timmerman & Platje 1987), presumably because the factory staff could afford to keep their houses warmer. Furthermore, there was some unexplained excessive use of fuel wood: in 1977 the factory used 32,000 kg of fuel wood to process 72,000 l of milk, over 3.5 times the estimated requirement of 9,120 kg of fuel wood (Durham University Himalayan Expedition 1977). The effective use of fuel wood at Sing Gompa is likely to be significantly higher still because, unlike at Kyangin, fuel wood is available near the grazing areas and milk is often preprocessed at the collection depots.

Cattle Grazing

The cheese factory operation has led to increased exploitation of the forest by large numbers of *chauri* from May through September. Each cheese factory collects milk from a 50 km² area by establishing as many as four collection and preliminary processing depots. Although it is difficult for factory staff, each depot is moved at least eight times annually to different elevations. This is necessary because herds are moved all over the area, through both forests and pastures between 3000–5000 m, in response to local depletion of fodder every few weeks. Moving the depots facilitates milk collection and encourages farmers to increase their livestock and graze over larger areas. Thus the grazing lands are degraded throughout most of the valley and there is ample evidence of pasture encroachment into the forest (Shrestha 1988).

In the Cholang-Dhokache area, which lies between 2900–3900 m, all accessible areas with slopes <50° were heavily grazed (Shrestha 1988). Larger trees are not affected by *chauri* grazing, but tree regeneration is inhibited by consumption of tree seedlings and browsing of saplings. For example, *chauri* browsed 92% of 200 fir (*Abies spectabilis*), mountain-ash (*Sorbus microphylla*), juniper (*Juniperus recurva*), and rhododendron (*R. campanulatum*) saplings that were monitored by Yonzon (1989) in the early monsoon season of 1987. Damage resulted not only from excessive grazing but also from trampling, which both crushed vegetation and compacted the soil. Another indication of overgrazing was the abundance of plant species associated with dry disturbed sites, such as *Berberis* spp. and *Rosa* spp., even on north-facing, predominantly moist slopes (Shrestha 1988).

Inadequate fodder during the dry season resulted in *chauri* living at a submaintenance level for about 7 months a year, causing reduced fertility and late maturity. When fodder availability is adequate, *chauri* calve every year, whereas in our survey of 297 adult female *chauri* of the Sing Gompa population, 13% of the adult females aborted calves (Yonzon 1988). Furthermore, several *chauri* have died in recent years because of a shortage of winter fodder. Farmers adapt to these limitations by thinning out young *zopkio*, hybrid males that are sterile. Stall feeding is becoming uncommon because the large number of animals makes it infeasible. In summary, the cheese factory has led to the valley having a high density of poorly nourished livestock grazing extensively throughout the forest and causing degradation.

Livestock, Red Pandas, and Other Mammals

The Langtang Valley has been subjected to grazing pressures from wild herbivores and livestock for many years. Currently, however, wild herbivores have been largely replaced by livestock during the monsoonal grazing period. For example, we observed only five serow (*Capricornis sumatraensis*) during the entire monsoons (June–September) of 1986 and 1987, while 25 musk deer (*Moschus chrysogaster*), 16 barking deer (*Muntiacus muntjak*), 6 black bear (*Selenarctos thibetanus*), 26 tahr (*Hemitragus jemlahicus*), 54 wild pigs (*Sus scrofa*), and 16 serow were observed during 12 months outside the monsoons (between March 1986–October 1987). Large carnivores such as leopard (*Panthera pardus*) and wild dog (*Cuon alpinus*) are still present, although discreetly persecuted, and prey on livestock in the absence of wild herbivores.

Effects of grazing disturbance on wildlife are particularly significant when the impacts fall on rare habitat specialists such as red pandas. Each summer, during the panda's birth season, about 500 *chauri* are brought to graze in the Cholang-Dhokache area, which is prime red panda habitat. This is an important problem because the red pandas of Langtang National Park appear to be in danger of extinction (Yonzon & Hunter, in press). Their preferred habitat of fir-bamboo forest between 2800 and 3900 m covers about 6% of the park, and their densities are so low (ca. 1 per 3 km²) that there are probably fewer than 40 individuals in the park (Yonzon et al., in preparation). Moreover, torrential rivers and other barriers probably isolate them into four or more populations. Their fecundity is also limited (usually one cub/female/year) and mortality of both cubs and adults is high: of 12–13 cubs born during the course of the field study, only three survived beyond six months of age and four of nine known adults died during the project (Yonzon & Hunter, in press). Most of the deaths from known causes (57%) were human-related; thus, the presence of *chauri*, their herders, and dogs was clearly detrimental to the pandas. Competition between *chauri* and red pandas for bamboo leaves (54–100% of the pandas' diet depending on season) may or may not be important (Yonzon & Hunter, in press). The red pandas fed higher on the bamboo than the *chauri*, but *chauri* may reduce bamboo abundance overall, particularly by trampling.

In sum, these observations strongly imply that Langtang's pandas are in a precarious state. Loss of any species is a tragedy, but the red panda is especially important because (1) it may be threatened throughout its range (Yonzon 1989); (2) taxonomically, it is the sole member of a monotypic subfamily; and (3) it is an exceptionally attractive animal that could easily function as a flagship species to catalyze worldwide public support for wise natural resource management throughout its range.

The Economics of Producing Cheese

Farmers are usually limited by insufficient capital resources and thus have a high demand for credit (Bow man & Hautman 1988). Therefore, development organizations such as the DDC have attempted to promote changes in agriculture by facilitating loans. In the Langtang Valley, banks provide loans for both crops and livestock, but livestock loans comprised 80% of all loans by the Agricultural Development Bank of Nepal (ADBN) to 71 farmers in the villages of Syabru and Bhargu during 1986 (Yonzon 1988). Livestock investments are favored because crop investments require a longer time for repayment and are riskier because of factors such as crop damage (e.g., losses of apples to the Himalayan palm civet, *Paguma larvata*). The usual requirement for a livestock loan is just a spot-check of the farmer's assets and a letter of recommendation from a cheese factory. Lending small sums to many farmers also brings risk to the banks because (1) bank overhead costs increase as the number of small loans for cattle increases; and (2) the likelihood of loan repayment diminishes as the environment is overused (Adams et al. 1984). Although the banks realize that agricultural loans are linked to the productivity of natural resources (Upadhyaya 1987), they have yet to act accordingly.

The DDC ensures its milk supply by providing the farmers one-year advance payments for their milk. The payments are the sole source of cash for many households, and the financial commitment precludes other options for the subsistence farmers. Moreover, the system encourages a few wealthy farmers who can afford to acquire large herds to overexploit the common grazing land, thus depriving small farmers of the full benefits of raising livestock (Tulachan 1985). For example, in Syabru village the average herd size is 10, but the range is 3–45 and only three farmers have over 30 *chauri*.

Targets and Production of a Cheese Factory

Despite support such as loans, advance payments, pasture development, and the availability of an unrestricted amount of fuel wood to process milk, the DDC has not been producing enough cheese to meet its target production. For example, production at the Sing Gompa cheese factory began in 1981, but production has never met the target except in 1987, when the target was lowered by 2000 kg (22%) and an additional milk depot was set up in a new area. From 1981–1987, the Sing Gompa factory produced 46,441 kg of cheese, averaging 6634 kg per year, but the target was 56,500 kg, on average 8071 kg a year.

The criteria used to set targets for cheese production are apparently arbitrary; they are certainly not based on the sustainable use of fodder and fuel. The amount of milk collected by the depots is largely a function of available fodder and reflects milk yield, which shows distinct seasonality. On average, it takes 10 l of milk to produce 1 kg of cheese and 100 gm of butter. Assuming four operational milk depots over a 5-month period (May–September), each milk depot should be collecting 135 l (80,714 liters/4 depots × 5 months × 30 days) of milk each day to fulfill its annual target. However, daily milk collection in all four depots averaged only 116 l (N = 560 total milk collection days) in 1987. This shortfall creates an incentive to further increase cattle herds and thus overgrazing problems are exacerbated.

A lesson from such an imbalance between targets and actual production is not difficult to learn. Production of a commodity such as cheese is not all economics; it is based on a renewable natural resource, and therefore its production limit should be defined with a commitment not to degrade the environment.

A Solution

A possible solution to this dilemma is to lower production targets and restrict the number of *chauri* while commensurately increasing the milk prices paid to farmers, such that the farmers' incomes remain the same. One may argue that the concomitant rise in the price of cheese will hurt the consumers, but cheese is not an item in the Nepali diet. Indeed, the proportion of calories from all milk products, such as butter and ghee, has been less than 5% for the Nepali populace in this decade (RAPA 1986). Cheese is sold in Kathmandu to Western tourists who come to Nepal and wish to consume cheese in their pizzas, cheese sandwiches, lasagna, etc. (Nembang 1987, Singh 1987). Mountain cheese (locally called yak cheese) sold in Kathmandu costs US $4.30 per kg, which makes it unaffordable for consumption by most Nepalis. On the other hand, the price is cheaper than most North American and European prices and could be increased significantly without damaging the market. Thus the DDC could, over time, establish a balance between ecology and economy that will sustain local people and improve their living conditions.

Conclusion

Sustainable economic development should provide a lasting and secure livelihood that minimizes resource depletion, environmental degradation, and social instability (Barbier 1987). Many economic developers believe the Nepal Himalayas can benefit from further development of dairy farming and marketing (Shah 1988), but high mountains have short growing seasons, season-specific rainfall, and extreme temperatures, so the vegetation is fragile. In Langtang National Park, alpine forests are more important to support sustainable agriculture and wildlife than to produce cheese to cater to the tourism industry. Tourism will inevitably become a major industry in the Langtang Valley and throughout much of Nepal because of its direct financial returns, but it brings with it many ecological and social problems that need to be managed (Yonzon 1988). In Langtang

National Park, tourism is having both a direct effect, primarily exacerbating the overharvest of fuel wood, and an indirect effect, supporting an unsustainable cheese production scheme. (For further information about the direct impacts of tourism on Himalayan environments, see Jefferies [1982] and McNeely et al. [1985].)

The consequences of excessive cheese production could include a slow deterioration in the quality of cattle health, loss of fuel wood, local extinction of some wildlife species such as the red panda, degradation of forest and pastures, and an increased risk of failure among small farmers. On a broader scale, these processes contribute to the larger problem of Himalayan deforestation, with environmental and social consequences felt throughout much of the Indian subcontinent (Myers 1986).

Farmers should have alternatives and a right to make decisions in their own long term best interests. The park managers should identify grazing lands and determine their carrying capacity while agricultural extension agents should inform farmers about optimal productivity. Present practices regarding bank loans and cheese production must be quantitatively evaluated and then corrected. With communication and cooperation among decision-makers from agriculture development, livestock extension, park management, and most importantly, the subsistence farmers, a balance between sustainable production of cheese and ecological integrity can be achieved.

The problems associated with commercial cheese production in Langtang National Park do not suggest that natural resource use by local people and wildland protection are fundamentally incompatible (McNeely & Miller 1984). There are many other examples of a reasonably successful marriage between these goals, such as exploitation of grass as thatch and canes in Chitwan National Park, Nepal (Lehmkuhl et al. 1988), harvesting large mammals for meat and hides in various protected areas in southern Africa (Child 1984; Lewis et al. 1990; Wolf & Wyckoff-Baird, in preparation), or culling vicunas (*Vicugna vicugna*) for wool and meat in Reserva Nacional de Pampa Galeras in Peru (Ponce del Prado 1984). The situation in Langtang simply highlights the manifest need to plan such programs very carefully with full participation by all interested parties. This is particularly important when the use of resources becomes commercial and involves development and financial institutions whose short term interests may conflict with sustainable resource use.

Acknowledgments

Financial and logistical support was provided by the World Wildlife Fund, the Department of National Parks and Wildlife (HMG), the East-West Center, the King Mahendra Trust for Nature Conservation, Tribhuvan University, and the University of Maine. For field assistance and support, we thank Keshav Chettri, Nima Sherpa, Ram Bir Rai, Parbat Rai, Bimal Gurung, Madan Oli, Mahendra Shrestha, Art Soukkala, Prasanna Yonzon, plus Warden Surya Panday, Subba Lama, and Mountain Travel. Kevin Boyle, Don Gilmour, Bill Glanz, Brad Griffith, Larry Hamilton, Ray Owen, Steve Sader, and Jim Sherburne critiqued early drafts. Maine Agricultural Experiment Station 1496.

Literature Cited

Adams, D. W., D. H. Graham, and J. D. Von Pischke. 1984. Undermining rural development with cheap credit. Westview Press, Boulder, Colorado.

Barbier, E. B. 1987. The concept of sustainable economic development. Environmental Conservation 14:101–110.

Bowman, F. J. A., and R. Houtman. 1988. Pawnbroking as an instrument of rural banking in the Third World. Economic Development and Cultural Change 37:69–89.

Child, G. 1984. Managing wildlife for people in Zimbabwe. Pages 119–121 in J. A. McNeely and K. R. Miller, editors. National parks, conservation and development. Smithsonian Institution Press, Washington, D.C. 825 pp.

Durham University Himalayan Expedition. 1977. Langtang National Park management plan. HMG/UNDP/FAO Project NEP/72/002. Kathmandu, Nepal.

Guillet, D. 1981. Land tenure, ecological zone, and agricultural regime in the central Andes. American Ethnologist 8:139–156.

Gurung, B. 1988. Socio-economics, development, and conservation in Syabru and Langtang, Langtang National Park, central Nepal. Tribhuvan University, Kathmandu, Nepal. Unpublished M.A. thesis.

Hagen, T. 1987. Wegen und Irrwette der Entwicklungshilfe. Das experimentieren an der Dritten Welt. Neue Zurcher Zeitung, Zurich, Switzerland.

Jefferies, B. 1982. Sagarmatha National Park: the impact of tourism in the Himalayas. Ambio 11:274–281.

Lehmkuhl, J. F., R. K. Upreti, and U. R. Sharma. 1988. National parks and local development: grasses and people in Royal Chitwan National Park, Nepal. Environmental Conservation 15:143–148.

Lewis, D. R., G. B. Kaweche, and A. Mwenya. 1990. Wildlife conservation outside protected areas: lessons from an experiment in Zambia. Conservation Biology 4:171–180.

McNeely, J. A., and K. R. Miller, editors. 1984. National parks, conservation, and development. Smithsonian Institution Press, Washington, D.C. 825 pp.

McNeely, J. A., J. Thorsell, and S. Chalise, editors. 1985. People and protected areas in the Hindu-Kush Himalaya. King Mahendra Trust for Nature Conservation and International Centre for Integrated Mountain Development, Kathmandu, Nepal.

Miller, K. R. 1984. The Bali Action Plan: a framework for the future of Roledi areas. Pages 756–764 in J. A. McNeely, and K. R. Miller, editors. National parks, conservation, and development. Smithsonian Institution Press, Washington, D.C. 825 pp.

Mishra, H. R. 1982. Balancing human needs and conservation in Nepal's Royal Chitwan Park. Ambio **11**:246–251.

Myers, N. 1986. Environmental repercussions of deforestation in the Himalayas. Journal of World Forest Resource Management **2**:63–72.

Nembang, L. B. 1987. Future prospects in cheese production and its marketing in Nepal. Pages 90–102 in D. D. Joshi, editor. Proceedings of the seminar on dairy development and management in Nepal. Dairy Development Corp., Kathmandu, Nepal.

Ponce del Prado, C. F. 1984. Inca technology and ecodevelopment: conservation of the vicunas in Pampa Galeras. Pages 575–580 in J. A. McNeely, and K. R. Miller, editors. 1984. National parks, conservation, and development. Smithsonian Institution Press, Washington, D.C. 825 pp.

Regional Office for Asia and the Pacific Area (RAPA). 1986. Selected indicators of food and agricultural development in Asia, Pacific region, 1975–85. FAO/SEAR, Pacific Office, Bangkok, Thailand.

Shah, S. 1988. Nepal's economic development: problems and prospects. Asian Survey **28**(9):945–957.

Sherpa, L. N. 1988. Conserving and managing biological resources in Sagarmahta (Mt. Everest) National Park, Nepal. Working Paper No. 8. EAPI, East-West Center, Honolulu, Hawaii.

Shrestha, M. K. 1988. Vegetation study of the red panda habitat in Langtang National Park, central Nepal. Unpublished M.SC. thesis, Tribhuvan University, Kathmandu, Nepal.

Singh, V. 1987. Integrated livestock development to provide dairy industry in Nepal. Pages 74–89 in D. D. Joshi, editor. Proceedings of the seminar on dairy development and management in Nepal. Dairy Development Corp., Kathmandu, Nepal.

Timmerman, C., and E. R. P. Platje. 1987. Environmental impact assessment of the energy requirements of the cheese factory in Kyangin (Langtang National Park). Department of National Parks and Wildlife Conservation, National Parks and Protected Areas Management Project NEP/85/011, Kathmandu, Nepal.

Tulachan, P. M. 1985. Socio-economic characteristics of livestock in Nepal. Research Report Series No. 1. HMG-USAID-GTZ-Winrock Project, Kathmandu, Nepal.

Upadhyaya, S. P. 1987. Role of credit in livestock and dairy development in Nepal. Pages 46–52 in D. D. Joshi, editor. Proceedings of the seminar on dairy development and management in Nepal. Dairy Development Corp., Kathmandu, Nepal.

Upreti, B. N. 1985. The park-people interface in Nepal: problems and new directions. Pages 19–24 in J. A. McNeely, J. Thorsell, and S. Chalise, editors. People and protected areas in the Hindu-Kush Himalaya. King Mahendra Trust for Nature Conservation and International Centre for Integrated Mountain Development, Kathmandu, Nepal.

Wolf, E. C., and B. Wyckoff-Baird, editors. Wildlife as a resource for development: a southern African perspective. Osborn Center, World Wildlife Fund, Washington, D.C. In preparation.

Yonzon, P. B. 1988. Conflicting issues of development infrastructures: a case study in relation to the red panda conservation. Theme paper. National Conference on Science and Technology, Royal Nepal Academy of Science and Technology, Kathmandu, Nepal.

Yonzon, P. B. 1989. Ecology and conservation of the red panda in the Nepal-Himalayas. University of Maine, Orono, Maine. Unpublished Ph.D. dissertation.

Yonzon, P. B., R. Jones, and J. Fox. Assessing habitat and estimating populations of red pandas in Langtang National Park, Nepal. In preparation.

Yonzon, P. B., and M. L. Hunter, Jr. Conservation of the red panda *Ailurus fulgens*. Biological Conservation. In press.

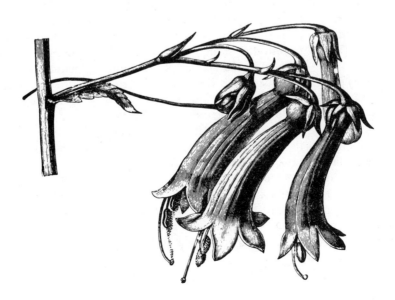

Conservation Education

Biological Diversity, Agriculture, and the Liberal Arts

> It is incumbent on us to take special pains . . . that all the people, or as many of them as possible, shall have contact with the earth and that the earth's righteousness shall be abundantly taught.
>
> Liberty Hyde Bailey

Until quite recently much of what people knew about the natural world they learned from growing up on farms or by periodic visits to nearby farms. For all of their flaws, farms were schools of a sort in natural history, ecology, soils, seasons, wildlife, animal husbandry, and land use. The decline of ecologically diverse farms and the experience of the natural world that they fostered explains in large part, I think, the increasing gap between the broad support for environmental causes evident in public opinion polls and the growing ignorance of how ecosystems work and how private consumption and economic growth destroy the environment. In other words, the sharp decline in the number of farms and the shift toward industrial farming has had serious consequences for society's ecological intelligence (see, for example, Tobin 1990:48–51).

To be sure, the experience of farm life varied greatly with the quality of the farm and the differences in individual perceptiveness, intelligence, and skill. Moreover, in the absence of vital rural communities, farm life was sometimes tedious, narrow, and parochial. On balance, however, I believe it was mostly otherwise. But in either case farms did what no other institution has ever done as well. They taught directly and sometimes painfully the relationship between our daily bread and soil, rainfall, animals, biological diversity, and natural cycles, which is to say land stewardship. They also taught the importance of human qualities of husbandry, patience, hard work, self-reliance, practical skill, and thrift. However imperfectly, farms served as a reality check about human possibilities in nature, a reality check presently lacking in urban societies.

The decline of small family farms and rural communities might still be justified as the necessary price for efficiency. But it is evident that unfair taxation and lavish subsidies for large scale had more to do with the demise of small farms than did their alleged failure to produce or to make a profit (see Strange 1988). On behalf of a short-term, devil-take-the-hindmost economics we destroyed farms and rural communities that have been the historical ballast for stable societies. We have good reason to believe with Aldo Leopold that social "stability seems to vary inversely to the mental distance from fields and woods" (Leopold 1991:286).

At its best traditional farming and rural life were, in Jacquetta Hawkes's words, "a creative, a patient and increasingly skillful love-making that had persuaded the land to flourish." (Hawkes 1951:202). In contrast, the industrialization of agriculture was "an upsurge of instinctive forces comparable to the barbarian invasions.... designed to satisfy man's vanity, his greed and possessiveness, his wish for domination" (202). As agriculture became more industrialized the number of farms declined, and with them, rural communities. The remaining farms became larger, ecologically less diverse, more expensive to operate, and more vulnerable to economic and ecological forces beyond the control of farmers. They also tended to become less interesting and less instructive places—hence the decline of knowledge of the land now evident throughout predominantly urban societies.

Farms were not only sources of instruction about the realities of the natural world. In many places they also protected biological diversity. Gary Nabhan's studies of the Papago and recent studies of peasant agriculture in Mexico and Central America show great ecological intelligence carefully and artfully woven into rural landscapes over a millennia or more (Nabham 1982; Wilken 1987). Traditional peasant farms were repositories of genetic diversity, often growing dozens or hundreds of varieties that are now disappearing in favor of a small number of hybrids purchased from multinational corporations.

In many parts of the Third World the imposition of high-yield agriculture helped to break apart the intimate relationship between cultures and ecosystems that had coevolved over millennia. As Angus Wright has shown in a brilliant study, *The Death of Ramon Gonzalez,* these systems and the biological diversity they preserved have been largely destroyed as part of a development strategy designed to "milk the soil and other natural resources of the nation and its poorest laborers by squeezing money out of agriculture" to pay for industrialization (Wright 1990: 227). In Mexico the result was to "concentrate earnings and investment in a few regions of agribusiness growth at the expense of the neglect, ruin, and abandonment of other regions" (Wright 1990:237).

What Wright calls "the modern agricultural dilemma" is simply that "the highly localized adaptations needed for ecologically healthy agriculture and healthy, stable rural communities are often in conflict with the apparent requirements of rapidly industrializing nations and an expanding international economy." (Wright 1990:227). From the perspective of a narrow science and economics this dilemma is hard to see, or more precisely it is difficult for some to want to see because it implies a failure of great consequence. From a wider perspective that includes a thorough knowledge of ecology and anthropology, it was in fact foreseen early on by people like Carl O. Sauer, Paul Sears, and Aldo Leopold (Wright 1990:247, 285).

The decline of the family farm in the United States and the destruction of traditional farming practices in what is called the "underdeveloped" world are the products of many forces, including the separation of the study of agriculture from its community, cultural, and ecological context. The modern agricultural dilemma described by Wright began when agricultural sciences were isolated in research institutions and from there evolved into technical disciplines whose purpose was to do one thing: increase production. Consequently they were not rooted in any coherent and sustainable social, philosophical, political, and ecological context. In this setting, a great many assumptions about nature, technology, farming, rural life, and the consequences of applying industrial techniques to complex biological and cultural systems went unchallenged.

It might have been very different had agriculture evolved instead within liberal arts colleges. Instead of becoming a series of disjointed technical specialties, agriculture might have come to be regarded, and rightly so, as a liberal art with technical aspects. In the context of liberal arts colleges, agriculturalists might have learned to see farming not as a production problem to be fixed, but as a more complex activity at once cultural, ethical, ecological, and political. This is not a new idea. Aldo Leopold once proposed that it was time to "swap ends ... to curtail sharply the output of professionals ... [in order to] tell the whole campus and ... the whole community what ... conservation is all about" (Leopold 1990, 301). By the same logic we have too many agricultural professionals and not nearly enough people who understand farming and its wider social and ecological context. The goal of liberal education (in "wildlife"), as Leopold described it, was "not merely a dilute dosage of technical education" but rather "to teach the student to see the land, to understand what he sees, and enjoy what he understands" (Leopold 1990: 302).

If agriculture would have evolved differently in a liberal arts setting, I think the inclusion of agriculture would have helped liberal arts colleges avoid the debilitating separation of abstract intellect and practical intelligence. Instead we have developed a version of the liberal arts in which it is assumed, without anyone ever quite saying as much, that learning is an indoor sport taking place exclusively in classrooms, libraries, laboratories, and computer labs, and that practical competence is to be avoided at all cost.

This leads me to propose that agriculture should be included as a part of a complete liberal arts education, first because it offers an important kind of experience no longer available to many young people from predominantly urban areas. Student responsibility for farm operations would teach the values of discipline, physical stamina, frugality, self-reliance, practical competence, hard work, cooperation, and ecological competence. Second, college farms properly used would be interdisciplinary laboratories for the study of sustainable agriculture, ecology, botany, zoology, animal husbandry, entomology, soil science, ornithology, landscape design, land restoration, mechanics, solar technology, business operations, philosophy, and rural sociology. Third, college farms could become catalysts in a larger effort to revitalize rural life in surrounding areas. Fourth, college farms could be used to preserve biological diversity jeopardized by development. Fifth, college farms could be a part of a global effort to reduce carbon emissions involved in the long-distance transport of food and to sequester carbon through agroforestry and tree cropping. Sixth, college farms could close waste loops by composting all campus organic wastes and incorporating them as soil admendments. Finally, by participating in the design and operation of college farms, students might learn that our problems are not beyond intelligent solution; that solutions are close by; and that institutions that often seem to be inflexible, unimaginative, and remote from the effort to build a sustainable society can be otherwise.

David Orr

Literature Cited

Hawkes, J. 1951. A land. Random House, New York.

Leopold, A. 1991. The river of the mother of God and other essays. S. Flader and J. B. Callicott, editors. University of Wisconsin Press, Madison, Wisconsin.

Nabhan, G. 1982. The desert smells like rain. North Point Press, San Francisco, California.

Strange, M. 1988. Family farming. University of Nebraska Press, Lincoln, Nebraska

Tobin, R. 1990. The expendable future: U.S. politics and the protection of biological diversity. Duke University Press, Durham, North Carolina. Pp. 48–51.

Wilken, G. 1987. Good farmers: traditional agricultural resource management in Mexico and Central America. University of California Press, Berkeley, California.

Wright, A. 1990. The death of Ramon Gonzalez: the modern agricultural dilemma. University of Texas, Austin, Texas.

Conservation Education

Professionalism and the Human Prospect

> "The mind can be permanently profaned by the habit of attending to trivial things, so that all our thought shall be tinged with triviality."
> Thoreau, "Life Without Principle"

> "I have always tried not to be a *professional* scientist."
> Erwin Chargaff, *Heraclitean Fire*

The tenure system was originally created to protect the right of professors to speak freely without fear of reprisal. One might have expected great and radical things to emanate from the safely tenured. With some notable exceptions, however, this has not happened often. Derek Bok, former President of Harvard University, has lamented the results:

> Armed with the security of tenure and the time to study the world with care, professors would appear to have a unique opportunity to act as society's scouts to signal impending problems. ... Yet rarely have members of the academy succeeded in discovering emerging issues and bringing them vividly to the attention of the public (Bok 1990).

Similarly, why have so few of the tenured joined the effort to preserve biological diversity and a habitable earth? Why are so few of the tenured willing to confront the large and portentous issues of human survival looming ahead?

The reason, I think, is that the professorate professionalized itself and professionalization has done what even the most flagrant college administrator would not dream of doing. The professionally induced fear of making a mistake or being thought to lack rigor has rendered much of the professorate toothless and confined to quibbles of great insignificance. One sure way for a young professor to risk being denied tenure is to practice what philosopher Mary Midgley calls "the virtue of controversial courage," the very reason for which tenure was created (Midgley 1989). For the consummate professional scholar, under professional and administrative pressures to secure large grants, the rule of thumb is that if it has no obvious and quick professional payoff, don't do it. A modern Charles Darwin, for example, would have gotten a large grant, flown to the Galapagos, returned to dash off a dozen articles to a dozen journals, and hired a publicity agent to get on TV and in the *Science Times,* all with the intention of getting a better job somewhere else. The real Darwin delayed publication of *Origin of the Species* for 20 years while he thought it over. Those with something to profess, in short, are being replaced by professionals with something to sell. This is not a new problem, but it is getting worse.

Professional scholars tend to think of themselves as a part of the established order, not as critics of it, let alone creators of something better. Knowledge has more and more become mastery of allegedly neutral techniques. In the process, however, neutrality has gotten confused with objectivity. In the words of historian Robert Proctor, "Neutrality refers to whether a science takes a stand; objectivity to whether a science merits certain claims to reliability" (Proctor 1991). The ideal of the broadly informed, renaissance mind has given way to the far smaller idea of the specialist. Even in the humanities where one might still expect dangerous thoughts about the plight and potentials of human kind, one often finds instead silliness, shrillness, and obscurity.

Furthermore, professionalization has Balkanized the intellectual landscape, each fiefdom having its own professional association, trade journals, and specialized jargon. Professing allegiance to one principality or another, few "professionals" know enough of the whole terrain to be dangerous to the established order. Narrowness, "methodolatry," and careerism have rendered many unfit and unwilling to ask large and searching questions. Where intellectuals once addressed the public, they now talk mostly to each other about matters of little or no consequence for the larger society. To the same degree that it is obscure, jargon-laden, and trivial, professionalized knowledge has come as a great windfall to the comfortable, serving to divert attention from behavior that is egregious, criminal, or merely embarrassing. When did an issue of the *American Political Science Re-*

view cause the comfortable in Congress to squirm? When did an issue of the *American Economic Review* ever cause the barons of Wall Street to tremble? When was the last time agribusiness felt itself threatened by the American Society of Agronomists? When was the last time the dispossessed felt befriended by an issue of the *American Sociological Review?* When did those now planning to re-engineer the fabric of life on the earth for profit feel their project threatened by any academic profession at all? And when did philosophy "cease to be 'the love of wisdom' and aspire to be a science?" (Solomon 1992).

The academy, in short, is a safer haven than it ought to be for the professionally comfortable, cool, and upwardly mobile. It is far less often that it should be a place for passionate and thoughtful critics. Professionalization has rendered knowledge safe for power thereby making it more dangerous than ever to the larger human prospect. And what are these dangers?

First, there is the danger that professionals will clone themselves, making their students knowledge technicians instead of broadly thoughtful, liberally educated persons willing to roll up their sleeves and join in the great struggles ahead to preserve a habitable planet. There is, second, the danger that a great deal of important "nonprofessional" knowledge will be dismissed or ignored altogether. I am referring to the kind of vernacular knowledge that people have always needed to live well in a place, to create enduring communities, to understand themselves, to be of service to those around them, and to find meaning amidst the mysteries of life. Third, despite all of the recent talk about multiculturalism, the plain fact is that the academy has become an agent of global homogenization. The great questions of human existence are being reduced to those amenable to narrow professionalized scholarship. The danger here is that a global monoculture, driven by the logic of a worldwide market economy and confined to the language of professional discourse, will not and, I think, cannot preserve a diverse biota for long. There is a fourth danger which is simply that higher education will mostly opt out of the great ecological issues of the 21st century because it could not summon enough vision and courage to take a stand.

What can be done? I do not propose to throw babies out with the bathwater. The tenure system continues to provide a defense against capricious administrators. It was never intended, however, to protect against the more subtle and powerful censorship of academic mandarins who are preoccupied with guarding disciplinary boundaries. For this I offer, accordingly, two ideas to curtail professionalism run amuck. The first is to suggest that all candidates for tenure appear before an institution-wide forum to answer questions such as:

1. Where does your field of knowledge fit in the larger landscape of learning?
2. Why is your particular expertise important? For what and for whom is it important?
3. What are its wider ecological implications and how do these affect the long-term human prospect?
4. Explain the ethical, social, and political implications of your scholarship.

The benefits of such a forum are clear. It would provide a great incentive for the candidate to think beyond the confines of his or her particular discipline. It would also provide considerable incentive to communicate plainly in commonly understood language, which would be something of a novelty. It would help to establish a balance between those with a sense of the larger issues of the time and those who are content to be narrowly specialized professionals. It would smoke out certain kinds of intellectual deficiencies such as ethical flaccidity or gross ignorance of the relation of ecological realities to one's scholarship. A tenure forum might even exert a moderately beneficial effect on the inquisitors, causing them to ponder the larger architecture and purpose of knowledge as well.

My second suggestion similarly aims to legitimize and encourage disciplinary boundary crossings and thereby weaken the hold of narrow professionalism. Educators might learn something from the way many prosperous business organizations have developed flexible teams that regularly reshuffle organizational responsibilities to achieve results not otherwise possible in a rigid or hierarchical structure. Colleges and universities might similarly pioneer new and daring ways to reorganize knowledge to attack problems of the day, but without the heavy baggage of disciplines and departments. I propose that all educational institutions create and amply fund regular forums, programs, courses, and projects that offer participants the opportunity to suspend their status as disciplinary specialists. University of California sociologist, Jeffrey Alexander and ten of his colleagues, for example, offer a course called "L.A. in Transition" (Alexander 1993). The course aims "to make Los Angeles into a guinea pig for broad study of contemporary racial, ethnic, and economic problems." In such settings scholars would have to step outside their particular disciplines in order to participate in a common dialogue, solve a problem, or cooperate in the creation of something new. Most colleges and universities have interdisciplinary programs that provide a setting for such endeavors, but participation is seldom seriously acknowledged, rewarded (for example, with tenure), or funded. And such programs are usually the first to be dropped when financial cuts are made. The remedy is straightforward: unequivocal institutional support and encouragement for boldness, breadth, and disciplinary boundary crossings.

For those willing to do it, the task of mastering a particular field in depth while acquiring a broad and contextual knowledge demands time, patience, intellectual skill, and great commitment. It demands scholars who pay attention to large issues and who have loyalties to things bigger than mere professionalism. These people deserve to be protected from both capricious administrators and from what Page Smith calls "academic fundamentalists" (Smith 1990). The world has always needed a dangerous professorate and now more than ever before. It needs a professorate with ideas that are dangerous to greed, shortsightedness, indulgence, exploitation, apathy, hi-tech pedantry, and narrowness.

Literature Cited

Alexander, J. C. 1993. The irrational discipinarity of undergraduate education. The Chronicle of Higher Education (December 1).

Bok, D. 1990. Universities and the future of America. Duke University Press, Durham, North Carolina.

Midgely, M. 1989. Wisdom, information, and wonder. Routledge, London.

Proctor, R. 1991. Value free science. Harvard University Press, Cambridge, Massachusetts.

Smith, P. 1990. Killing the spirit. Viking, New York.

Solomon, R. 1992. Beyond reason: the importance of emotion in philosophy. In Ogilvy, J., editor. Revisioning philosophy. State University of New York Press, Albany, New York.

David W. Orr

Conservation Education

Architecture as Pedagogy

It is paradoxical that buildings on college and university campuses, places of intellect, characteristically show so little thought, imagination, sense of place, ecological awareness, and relation to any larger pedagogical intent. The typical academic building would seem to have the architectural elegance and performance standards common to shopping malls, motels, and drive-through funeral parlors; places where, one might infer, considerations of "throughput" are uppermost in the minds of designers. How has this come to be? Some believe it is the result of a conspiracy of sorts between a wealthy donor wishing to make an end-run around mortality by having his or her name on a building, a college president wishing to enhance a reputation for getting things done, an architect seeking a professional reputation by designing showy buildings that don't work very well, and a financial officer whose job it is to economize on beauty, humanity, and common sense in the name of fiscal integrity.

Personally, I do not believe that the design of academic buildings is the result of a conspiracy at all. Most academic administrators and trustees are fully capable of doing all of this on their own in broad daylight. They have not conspired because they didn't need to—faculty and students have been effectively excluded from the process whereby ambition tempered by dullness and tortured by utility is rendered into architectural form.

The problem isn't just that many academic buildings are unsightly, don't work very well, or that they don't fit their place or region. The deeper problem is that we've assumed, wrongly I think, that learning takes place in buildings but that none occurs as a result of how they are designed or by whom, how they are constructed and from what materials, how they fit their location, and how they operate and how well. My point is that academic architecture is a kind of crystallized pedagogy and that buildings have their own hidden curriculum that teaches as effectively as any course taught in them. And what lessons are taught by the way we currently design, build, and operate academic buildings?

The first lesson is that architecture is the prerogative of power and not that of those who teach or learn. Implicit in this view is the assumption that architecture does not influence the flow of ideas and the quality of learning and the human relationships in which learning is embedded. So, faculty and students are rarely consulted on whether or what to build or where. From this they learn that power can impose what it wishes on the academic landscape without having to explain much.

The second lesson is that architecture and building design is merely technical and is thus best left to people with technical competence. It follows that ethical, ecological, or aesthetic aspects of buildings don't matter nearly as much as technique and technology. In deference to expertise then, we learn passivity toward the "built environment." This may explain our subsequent failure to protest the spread of ugliness and banality across the landscape as well as our apparent obliviousness to how these blights cheapen our lives and diminish our prospects.

From the design and materials used in construction a third lesson is learned: the environmental and energy costs of buildings do not matter much. Academic buildings are seldom designed to maximize solar gain or energy efficiency, or minimize unpriced environmental costs of materials, or utilize local materials. Thus, we learn carelessness that accompanies waste and inefficiency, as well as callousness to the degradation of other places where materials and energy originate.

Fourth, a "successful" building is one that quietly serves the educational process but requires no care and interaction of those who use it. From this we learn passivity and disengagement from our surroundings and the irresponsibility invited by never having to know how things work, or why, or what alternatives there might have been. The same building in which sophisticated theories are propounded, unobtrusively teaches its occupants that it's O.K. to be oblivious to the most basic aspects of life support.

Fifth, modern academic architecture teaches us about the limits of imagination. It is assumed, without anyone ever saying as much, that intellect can be nurtured in sterile places largely devoid of imagination. So, creativity in academic architecture is mostly confined to facades replete with lots of glassy flourishes of form disengaged from any purpose beyond that of impressing the easily impressible. The use of imagination mostly stops short of the places where learning is supposed to happen, the design of which is still the cubical classroom, or the lecture hall (a cavernous space with audio-visual equipment), both of which reached near state-of-the-art sometime before the Dark Ages. Such spaces do little to lift the spirit, stir the imagination, fuel the intellect, or remind us that we are citizens of ecological communities.

We've not thought of academic buildings as pedagogical, but they are. We've not exercised much imagination about the design of academic buildings, and it shows in a manifest decline in our capacity to envision alternatives to the urban and suburban excrescence oozing all around us. We've assumed that people who know little about learning and pedagogy were competent to design places where learning is supposed to occur. They aren't, not alone anyway. What do I propose?

Let's begin by asking what might be learned from the design, construction, and operation of the places where formal education occurs? First, the process of design and construction is an opportunity for a community to deliberate over the ideas and ideals it wishes to express and how these are rendered into architectural form. What do we want our buildings to say about us? What will they say about our ecological prospects? To what large issues and causes do they direct our attention? What problems do they resolve? What kind of human relationships do they encourage? These are not technical details, but first and foremost issues of common concern that should be decided by the entire campus community. When they are so decided, the design of buildings fosters civic competence and extends the idea of citizenship.

Second, the architectural process is an opportunity to learn something about the relationship between ecology and economics. For example, how much energy will a building consume over its lifetime? How much of what kinds of materials will be required for its upkeep? What unpriced costs do construction materials impose on the environment? Are they toxic to manufacture, install, or, later, to discard? How are these costs paid? What is the total energy embodied in materials used in the structure? It is possible to design buildings that repay those costs by being net energy exporters? If not, are there other ways to balance ecological accounts? Can buildings and the surrounding landscape be designed to generate a positive cash flow?

These questions cannot be answered without engaging issues of ethics. How are building materials extracted, processed, manufactured, and transported? What ecological and human costs do various materials impose where and on whom? What in our ethical theories justifies the use of materials that degrade ecosystems, jeopardize other species, or risk human lives and health? Where those costs are deemed unavoidable to accomplish a larger good, how can we balance ethical accounts?

Fourth, within the design, construction, and operation of buildings is a curriculum in applied ecology. Buildings can be designed to recycle organic wastes through miniature ecosystems which can be studied and maintained by the users. Buildings can be designed to heat and cool themselves using solar energy and natural air flows. They can be designed to inform occupants of energy and resource use. They can be landscaped to provide shade, break winter winds, propagate rare plants, provide habitat for animals, and restore bits of vanished ecosystems. Buildings and landscapes, in other words, can extend our ecological imagination.

Fifth, they can also extend our ecological competence. The design and operation of buildings is an opportunity to teach students the basics of architecture, landscape architecture, ecological engineering for cleaning wastewater, aquaculture, gardening, and solar engineering. Buildings that invite participation can help students acquire knowledge, discipline, and useful skills that cannot be acquired other than by doing.

Finally, good design can extend our imagination about the psychology of learning. The typical classroom empties quickly when not required to be used. Why? The answer is unavoidable: it is most often an uninteresting and unpleasant place, designed to be functional and nothing more. And the same features that make it unpleasant make it an inadequate place in which to learn. What makes a place a good educational environment? How might the typical "classroom" be altered to encourage ecological awareness? creativity? responsiveness? civility? How might materials, light, sounds, water, spatial configuration, openness, scenery, colors, textures, plants, and animals be combined to enhance the range and depth of learning? My hunch is that good learning places are places that feel good to us—human scaled places that combine nature, interesting architecture, materials, natural lighting, and "white" sounds (e.g., running water) in interesting ways that resonate with our innate affinity for life.

My point is that the design, construction, and operation of academic buildings can be a liberal education in a microcosm that includes virtually every discipline in the catalog. The act of building is an opportunity to stretch the educational experience across disciplinary boundaries and across those dividing the realm of thought from that of application. It is an opportunity to work together on projects with practical import and to teach the art of "good work." It is also an opportunity to lower life-cycle costs of buildings and to reduce a large amount of unnecessary damage to the natural world incurred by careless design.

As a test of these ideas I have been working with 25 students for the past year on the design of a modestly sized, aesthetically attractive building (~5000 ft^2) that uses no fossil fuels, monitors its energy production and use, purifies its wastewater on site, is built from nontoxic materials, recovers its initial energy investment, generates a positive cash flow, is landscaped to include rare species, and expands its occupants' sense of ecological possibilities. A tall, but not impossible, order. Mark DeKay at Virginia Tech is asking similar questions and is forming a "Society of Building Science Educators" to address ecological issues in architecture (Professor Mark DeKay, Virginia Polytechnic and State University, College of Architecture, 202 Cowgill Hall, Blacksburg, VA 24061).

Among the newer, U.S. sources we've found useful I recommend the following: Richard Crowther, *Ecologic Architecture* (Boston: Butterworth, 1992); William McDonough, *The Hannover Principles: Design for Sustainability* (Bill McDonough Architects, 116 East 27th St., New York, NY 10017); David Seamon, *Dwelling, Seeing, and Designing.* (SUNY Press, 1993); Andrew St. John, *The Sourcebook for Sustainable Design* (Boston: Society of Architects, 52 Broad St., Boston, MA 02109); Stephen Strong, *The Solar Electric House* (Still River, 1991); and *Environmental Building News* (RR1, Box 161, Brattleboro, VT 05301).

David W. Orr

Conservation Education

The Virtue of Conservation Education

At a recent gathering of D.C. environmentalists the prevailing wisdom held that the public could not be led to a survivable future with moral arguments, but only by those that appealed to short-term economic self-interest. This is a widely held opinion and one that raises a serious question for conservation educators. Is conservation primarily a technical subject with minor moral implications, or is it fundamentally about morality with technical aspects? If the former, then having equipped our students with a thorough grasp of the pertinent scientific disciplines, the technological basis of efficient resource use, and a bit of economics we may regard our duties as educators adequately discharged. If the latter we must do all of the above *and* enable students to think clearly about (what was once without apology called) virtue and motivate them to live accordingly. The difference between the two is partly that between reform and perestroika. It is a difference in whether one thinks that with the right technologies and prices we will make an orderly transition to the condition of sustainability, or whether we will make it, if we make it, by the margin of a gnat's eyebrow with the four horsemen in hot pursuit. On grounds of prudence and my reading of the evidence I am persuaded of the latter and hence of the need to think seriously about the relationship between sustainability and the human qualities subsumed in the word "virtue."

But what is virtue? Philosopher Alasdair MacIntyre (1981) believes that the modern world suffers from moral amnesia, the vague awareness of a deficiency of virtue that we can no longer describe. To understand virtue he argues that we must return to its ancient roots for "the tradition of the virtues is at variance with central features of the modern economic order and more especially its individualism, its acquisitiveness and its elevation of the values of the market to a central social place."

As it was understood in the ancient world, virtue was founded on the bedrock of community. One's virtue was inseparable from one's life within a community. From this perspective, in MacIntyre's words, "The egoist is thus ... someone who has made a fundamental mistake about where his own good lies." Robert Proctor (1988) has made the same point in a remarkable book, *Education's Great Amnesia*: "The ancients ... conceptualized and experienced their humanity not as separation, but as participation in the whole order of being." Virtue was regarded, first, as an exercise in participation and fulfillment of the obligations of membership in a community that was embedded in a larger cosmic order.

A second aspect of ancient virtue was the quality of moderation, in Cicero's words, "the ability to restrain the passions and to make the appetites amenable to reason." Moderation as Aristotle defined it was the mean between extremes of excess and deficiency that could be defined by a person of practical wisdom. Virtue, for Aristotle, is chosen through the exercise of reason. "It is not possible," in his words, "to be good in the strict sense without practical wisdom, nor practically wise without moral virtue." In other words, virtue is the result of choosing intelligently between extremes.

Third, for the Greeks and Romans virtue was never separated from politics and from participation in the civic life of the community. For Aristotle the cultivation of virtue was both a goal of politics ("to engender a certain character in the citizens and to make them good and disposed to perform noble actions") and a prerequisite for civic order, because no good community could be built by people without virtue. Modern politics has rejected that tradition, replacing authority based on virtue with scientific management and public relations.

In the ancient world virtue also meant the cultivation of qualities of courage, fortitude, honesty, restraint, charity, chastity, family, personal rectitude, integrity, and reverence. However imperfectly these were realized in practice, they provided the standard by which people judged themselves and the social order. The fact that this list sounds archaic to the modern ear is an indication of how far we have gone in the contrary direction. Modern societies are increasingly operated by and for that subsystem called the "economy," the same economy that, as Lewis Mumford once observed, converted the seven deadly sins (pride, envy, anger, sloth, avarice, gluttony, lust) into virtues after a fashion, and the seven virtues (faith, charity, hope, prudence, religion, fortitude, and temperance) into sins against gross national product. The dependence of the economy on sin is a fact only infrequently studied by economists. Sin, a contentious subject, has been replaced with the more socially agreeable doctrine that all things are relative so that anyone's opinions or behavior are as good as those of any other, or at least not much worse. But lacking the qualities of virtue, can we do the hard things that will be necessary to live within the boundaries of the earth?

I think not, first, because people lacking a sense of community that undergirds the practice of virtue are not likely to care how their actions affect the larger world in any but the most superficial way. Can we expect rational maximizers of self-interest, who discount the future interests of their own children and grandchildren, to be moved by their kinship to bugs and biota? Not likely. Virtue as Aristotle and Cicero described it was founded on a kind of moral ecology (albeit one that excluded lots of people); an awareness of mutual dependence. Lacking this sense, people are not likely to care deeply enough to join the constituency for change that must finally think, live, and vote differently. People who regard their welfare narrowly are unlikely to support large scale social change when it costs something. Hence without a virtuous public that cares deeply about the protection and enhancement of life, there will be no constituency for hard choices ahead and for the policy changes necessary for sustainability.

Second, sustainability will require a reduction in consumption in wealthy societies and changes in the kinds of things consumed toward products that are durable, reusable, useful, efficient, and sufficient. This will come about by enough people choosing to consume less or by scarcity imposed by circumstances and enforced by authoritarian governments as Robert Heilbroner once predicted. It will not come about by putting bandaids on potentially terminal wounds, making plastics that are biodegradable for example. If we are not to turn the earth into a toxic dump or bankrupt ourselves by expensively undoing what should not have been done in the first place, moderation must replace self-indulgence. The appetites, as Cicero put it, must be made "amenable to reason," which for us means making them less amenable to advertising and television.

Third, a great deal has been said about the potential for least-cost end-use analysis that hitches narrow economic rationality to the efficient use of resources with better technology. This is all to the good. But problems arise when that same economic rationality causes consumers to observe that least-cost is not the same as full-cost. For example, the fully informed consumer armed with least-cost reasoning would certainly choose to buy compact fluorescent light bulbs that have lower life-cycle costs than incandescent bulbs. But the same narrow economic rationality would cause them to refuse to pay higher utility costs to clean up nuclear wastes and decommission reactors used to generate the electricity that is used with greater efficiency. At this point economic rationality stops and virtue begins. Least-cost reasoning applies to those costs that must be paid now; full cost applies to those that can be put onto others or deferred to future generations. Only people who take their obligations seriously, people of virtue, would willingly pay the full costs of their actions or even demand to do so.

Fourth, it is implausible, as E. F. Schumacher once noted, that we can systematically cultivate pride, gluttony, lust, avarice, sloth, envy, and anger and remain intelligent. The seven deadly sins are sins in large part because they corrode the intellect. Virtue is a product of reason not of impulse, whim, and fantasy. Anything that destroys the capacity for reasoned choice promotes sin and a grosser national product. On a larger scale, does the deliberate cultivation of sin make us a dumber society? Aristotle would have thought so. And as we become dumber, more passive, and less morally adept, do we also become more tolerant of (or less capable of recognizing and being outraged by) malfeasance, arrogance, stupidity, and vacuousness by public officials? As officialdom becomes more corrupt, inept, and shortsighted can its management of the environment become better? Hardly.

All of this is only to say that in the struggle to restore a decent world, and one that is humane and just, virtue will count a great deal and utilitarians notwithstanding, people in the main are moved as much by considerations of right and wrong as by self-interest. Most people want to do good and given the chance will do it. At the same time the idea of virtue has been corrupted by political charlatans, electronic evangelists, and by our means of livelihood. The ancient concept of virtue accordingly needs to be dusted off, updated, broadened, ecologized, feminized, and reintroduced into the contemporary curriculum and from there into the mainstream of an increasingly cynical society. The conservation of nature is not just a technical subject. It is about morality, the distinction between right and wrong with room for subtleties in between. Clear thought about these categories of thought and behavior should be a primary aim of conservation educators.

Send comments and news items to David Orr at the address below.

David W. Orr
Environmental Studies
Oberlin College
Oberlin, Ohio 44074

Literature Cited

MacIntyre, Alasdair. 1981. After Virtue. Notre Dame University Press, Notre Dame. p. 213, 237.

Proctor, Robert. 1988. Education's Great Amnesia. Indiana University Press, Bloomington. p. 166.

Cicero. On Duties. In: Michael Grant (trans) 1987. On the Good Life. Penguin Books, London. p. 128.

Aristotle. Nichomachean Ethics. In: Martin Ostwald (trans) 1962. Bobbs Merrill, Indianapolis. p. 172.

The Social Dimension—Ethics, Policy, Law, Management, Development, Economics, Education

References

Alcorn, J. B. 1993. Indigenous peoples and conservation. Conservation Biology **7(2)**:424–426.

Anunsen, C. S. and R. Anunsen. 1993. Response to Scheffer. Conservation Biology **7(4)**:954–957.

Ayres, J. M., R. E. Bodmer, R. A. Mittermeier. 1991. Financial considerations of reserve design in countries with high primate densities. Conservation Biology **5(1)**:109–114.

Balick, M. J. and R. Mendelsohn. 1992. Assessing the economic value of traditional medicines from tropical rain forests. Conservation Biology **6(1)**:128–130.

Berger, J., C. Cunningham, A. A. Gawuseb, and M. Lindeque. 1993. "Costs" and short-term survivorship of hornless black rhinos. Conservation Biology **7(4)**:920–924.

Costanza, R. and H. E. Daly. 1992. Natural capital and sustainable development. Conservation Biology **6(1)**:37–46.

Contreras-B, S. and M. Lourdes Lozano-V. 1994. Water, endangered fishes, and development perspectives in arid lands of Mexico. Conservation Biology **8(2)**:379–387.

Diamond, J. M. 1987. Extant unless proven extinct? Or, extinct unless proven extant? Conservation Biology **1(1)**:77–79.

Dudley, J. P. 1992. Rejoinder to Rohlf and O'Connell: biodiversity as a regulatory criterion. Conservation Biology **6(4)**:587–589.

Ehrenfeld, D. 1992. The business of conservation. Conservation Biology **6(1)**:1–3.

Gainnecchini, J. 1993. Ecotourism: new partners, new relationships. Conservation Biology **7(2)**:429–432.

Gullison, R. E. and E. C. Losos. 1993. The role of foreign debt in deforestation in Latin America. Conservation Biology **7(1)**:140–147.

Hall, S. J. G. and J. Ruane. 1993. Livestock breeds and their conservation: a global overview. Conservation Biology **7(4)**:815–825.

Holdgate, M. and D. A. Munro. 1993. Limits to caring: a response. Conservation Biology **7(4)**:938–940.

Karr, J. R. 1990. Biological integrity and the goal of environmental legislation: lessons for conservation biology. Conservation Biology **4(3)**:244–250.

Kellert, S. R. 1993. Values and perceptions of invertebrates. Conservation Biology **7(4)**:845–855.

King, F. W. 1988. Extant unless proven extinct: the international legal precedent. Conservation Biology **2(4)**:395–397.

Mace, G. M. and R. Lande. 1991. Assessing extinction threats: toward a reevaluation of IUCN threatened species categories. Conservation Biology **5(2)**:148–157.

Meffe, G. K. 1992. Techno-arrogance and halfway technologies: salmon hatcheries on the Pacific coast of North America. Conservation Biology **6(3)**:350–354.

Murphy, D., D. Wilcove, R. Noss, J. Harte, C. Safina, J. Lubchenco, T. Root, V. Sher, L. Kaufman, M. Bean, and S. Pimm. 1994. On reauthorization of the Endangered Species Act. Conservation Biology **8(1)**:1–3.

O'Connell, M. 1992. Response to :"Six biological reasons why the Endangered Species Act doesn't work and what to do about it." Conservation Biology **6(1)**:140–143.

Orr, D. W. 1990. The question of management. Conservation Biology **4(1)**:8–9.

Orr, D. W. 1990. The virtue of conservation education. Conservation Biology **4(3)**:219–220.

Orr, D. W. 1991. Biological diversity, agriculture, and the liberal arts. Conservation Biology **5(3)**:268–270.

Orr, D.W. 1993. Architecture as pedagogy. Conservation Biology **7(2)**:226–228.

Orr, D. W. 1994. Professionalism and the human prospect. Conservation Biology **8(1)**:9–11.

Peres, C. A. 1993. Indigenous reserves and nature conservation in Amazonian forests. Conservation Biology **8(2)**:586–588.

Redford, K. H. and A. M. Stearman. 1993. Forest-dwelling native Amazonians and the conservation of biodiversity: interests in common or in collision? Conservation Biology **7(2)**:248–255.

Redford, K. H. and A. M. Stearman. 1993. On common ground? Response to Alcorn. Conservation Biology **7(2)**:427–428.

Robinson, J. G. 1993. "Believing what you know ain't so": response to Holdgate and Munro. Conservation Biology **7(4)**:941–942.

Robinson, J. G. 1993. The limits to caring: sustainable living and the loss of biodiversity. Conservation Biology **7(1)**:20–28.

Rohlf, D. J. 1991. Six biological reasons why the Endangered Species Act doesn't work—and what to do about it. Conservation Biology **5(3)**:273–282.

Rohlf, D. J. 1992. Response to O'Connell. Conservation Biology 6(1):144–145.

Saberwal, V. K., J. P. Gibbs, R. Chellam, and A. J. T. Johnsingh. 1994. Lion-human conflict in the Gir Forest, India. Conservation Biology 8(2):501–507.

Salafsky, N., B. L. Dugelby, and J. W. Terborgh. 1993. Can extractive reserves save the rain forests? An ecological and socioeconomic comparison of nontimber forest product extraction systems in Petén, Guatemala, and West Kalimantan, Indonesia. Conservation Biology 7(1):39–52.

Salzman, J. E. 1989. Scientists as advocates: The Point Reyes Bird Observatory and gill netting in central California. Conservation Biology 3(2):170–180.

Scheffer, V. B. 1993. Reply to the Anunsens. Conservation Biology 7(4):958.

Scheffer, V. B. 1993. The Olympic goat controversy: a perspective. Conservation Biology 7(4):916–919.

Sidle, J. C. and D. B. Bowman. 1988. Habitat protection under the Endangered Species Act. Conservation Biology 2(1):116–118.

Turner, R. E. 1993. Brazilian crocodilian tears. Conservation Biology 7(2):225.

Weir, J. 1992. The Sweeteater Rattlesnake round-up: a case study in environmental ethics. Conservation Biology 6(1):116–127.

Wilcove, D. S., M. McMillan, and K. C. Winston. 1993. What exactly is an endangered species? An analysis of the U.S. endangered species list: 1985–1991. Conservation Biology 7(1):87–93.

Yonzon, P. B. and M. L. Hunter, Jr. 1991. Cheese, tourists, and red pandas in the Nepal Himalayas. Conservation Biology 5(2):196–202.